Benjamin Lewis Robinson
**Bureaucratic Fanatics**

# Paradigms

---
Literature and the Human Sciences

Edited by
Rüdiger Campe · Paul Fleming

Editorial Board
Eva Geulen · Rüdiger Görner · Barbara Hahn
Daniel Heller-Roazen · Helmut Müller-Sievers
William Rasch · Joseph Vogl · Elisabeth Weber

# Volume 8

Benjamin Lewis Robinson

# Bureaucratic Fanatics

—

Modern Literature and the Passions of Rationalization

**DE GRUYTER**

ISBN 978-3-11-073669-4
e-ISBN (PDF) 978-3-11-060693-5
e-ISBN (EPUB) 978-3-11-060604-1
ISSN 2195-2205

**Library of Congress Control Number 2019933420**

**Bibliographic information published by the Deutsche Nationalbibliothek**
The Deutsche Nationalbibliothek lists this publication in the Deutsche Nationalbibliografie;
detailed bibliographic data are available on the Internet at http://dnb.dnb.de.

© 2020 Walter de Gruyter GmbH, Berlin/Boston
This volume is text- and page-identical with the hardback published in 2019.
Cover image: Francisco de Goya: El sueño de la razón produce monstruos
Printing and binding: CPI books GmbH, Leck

www.degruyter.com

For FEL

# Contents

**Acknowledgments** —— XI

**Introduction** —— 1
    *The Dream of Reason produces Monsters* —— 1
    Bureaucracy and *Schwärmerei* —— 1
    Fanatical Bureaucrats – Peculiar Fanatics —— 3
    Literature and the Political —— 7

**1 Bureaucratic Fanaticism** —— 9
    Exceptions (Agamben, Schmitt, Kierkegaard) —— 9
    Bureaucracy as a Vocation (Kant, Weber) —— 12
    Abstraction (Hegel, Marx) —— 19
    The Confusion of the Present Age (Kierkegaard I) —— 26
    1848 – More Confusion (Kierkegaard II) —— 29
    Another Kind of Despair (Kierkegaard III) —— 32

**2 A New Kind of *Schwärmer*: Kleist's *Michael Kohlhaas*** —— 35
    Wolves and Mad Dogs —— 35
    *Schwärmerei* —— 41
    Bear Nature and Bureaucracy —— 47
    Humanity? —— 50
    Nags – and other negligible things —— 56
    Bureaucrat against Bureaucracy —— 61
    Doubt, Forgiveness, and the Givenness of *das Vergebliche* —— 63
    The Gypsy Soothsayer —— 67

**3 Architecture of the Office: Melville's *Bartleby*** —— 75
    Of Offices —— 75
    In the Office —— 79
    Insecurities —— 83
    Easiness —— 86
    What is to be done? —— 90
    "Ah Bartleby! Ah humanity!" —— 95
    Vagrancy —— 98

## 4 Abstraction of the Earth: Conrad's *Heart of Darkness* —— 102
Disfigurements —— 102
Ivory, Ideas, Rivets —— 108
The Outraged Law —— 115
Voice, Voices —— 124

## 5 Poverty of Agency: Conrad's *The Secret Agent* —— 130
Temperature —— 130
A Period of Reaction —— 133
Dulling Agents —— 136
Outcast London —— 140
Institutional Agencies and Dependencies —— 142
"Poor Stevie" —— 147
Actions and Accidents —— 150
Winnie, "the free woman" —— 152
Blank Wall —— 154

## 6 State of Embarrassment: Kafka's *In the Penal Colony* —— 157
The Postcolonial Moment —— 157
Pain after Punishment —— 159
Beyond Doubt – The Officer —— 164
Beyond Doubt – The Research Traveler —— 171
Postcolony —— 174
The Harbor —— 182

## 7 Society of Security: Kafka's *The Metamorphosis* —— 185
Security and its Discontents —— 185
Authenticity —— 190
Who cares? / *Wen juckts?* —— 193
Indiscreet Disclosures —— 198
Carefree Compartments —— 201
Bites of Conscience: Guilt, Debt, Hunger —— 205
Family Ties —— 207
Dying from Taste —— 210

**Epilogue: Guerrilla / Gardener Coetzee's *Life & Times of Michael K* —— 218**
        Of Monstrous Pictures of Michaels —— 219
        Translating Michael Kohlhaas —— 223
        The Savor of the Earth —— **228**
        Sparrows —— **234**
        Justice —— **240**

**Bibliography —— 245**

**Index —— 262**

# Acknowledgments

This project began as I waited with many others in the sunbaked courtyard of the temporary offices – long since de facto permanent – of the Department of the Registrar General in Harare. We were waiting to be called. I had temporarily lost my citizenship on a technicality, and was reading *Moby Dick* to try to kill the time and assuage my frayed nerves. What were we all doing there? How had it come to this? One elderly woman was told to return to the town of her birth on the other side of the country to have her birth certificate changed as it gave the wrong gender. Each infuriating encounter with the bureaucracy gave way to a disciplined kind of apathy. No less remarkable were the bureaucrats themselves, who demonstrated a painstaking patience in explaining and executing policies that were in not a few cases – at least to those of us languishing in the courtyard – patently absurd. I was inspired, if that is the right word, to look into the history of the experience of bureaucracy. Whatever else I may have been able to discover in my investigations of the literature of bureaucratization, I have to acknowledge that this book fails to offer the one thing that may have made it worthwhile, namely, a therapy for the passions of bureaucratization. Much as I would like to, I cannot recommend taking *Bureaucratic Fanatics* to the passport office – best stick with *Moby Dick*.

At a time when university life has become so bureaucratic, I count myself lucky to have begun this project at Northwestern in the invigorating intellectual company afforded by my dissertation committee: Peter Fenves, Samuel Weber, Jörg Kreienbrock, and Susannah Gottlieb. For their always thoughtful and thought-provoking guidance, my thanks. The friendship and collegiality I enjoyed at Northwestern was complemented by the generosity of students and faculty at the various institutions with which I was associated in Europe. Eva Geulen at the Goethe University, Frankfurt sponsored my stay in Germany 2014 – 2015 on a DAAD scholarship and I am very grateful for her continuing support of my work. Thanks to Sigrid Weigel and Daniel Weidner, I was able to write in spring 2015 among the welcoming scholarly community at the Center for Literary and Cultural Research (ZfL) in Berlin. In my year as a postdoc in Berlin, the fellows and staff of the Institute for Cultural Inquiry (ICI) provided an incomparable environment for the rethinking and revision of the dissertation into a book. My thanks go to Stanley Corngold, Christoph Menke, Yv Nay, Nathan Taylor, and Arnd Wedemeyer, for reading and remarking on various chapters, and to Paul North, Chad Wellmon, and Caroline Lillian Schopp for their generous comments on the whole manuscript.

Research for the epilogue on Coetzee was conducted in March 2016 at the Harry Ransom Center at the University of Texas at Austin with a Northwestern Graduate Research Grant as well as research funding from the Kaplan Institute for the Humanities. Versions of chapters of this book have appeared elsewhere and I am grateful to the publishers for allowing their publication here: "Passions for Justice: Kleist's Michael Kohlhaas and Coetzee's Michael K," *Comparative Literature* (2018); "State of Embarrassment: Kafka's *In der Strafkolonie*," *The Germanic Review* (2015); and "The Poverty of Agency in Conrad's *The Secret Agent*," *Conradiana* (2014). References to Coetzee's unpublished papers are cited with permission of the Harry Ransom Center at the University of Texas at Austin.

Paul Fleming and Rüdiger Campe showed faith in the project and I am particularly pleased that *Bureaucratic Fanatics* is part of the *Paradigms* series. I am grateful to Manuela Gerlof and Stella Diedrich at De Gruyter for their attention and support. Eva Horn, who advised me on some of the earliest drafts, has been a continuous source of intellectual stimulation and encouragement. I would like to thank my colleagues in the Department of German at the University of Vienna for furnishing me with "office-room" – and time – to complete this book. Finally, to CLS and FEL without whom this undertaking would doubtless have ended in despair – thank you!

# Introduction

## *The Dream of Reason produces Monsters*

Francisco Goya's *Los Caprichos* (1799), a series of eighty etchings, present an archive of all-too-human caprices and depravities from the mundane to the metaphysical. In the middle of the series, plate 43 depicts a singular figure, usually taken to be the artist himself. Dressed in formal attire, he sits slumped over the papers and writing utensils on his desk, while a multitude of night creatures swarms around him. The title is usually translated as "sleep of reason produces monsters" in order to insist on the univocality of Goya's Enlightenment message. But the ambiguity of the Spanish word *sueño*, no less than the ambiguity of Goya's images, brings into focus a dialectic of Enlightenment that was already legible on the brink of the nineteenth century – and that cannot be missed by a viewer of Goya's work today. Indeed, I wonder, could this dreaming figure present a type so central to the aspirations and the caprices of the century to follow, a type otherwise missing from Goya's *Caprichos*, namely, the bureaucrat?

Between the asses in the plates immediately preceding and the witches or fates in those following, the bureaucrat slumbers at his office desk. Does his "sleep" – the sleep of reason – solicit the terrifying swarm in which the earthly and the animal are confused with the mythical and the magical? Or is it rather his "dream" – the dream of reason – that produces, as an unconscious or unconscionable accompaniment, a capricious swarming that facilitates the atrocities, injustices, and indignities to which the rest of the series is testament? If the dream of reason is the edifying idealistic response to the corruption and injustices of real life, Goya presents this dream itself as monstrous. For the worldly corruption that bureaucracy is supposed to set in order may also prove to be the effect of the bureaucratic fanaticism to which the dream of reason gives rise.

## Bureaucracy and *Schwärmerei*

At a conference organized by the Association for Social Politics in Vienna in 1909, Max Weber, in an uncharacteristic expression of affect, declared: "This passion for bureaucratization [...] is enough to drive one to despair."[1] To be

---

[1] Max Weber, "Diskussionsbeitrag auf der Generalversammlung des Vereins für Sozialpolitik, 1909: Über die wirtschaftlichen Unternehmungen der Gemeinden," in *Wirtschaft, Staat und So-*

sure, he was in the first place referring to the enthusiastic endorsement of bureaucratization among the social scientists he encountered at the Vienna meeting. He was also, however, giving vent to an everyday experience of bureaucratization that seemed to fly in the face of everything that, technically and scientifically speaking, "bureaucracy" was meant to mean. Far from being "dispassionate, upright and mild," as Hegel had written in 1820, bureaucracy turned out to be motivated by powerful and escalatory passions – and to be generative of them as well.[2] Doubtless the passions of bureaucratization contributed to its unprecedented global expansion but also to the excesses of officiousness and violence that have today brought the word "bureaucracy" into disrepute – without by any means halting the preponderance of bureaucratization.

If bureaucratization inspired new passions, these proved to be disposed to extremist escalation evocative of the history of religious enthusiasm. This study uncovers a genealogy that relates two supposedly distinct and opposing currents in the history of modern politics: the history of rationalization and the history of the distinctly modern form of religious fervor captured in the German word, *Schwärmerei*. Usually translated into English with the older Latinate term "fanaticism," *Schwärmerei* was coined by Luther to refer polemically to figures who claimed a specious kind of transcendence that he related, drawing on the literal meaning of the word, to the *swarming* of certain insects and animals and the dangerous mass violence with which they were associated. *Schwärmerei* thus names both a political-theological and a political-zoological phenomenon – or rather, it expresses the confusion of the two. In the late eighteenth century, in the wake of the French Revolution and Kant's revolution in philosophy, the question of *Schwärmerei* – so long synonymous with irrationality – would emerge with renewed urgency as an expression of the crisis of Enlightenment reason. For reason showed itself to be predisposed to transgress its own limits and perpetrate inhuman excesses of rationality. And if the term *Schwärmerei* lost its polemical force over the course of the nineteenth century, this is because the kind of swarming that it evoked became the subject of the biopolitical discourses and governmental practices of bureaucratization.

The figures I call "bureaucratic fanatics" exhibit a paradoxical relation to an instance of transcendence that bureaucracy, in principle, denies or shuts out. As such, these figures are exponents of what Kierkegaard, writing in and around 1848, would refer to as the religious and political "confusion" of the present

---

*zialpolitik: Schriften und Reden 1900–1912, Gesamtausgabe*, ed. Wolfgang Schluchter (Tübingen: J. C. B. Mohr, 1998), I/8:128. Translation mine.
[2] Georg Wilhelm Friedrich Hegel, *Elements of the Philosophy of Right* (1820), ed. Allen Wood, trans. H. B. Nisbet (Cambridge: Cambridge University Press, 1991), § 296; translation modified.

age.³ A confusion is more than a crisis: for in confusion, the critical discourse is entangled in the crisis it seeks to diagnose. If *Schwärmer* refers to someone who makes a problematic claim to authoritative or revelatory communication, then the confusion of the present age is that, under conditions of bureaucratization, every claim to "authority" or "revelation" has become problematic. So as not to fall victim to the *Schwärmerei* one seeks to identify, it is necessary to resort to a mode of expression that is more circumspect about its authoritative and revelatory character. Such "indirect communication," to use Kierkegaard's phrase, is generated by techniques that are usually considered literary. For literature does not make descriptive or normative assertions about reality but rather reflects on what it means to suspend such claims. The historical and theoretical contextualization of bureaucratic fanaticism from Kant to Weber, to which Chapter One is devoted, leads, therefore, to a reflection on the problem of form. For even the likes of Kant and Weber, who are hardly considered great stylists, seem compelled to turn to ironic, if not actually literary, modes when addressing the problem of fanatical rationality.

Bureaucratic fanaticism proves to be a distinctly literary problem, while remaining, to be sure, a political one. As such it presents peculiar challenges to political theory, which is compelled to address the sort of epistemological confusions that are usually confined to the domain of literary theory. It may be for this reason – and not only due to the complicity between these disciplines and the bureaucracies they studied – that the fanaticism to which bureaucracy was disposed was only acknowledged late, if at all, in political and social theory. In contrast, a distinctive kind of literature emerged early, fascinated by the exaltation and agitation bureaucracy provokes. In the work of Kleist, Melville, Conrad, and Kafka, the literary engagement with "bureaucratic fanaticism" produced probing explorations of the bureaucratic transformation of political life and of the global political landscape in the long nineteenth century. It is not by chance that these writings, which afford exemplary presentations of the "confusion" of the age, are also some of the most emblematic works of modern literature.

## Fanatical Bureaucrats – Peculiar Fanatics

Bureaucratization emerges in the course of this book as the "humanitarian" project to secure the earth for humanity. Each bureaucratic fanatic presents occasion

---

3 Søren Kierkegaard, *The Book on Adler*, ed. and trans. Edna Hong and Howard Hong (Princeton: Princeton University Press, 1998).

for an interrogation of the humanitarian determination of the human and the bureaucratic measure of the earth. These exceptional literary figures are of consequence not as political examples or exemplars, but as markers of political and theoretical problems that might not otherwise appear as such. Testing even the limits of literary representation, the literature of bureaucratic fanaticism excavates forms of life and aspects of the earth that, on account of bureaucratization, may have been overlooked or disfigured.

It is in Kleist's *Michael Kohlhaas* (1808/10) that the new kind of *Schwärmerei* – bureaucratic fanaticism – makes its literary appearance. Kohlhaas's fanaticism is a function of his injured *Rechtgefühl*, an ambiguous term that may be translated as righteous feeling or passion for justice but also, crucially, as passion for the law (*Recht*). In response to a corrupt bureaucrat's appropriation and abuse of a pair of his horses and the intransigence of the legal system from which he seeks remedy, Kohlhaas leads a partisan revolt against the entire bureaucratic order. Throughout, however, he presents himself not as an outlaw or freedom fighter but as a superior kind of bureaucrat – and ultimately as the representative of the archangel Michael. He thus emerges, I argue in Chapter Two, as the paradoxical, but perhaps nonetheless paradigmatic, figure of a fanatical bureaucrat who wants to do away with bureaucracy once and for all.

Kleist's bureaucracy novella is also a partisan one. This is no coincidence: the colonization and neutralization of political life under bureaucratization solicits partisan warfare as politics by other means. As a partisan, Kohlhaas feels compelled to engage the bureaucratic dispensation on its own legalistic terms, the inadequacy of which only fuel his righteous frustration. Only in the final part of the novella, once he has handed himself over in despair to the authorities, does he begin to consider the possibility that the apparently futile cause of justice may require a peculiar mode of engagement that would have nothing to do with bureaucracy.

Subsequent chapters articulate the ambivalence legible in the figure of Kohlhaas and his *Rechtgefühl* in two opposing modalities of bureaucratic fanaticism: the "fanatical bureaucrat," on the one hand, and the "peculiar fanatic," on the other. One thread of chapters traces the escalation from the exemplary figure of the lawyer in Melville's *Bartleby* (1853), to such "fanatical bureaucrats" as Kurtz in Conrad's *Heart of Darkness* (1899/1902) and the officer and the research traveler in Kafka's *In the Penal Colony* (1914–1919), who find themselves propelled to the ends of the earth by the intoxicating "rationality" of the colonizing structures they represent. These figures are complemented at the other extreme by a series of "peculiar fanatics" like Bartleby, Stevie in Conrad's *The Secret Agent* (1907), and Gregor in Kafka's *The Metamorphosis* (1915). Theirs is a bureaucratic fanaticism that passionately reacts against the encroachments of bureaucratization. In another time and place, they might have become exalted fanatics, but under conditions of bureaucratization they are shut off from the prospect of enthusiastic transcendence. Their aversion to the incursions of bureaucracy gives rise to passionate, if apparently passive, behavior that is "peculiar" insofar as it is not, and cannot be, subsumed by bureaucratic measures.

Already in Kleist the figure of the fanatical bureaucrat evokes Fichte's radicalization – or bureaucratization – of the Kantian principle of the "primacy of the practical." In this scheme, it is one's duty, one's office, to "rationalize" the world according to the imperative to act, and in accordance with one's capacity to act. The imperative of thoroughgoing activity over and against the "natural" world emerges as the principle of bureaucratization – with drastic and sometimes devastating ecological as well as human consequences. Under conditions of bureaucratization, the primacy of *what ought to be* over *what is* becomes the privileging of the official world over the world as it is given. The perplexities that arise from this bureaucratic doubling of the world are succinctly captured in the ambiguity of the word "oversight." Much of the fanatical passion provoked by bureaucratization emerges out of the tension between what is administered by bureaucratic oversight and what is over-seen by it.

Bureaucratization, understood as the transformation of the world according to the imperatives of the office, was accomplished with the help of a particular architectural innovation that separated the officiator from the officiated, the officer from humanity. If "humanity" names the ultimate office-duty and presents therefore the ultimate legitimacy of all bureaucratization, it is also, for the sake of expediency, what must be excluded from the office and from the person of the bureaucrat. This dispositive of the office is laid out in *Bartleby* to which Chapter Three is devoted. The lawyer's snug office is thrown into confusion when an impassive figure appears who, insofar as he neither fulfills nor demands any offices, undermines the ostensibly humanist principle of the office. This confusion is captured in the lawyer's despairing final line: "Ah Bartleby! Ah humanity!"

Characteristic of the experience of the age of bureaucratization is the constraining sense that there is nowhere else to go. In this regard *Michael Kohlhaas* presents a threshold in the history of the bureaucratization of space. Henceforth, those who oppose the increasingly "global" systems of governance in which they find themselves, and of which there is no outside, will have to choose, like Kohlhaas, between partisan warfare and the apparently passive resistance of peculiar fanaticism. Around 1850 in New York, Bartleby can only find a space free of the demands and imperatives of official life in the "office room" he is afforded, for a time, by the lawyer.

The implication of the imperatives of the office with the official articulation of space is at the center of Conrad's writings around 1900 in the form of an interrogation of agency and geography. The culmination of the European epistemological-political project to dominate the earth, which Conrad called "geography triumphant,"[4] is emblematized in the colonization of the "Congo" on the one hand, and in the establishment of Greenwich as the prime meridian on the other. Far from triumph, however, the turn of the century was pervaded by a sense of displacement or disorientation. Conrad himself would refer to the moment presented in *Heart of Darkness* as a "disfigurement" of the human conscience and of geography. This disfigurement is explored in Chapter Four with attention to the first-class agent, Kurtz, whose charismatic voice conveys an imperial attitude of abstraction that operates without regard to the reality it is supposed to articulate. When the earth fails to map onto the topology expressed in his voice, measureless violence ensues.

While *Heart of Darkness* presents the contours of agency at the outer limits of the earth and of experience, *The Secret Agent* addresses the crisis of agency in the midst of the modern metropolis. The "outrage" in which the "peculiar Stevie" is involved solicits an interrogation in Chapter Five of the liberal-bureaucratic paradigm of the agent as a responsible, self-conscious, and conscientious actor. What does it mean to be an "agent" in the context of the emerging and proliferating impersonal agencies – institutional, social, and biopolitical – that define modern urban life? The literary treatment of bureaucratic fanaticism thus brings into focus the question of what it means to be an agent or a patient under bureaucratization – and how it feels. The pervasive anxiety about action and passion exhibited and occasioned in all the texts I consider betrays a fundamental mood that, drawing on Kierkegaard, I call despair.

The tension between the necropolitical excesses of European colonialism and the increasingly invasive biopolitical and governmental practices of European metropoles and metropolises culminates in the writing of Kafka. *In the Penal Colony*, which was composed on the eve of, and finally published following, the First World War, presents, in the figures of the officer and the research traveler, an unfolding crisis of bureaucratic authority. The end of the penal colony discloses, I argue in Chapter Six, a postcolonial site of excruciating political and epistemological uncertainty – or embarrassment. Meanwhile, Gregor Samsa, presents, as I suggest in Chapter Seven, a radical rejection of the emergent society of security. In his bourgeois family home, he literally turns into a *Schwärmer*, albeit an abject, fallen, and flightless one. While the security imperative governing everyday life reduces the sphere of recognizable concerns to those that define a "normal" human life, Gregor's transformation, which I

---

4 Joseph Conrad, "Geography and some Explorers," *National Geographic*, March 1924, 250.

argue takes the form of a monstrous indiscretion, facilitates a kind of underground transcendence of security. At stake is not a sphere of more authentic, edifying, or substantial cares but, on the contrary, an opening in which one might cultivate a more discriminating, if altogether improper, taste for peculiarity.

The Epilogue takes up J. M. Coetzee's *Life & Times of Michael K* (1983), which presents a literary reflection on the political and literary history of bureaucratic fanaticism addressed in this study. The book was originally conceived, as is revealed in Coetzee's notes and manuscripts, as an "interpretative translation" of Kleist's *Michael Kohlhaas*, meant to evoke the "urgency and passion" of his partisan uprising in the context of apartheid South Africa. Coetzee ended up, however, presenting in Michael K a figure who belongs instead in the literary genealogy of peculiar fanatics like Bartleby, Stevie, and Gregor. Rather than joining the guerillas, Michael K's "passion for justice" goes into tending an obscure patch of earth in the desolate landscape of the Karoo. Between the guerilla and the gardener Coetzee excavates an antinomy of modern political life under conditions of bureaucratization that, I argue, is already legible in *Michael Kohlhaas*. Michael K does not care to engage the colonial system, neither as a subject nor as an antagonist. He furthermore rejects, or attempts to reject, the solicitous attentions of literature and frustrates even the subtle, and ostensibly liberal, norms of literary analysis. In so doing, Coetzee's novella stages in the figure of Michael K a critical reflection on the complicities of the institution of literature in the political history of the modern institutionalization of life. Indeed, all the bureaucracy literature traced in this book is profoundly suspicious of, and tends to ridicule, the pretention of literature to coherently articulate a world of significance. This is, after all, a bureaucratic aspiration and, indeed, a fanatical one.

## Literature and the Political

It has become the norm to treat bureaucracy with contempt – in everyday life and in public discourse. Nonetheless, it remains all but inconceivable to address the most pressing political problems facing societies around the globe without a bureaucratic program. This familiar conundrum of everyday life has given rise to an impasse in contemporary political theory. Either "activists" have to embrace some form of bureaucratization as an inherent aspect of programmatic political action, or else reject it altogether by adopting a so-called "politics of refusal." *Bureaucratic Fanatics* excavates the origin of this tension – a tension inherent to bureaucratization and the "passion for justice" it inspires – by turning to the literature of the long nineteenth century in which its contours are already to be found. This relation is indicated in the current fascination and frustration in political theory with Bartleby, who is treated as if he were a contemporary rather than a nineteenth-century literary figure.[5] By tracing the interwoven genealogy, as it is rendered legible in literature, of the politics of refusal and of fanatical bureaucratization to their common political-theological roots in the problem of *Schwärmerei*, this book provides a historical and theoretical perspective with which to assess contemporary political thought and practice – whether it favors increased governmental intervention for the protection and institutionalization of human (and not only human) rights, or seeks radical alternatives to the modern administration of life.

---

[5] See esp. Giorgio Agamben, "Bartleby, or On Contingency" (1993), in *Potentialities: Collected Essays in Philosophy*, ed. and trans. Daniel Heller-Roazen (Stanford: Stanford University Press, 1999); Gilles Deleuze, "Bartleby; or, The Formula" (1989), in *Essays Critical and Clinical*, trans. Daniel Smith and Michael Greco (Minneapolis: Minnesota University Press, 1997); also Lee Edelman, "Occupy Wall Street: 'Bartleby' Against the Humanities," *History of the Present* 3.1 (Spring 2013): 99–118. In a critical but sympathetic assessment, Bonnie Honig refers to the "Bartleby-love" in the work of Agamben, Deleuze and others in her "Charged: Debt, Power, and the Politics of the Flesh in Shakespeare's *Merchant*, Melville's *Moby-Dick*, and Eric Santner's *The Weight of All Flesh*," in *The Weight of All Flesh: On the Subject-Matter of Political Theology*, ed. Kevis Goodman (Oxford: Oxford University Press, 2016). It would be equally illuminating to consider the indignation with which "Bartleby-haters" respond.

# 1 Bureaucratic Fanaticism

The critical analysis of bureaucratic fanaticism runs the risk of becoming entangled in the very "confusion" it seeks to diagnose. For if bureaucratic fanaticism is, in bureaucratic terms at least, unaccountable, it is also, as Kierkegaard had already argued around 1848, characteristic of a bureaucratic age that other-than-bureaucratic communications, namely exceptional ones, are no longer credited or creditable. Indeed, they will just be dismissed as "fanatical." The attempt to make this confusion explicit involves an acknowledgement of, and engagement with, form. To provide a theoretical and historical analysis of bureaucratic fanaticism, other less credible and more insignificant forms of communication, of the sort usually consigned to literature, have to be taken into account.

This book is devoted to literary presentations of exceptional figures who emerge at the extremes of the nineteenth-century bureaucratic project. Despairing of, or over, bureaucracy, these bureaucratic fanatics draw attention to the limits, the internal contradictions, the points of breakdown, excess and dislocation, in short to those moments at which bureaucratization runs into perplexities it can neither resolve nor circumvent. This literature of bureaucratic fanaticism proves definitive for a peculiar tradition of modern literature, which may itself betray fanatical-bureaucratic tendencies, for it takes upon itself to account for what otherwise cannot be accounted for or, failing that, for the limits of accountability.

## Exceptions (Agamben, Schmitt, Kierkegaard)

On account of their exceptionality, I take bureaucratic fanatics to be revelatory of more general tendencies in the bureaucratic transformation of modern life. This approach draws on a significant strand of contemporary political theory that departs from the exemplary character of the exception. Taking up Carl Schmitt's theory of sovereignty: "Sovereign is he who decides on the state of exception,"[1] Giorgio Agamben has influentially sought to explain the history of Western politics with reference to the exceptional figure of *homo sacer*.[2] In developing – and

---

[1] Carl Schmitt, *Political Theology*, trans. George Schwab (Chicago: University of Chicago Press, 2005), 5.
[2] Giorgio Agamben, *Homo Sacer: Sovereign Power and Bare Life*, trans. Daniel Heller-Roazen (Stanford: Stanford University Press, 1998).

legitimizing – the significance of the exception for political theory, Agamben cites Schmitt citing another author:

> The exception is more interesting than the regular case. The latter proves nothing; the exception proves everything. The exception does not only confirm the rule; the rule as such lives off the exception alone. A Protestant theologian who demonstrated the vital intensity of which theological reflection was still capable in the nineteenth century said: The exception explains the general and itself. And when one really wants to study the general, one need only look around for a real exception. It brings everything to light more clearly than the general itself. After a while, one becomes disgusted with the endless talk about the general – there are exceptions [*es gibt Ausnahmen*]. If they cannot be explained, then neither can the general be explained. Usually the difficulty is not noticed, since the general is thought about not with passion but only with comfortable superficiality. The exception, on the other hand, thinks the general with intense passion [*mit energischer Leidenschaft*].³

Agamben parenthetically names the "protestant theologian" to whom Schmitt refers: he is "none other than Søren Kierkegaard."⁴ It is significant, however, that the citation is not written in Kierkegaard's own voice, but in that of a pseudonym, Constantin Constantius, and that the book in question, *Repetition* (1846), is explicitly *not* a theological work but an "esthetic" one.⁵ Indeed, it is arguably the most literary and, given that its literary character is persistently underlined in the text, certainly the most metaliterary of Kierkegaard's writings. In the "Concluding Letter," for example, from which the passage on the exception is taken, the pseudonymous author, addressing himself to "my dear reader," goes so far as to admit that not only had the young man, whose case and letters to Constantius take up much of the book, been a literary invention but that he even considers his reader to be a fiction (*en poetisk Person*).⁶

It is not clear what it means to cite *Repetition* in the way that Agamben and Schmitt do – especially given that the possibility of repetition is itself the ostensible subject of the book. In taking a passage that, treated on the level of statement as a "direct communication," seems to expound a theory of the exception,

---

**3** Agamben, *Homo Sacer*, 17; quoting Schmitt, *Politische Theologie* (Berlin: Duncker & Humblot, 1922), 21; citing Søren Kierkegaard, *Wiederholung*, in *Gesammelte Werke*, vol. 3, trans. H. C. Ketels and Hermann Gottsched (Jena: Eugen Diederichs Verlag, 1909), 200.
**4** Agamben, *Homo Sacer*, 18.
**5** Søren Kierkegaard, *Repetition: A Venture in Experimenting Psychology by Constantin Constantius* (1843), in *Fear and Trembling; Repetition*, ed. and trans. Edna Hong and Howard Hong (Princeton: Princeton University Press, 1983), 227. For the Danish: Søren Kierkegaard, *Gjentagelsen: Et Forsøg i den Experimenterende Psychologi af Constantin Constantius* in *Søren Kierkegaards Skrifter*, ed. Niels Jørgen Cappelørn et al. (Copenhagen: Forskningscentret and Gads Forlag, 1997), 4:6–96.
**6** Kierkegaard, *Repetition*, 225.

and in lifting it out of a context that foregrounds its indirectness, is it possible that something essential about exceptions, or at least what is said about them in *Repetition*, is irretrievably lost? Here an impasse between political theory and literary criticism is reached that is not without consequence for the study of politics proposing to depart from the thought of the exception.

The impasse can be indicated in the dialectic concentrated in the ambivalent status of the line: "There are exceptions." Schmitt and Agamben take up the phrase as if it referred immediately to the concrete political situation that their own theories are supposed to describe. In other words, they do not find it necessary, in this case at least, to problematize the relation between discourse and reality and so, on the assumption of the transparent disclosure of reality it affords, are free to address themselves simply to the concept of the exception or the exception as a concept. The elaboration of the political departing from the exception thus becomes the elaboration of a political *logic* of the exception. It becomes indeed, as the "logic of sovereignty" that Agamben develops following Schmitt shows, a *generalized* logic of the exception, of the exception *in general*. It thus obscures, rather than bringing into focus, the singularity of the exception – if such a thing exists.

The political theorist thus falls into the position of an "ordinary reviewer" (*en almindelig Recensent*) who, Constantius concedes, cannot be expected "to be interested in the dialectical battle in which the exception arises in the midst of the universal."[7] And this remark is all the more significant since Constantius, the self-proclaimed and hyper-reflexive observer, is a theorist par excellence. By reading the book in a manner that seeks to reduce it to a set of generalizable assertions, the dialectical battle staged in *Repetition* is overlooked. In a manner characteristic of Kierkegaard's indirect treatment of "existential" concepts, the passage on the problematic exposition of the exception in fact rehearses the dialectical tension between the problem and its exposition with which the pseudonymous author claims to have been concerned all along.

Read with attention to the "literary" markers that characterize its context, the phrase "there are exceptions" does not describe an ontological situation – no more than any other line of *Repetition*. One cannot ascertain from the text the existence of exceptions (any more than of repetitions) nor indeed arrive at the indisputable conditions in which it would even be possible to say, "There are exceptions." On the basis of what is said in *Repetition*, there can be no (political) science of exceptions, if such a science depends on the assumption of their existence, but only a literary-critical "venture in experimenting," to cite

---

7 Kierkegaard, *Repetition*, 226.

the subtitle of the book, that acknowledges and operates with the ontological uncertainty of the category it examines. Such a literary approach is able to elaborate the wrestling, or the rupture, that the thought of the exception involves: "The whole thing is a wrestling match in which the universal breaks with the exception, wrestles with it in conflict, and strengthens it through this wrestling."[8]

The relevance of literature for political thought is not limited to providing compelling examples in the form of literary types or prescient scenarios. It concerns the exception – the exception to political theory. And if literature concerns what is excepted, or excepts itself, from the domain of political theory, it is not by that means rendered apolitical nor absolved of complicity in the political context out of which it emerges. Rather it becomes, on account of its exception to theory, all the more relevant and indeed revelatory for thinking through politics and interrogating concepts of the political. Kierkegaard's Constantius compares the "very dialectical and infinitely nuanced" attempt to think the exception to a curious, in fact impossible, biopolitical operation: "it is just as difficult as to kill a man and let him live."[9] Literature is the site in which such a difficult undertaking, one that is not without political consequence, can plausibly be ventured.

This study in literary criticism engages political theory not in order to contribute to or criticize a given theory, but with an eye to the ruptures that emerge when theory is put to the test of the exception in literature. In this way certain otherwise unseen, or overlooked, limits and aporia of what is called the political, and what is taken to constitute political practice, come to light. For bureaucratic fanatics are distinguished from the "exceptional" figures in political theory insofar as they are not exhaustively defined by the norms that exclude them. They exhibit a singular mode of existence of their own, albeit only a "literary" one – or only in literature.

## Bureaucracy as a Vocation (Kant, Weber)

Towards the end of the eighteenth century, and with increasing intensity in the wake of the French Revolution, the question of fanaticism, of that distinctly modern type of fanaticism that in German is called *Schwärmerei*, was at the center of heated debates in the German public sphere. The term, coined by Luther, had, since the Reformation, referred to suspicious and specious claims to transcendence. Throughout the Enlightenment it was used polemically to refer to its in-

---

[8] Kierkegaard, *Repetition*, 226–227; translation modified.
[9] Kierkegaard, *Repetition*, 226.

tellectual and political enemies, disparaged for their excessive and irrational behavior. By the end of the eighteenth century, however, it was no longer possible to ignore reason's own propensity for a certain kind of fanaticism. The heated claims raised by figures opposed to the Enlightenment about the inhumanity of reason on account of its cold, calculated, heartlessness, were, paradoxically, supplemented by the preponderance of an exalted and infectious passion for rationality that was disposed to indulge in inhuman acts and ultimately presented a threat to humanity as such.

Kant's critical project can be broadly construed as responding to this atmosphere of disappointment and disaffection with Enlightenment reason. For it is concerned with reconciling the finitude of human existence with the dangers of a capacity for reason that at once defines and transcends it. Although Kant does not systematically take up the question of *Schwärmerei*, the term in his critical work refers persistently to a pathological affection for rationality, a desire to fulfill or realize in the world – in what would be an inhuman or superhuman project – a dream of reason. In the *Critique of the Power of Judgment* (1790), Kant defines *Schwärmerei* as "a delusion of being able to *see* something beyond all bounds of sensibility i.e., to dream in accordance with principles (to rave with reason)" (*ein Wahn [...], über alle Grenze der Sinnlichkeit hinaus etwas sehen, d.i. nach Grundsätzen träumen (mit Vernunft rasen) zu wollen*).[10]

It is not by chance, therefore, that the earliest published traces of the critical revolution in Kant's thought are to be found in an attempt to engage the problem of *Schwärmerei* in a curious book entitled *Dreams of a Spirit Seer, Elucidated by Dreams of Metaphysics* (1766).[11] The method proposed in the title, to elucidate dreams of the visionary variety by means of the dreams of metaphysics, in fact serves to amplify the perilous proximity that Kant is interested in negotiating between the rationalist and the visionary accounts of the world. The ambivalence of this approach would be remarked upon by Mendelssohn in his brief review of the book: "the jocular profundity with which this little work is written leaves the reader at times in doubt as to whether Herr Kant wants to make meta-

---

[10] Immanuel Kant, *Critique of the Power of Judgment*, ed. Paul Guyer, trans. Paul Guyer and Eric Matthews (Cambridge: Cambridge University Press, 2001), 156. German references to Kant from Immanuel Kant, *Gesammelte Schriften*, ed. Preussische [later Deutsche] Akademie der Wissenschaften (Berlin: Reimer, later de Gruyter, 1900–). References given as AA for *Akademie-Ausgabe* with volume and page number. AA, 5:275.

[11] Immanuel Kant, *Träume eines Geistersehers: erläutert durch Träume der Metaphysik*, AA, 2:315–373. Translations mine. For an English translation, see *Kant on Swedenborg: Dreams of a Spirit-Seer and other Writings*, ed. Gregory R Johnson, trans. Gregory R. Johnson and Glenn Alexander Magee (West Chester: Swedenborg Foundation Publishers, 2007).

physics laughable or spirit-seeing credible."[12] This ironic procedure enables the writer to avoid committing to either dogmatism or skepticism in a manner that anticipates the alternative approach that would find a more austere articulation in the *Critique of Pure Reason*.[13] It is as if the philosophical attempt to treat the question of *Schwärmerei* head-on necessitated, in order to avoid the seduction of substituting his own metaphysical fancies for those he condemns, recourse to a method that strenuously, in a sometimes forced show of wit, emphasizes that he does not earnestly mean what he says.[14]

If Kant's text is bemusing when considered as a philosophical treatise, it is because it displays a critical method that is concerned more with its performance than with the presentation of knowledge. The argument which, as already emphasized in the preface, "promises very little for the discussion," is construed not with regard to its truth content (by his own confession, "he found nothing"), but to its *use*.[15] If "properly used" (*gehörig genutzt*), the study promises nothing less than to exhaust all "philosophical insight into metaphysics."[16] With an eye to its use, metaphysics emerges, on the one hand, as "the science of the limits of human reason," and, on the other, as the expression of "hope for the future" that motivates action beyond the world as it is determined and disclosed to human knowledge.[17]

The occasion for the text and the particular visionary in question, was Emanuel Swedenborg, whom Kant introduces as a man who lives in Stockholm, "without office or employment," whose entire occupation consists in consorting with and reporting on spirits and departed souls.[18] Kant, whose text insists on the category of usefulness, is quick to insist on Swedenborg's unemployment, which leaves him the time for such idle, although possibly also lucrative, rav-

---

12 Moses Mendelssohn, "Rezension der Träume," in *Allgemeine deutsche Bibliothek* 4:2 (Berlin: Friedrich Nicolai, 1767), 281; reprinted in Immanuel Kant, *Träume eines Geistersehers: erläutert durch Träume der Metaphysik*, ed. Rudolf Malter (Stuttgart: Reclam, 1976), 118. Translation mine.
13 On the significance of *Träume* for Kant's critical project, see Susan Meld Shell, *The Embodiment of Reason: Kant on Spirit, Generation, and Community* (Chicago: University of Chicago Press, 1996), chap. 5; also Monique David-Ménard, *La folie dans la raison pure: Kant lecteur de Swedenborg* (Paris: Vrin, 1990).
14 There is a long history of using humor for treating, or as an antidote to, *Schwärmerei*, notably, the conclusion of the Earl of Shaftesbury's "Letter Concerning Enthusiasm," in *Characteristics of Men, Manners, Opinions, Times* (1711), ed. Lawrence E. Klein (Cambridge: Cambridge University Press, 1999).
15 Kant, *Träume*, AA, 2:318.
16 Kant, *Träume*, AA, 2:351.
17 Kant, *Träume*, AA, 2:368, 349.
18 Kant, *Träume*, AA, 2:354.

ings. In fact, Swedenborg lived off a generous state pension proceeding from the post he had held as assessor of the board of mines. It was this aspect of his biography that drove Paul Valéry, in a 1936 preface he wrote to the French translation of a Swedenborg biography, to ask: *"How is a Swedenborg possible?"*[19] How, he proceeded, could "a learned engineer, an eminent functionary, a man wise in practical matters and knowledgeable in all things" become a visionary, who, furthermore, did not hesitate to publicize the fact that he is "visited by inhabitants of another world, taught by them, and living a part of his life in their mysterious company?"[20]

While Kant concerned himself with the parallels between the psychology of the metaphysician and that of the seer, Valéry was struck by the fact that one and the same person could be an esteemed scientist and functionary, today we might call him a technocrat, and, apparently without any qualms, present himself as a seer as well. Swedenborg was a bureaucrat and a *Schwärmer* – and if he was not yet a "bureaucratic fanatic," it is because he did not immediately identify his visions with his office. Valéry, himself an indefatigable rationalist, attempts to make sense of the "sign" that Swedenborg takes to be the assurance of the veracity of his visions, wondering whether it may not simply be "the sensation of energy, of happy plenitude, of well-being that he always felt when giving himself up to the production and organization of his spiritual world."[21] If his conviction were generated by the satisfaction of seeing a world in accord with his spirit, such "creative delight" would, Valéry speculates, be enough to defer his doubts and defuse his critical sensibility.[22] Nonetheless, Valéry concedes, there remain certain aspects of the biography that challenge the plausibility of such a thesis, including the astonishing lack of inventiveness of Swedenborg's visions.[23] But perhaps this is explained by the very juxtaposition of the bureaucrat and the visionary that he wonders at: the delirious security and serenity afforded Swedenborg by his visions is the sense of fulfillment of his dreams *as a bureaucrat*. The answer to how "a Swedenborg" is possible would not then be in spite of the technical-scientific knowledge that qualified him to be a respected functionary of a state association, but a function of it.

In the section of *Dreams of a Spirit Seer* entitled "A *Schwärmer*'s Ecstatic Journey through the Spirit World," Kant too remarks on the dullness of Sweden-

---

[19] Paul Valéry, "Svedenborg" (1936), in *Oeuvres*, ed. Jean Hytier (Paris: Gallimard, 1957), 1:878. Translation mine.
[20] Valéry, "Svedenborg," 878.
[21] Valéry, "Svedenborg," 880.
[22] Valéry, "Svedenborg," 880.
[23] Valéry, "Svedenborg," 882.

borg's writing.[24] It is in part what leads him to suggest that his madness is due to a fanatical intuition, a systematic delusion of the senses (*Wahn-sinn*) as opposed to an insanity of reason (*Wahn-witz*). While reason can, in principle, "constrain its own empty curiosity" (*Bändigung eines leeren Vorwitzes*), the confusion of the senses strikes at the "fundament of all judgments."[25] Nonetheless, the point of the exercise is ultimately to bring the "butterfly-wings of metaphysics," that is, of the rationalist not the visionary, back down to earth, "on the base ground of experience and of common understanding."[26] For the ecstatic domain that the *Schwärmer* experiences as a spirit-world, and through which the inquisitive reader journeys with Kant to come out with nothing but exhaustion, is the very domain that metaphysics presumes to articulate and understand in the absence of the least data of common-sense experience. And it is this willful desire to overlook, in principle, the limits of sensible experience and to race off, or rave, with reason that Kant will henceforth call *Schwärmerei*.

As already shown in the case of Swedenborg, it was not only self-proclaimed seers and metaphysicians who were disposed to fanaticism, but also – and above all – bureaucrats. For while Kant conceived of the limits of human reason as being fundamentally defined by the fact of embodiment and the human faculty of sensibility, the bureaucrat was increasingly absolved, in principle at least, of such sensibility by the apparatus – "the office" – in which he operated. Bureaucratization is the dream of reason freed from the constraints of human sensibility and possessed by the aspiration, possibly also the technical-scientific means, to institute the organization of life and of the earth according to its principles. Bureaucratization, left to its own devices, is the institutionalization of *Schwärmerei*.

The metaphysics of bureaucracy is anything but idle. On the contrary, it is all about the implementation of its visions in reality. For this reason the European or Eurocentric transformation of the world that Weber called "rationalization" proved to have less to do with the critical project of circumscribing the field of the legitimate application of reason than with a visionary project of limitless expansion and intensity concerned with discovering and dominating new grounds. The domination, or the dream of domination, implied in rationalization was central to Weber's "Science as a Vocation," the lecture he gave in Munich towards the end of the First World War:

> The growing process of intellectualization and rationalization does *not* imply a growing understanding of the conditions under which we live. It means something quite different. It is

---

24 Kant, *Träume*, AA, 2:360.
25 Kant, *Träume*, AA, 2:361.
26 Kant, *Träume*, AA, 2:367.

the knowledge or the conviction that if *only we wished* to understand them we *could* do so at any time. It means that in principle, then, we are not ruled by mysterious, unpredictable forces, but that, on the contrary, we can in principle *control everything by means of calculation.*[27]

The violence of rationalization is first of all the violence of an attitude that is at once theoretical and practical, a bureaucratic attitude in short, that can consider the world only under the aspect of its accessibility to calculation and control. Although it promises such mastery, this attitude is always one of disappointment, for it brings about, famously, the "disenchantment of the world" (*Entzauberung der Welt*).[28] If rationalization did indeed incline to *Schwärmerei*, by Weber's time it exalted over an increasingly barren realm of depleted experience, which, however, seemed only to fuel the "passion for bureaucratization" and exacerbate its extremism.

This dialectic of rationalization – its peculiar and persisting force and attraction even in the face of the disenchantment and alienation to which it led – concealed an even more troubling conundrum for someone who considered himself to be a scientist by vocation. For Weber's groundbreaking sociological studies of rationalization were the outcome and expression of a period in which reason had undergone an unprecedented relativization and historicization, to the point, indeed, of fundamentally discrediting its very rationality. Profoundly informed by Nietzsche's critique of modern science in *On the Genealogy of Morality* (1887), Weber found himself engaged in a scientific project that sought to rationally study reason and rationalization in their social and historical dimensions, where they proved to be anything but rational. To formulate the paradox more explicitly: the sociologist found himself in the position of having to rationally concede that there may be no such thing as reason, even as "rationalization" seemed to be imposing itself ever more forcefully and seemingly irrevocably across the globe.

Herbert Marcuse addressed the ambivalence towards reason found in Weber in his essay, "Industrialization and Capitalism" (1965), which preliminarily sought to understand and diagnose Weber's "passionate and [...] malevolent attack on the socialist attempts of 1918."[29] While the Weber of "Science as a Vocation" is critical of the rationalization characteristic of capitalist-industrial society, he reverts, in a speech on socialism given in Vienna in 1918 for example, to an

---

[27] Max Weber, "Science as a Vocation" (1917, published 1919), in *The Vocation Lectures*, ed. David Owen and Tracy B. Strong, trans. Rodney Livingstone (Indianapolis: Hackett, 2004), 13.
[28] Weber, "Science as a Vocation," 13.
[29] Herbert Marcuse, "Industrialization and Capitalism," *New Left Review* I.30 (April 1965): 3.

apologia for capitalist reason insofar as he denies, or fails to acknowledge, the possibility of "a qualitatively *different* historical rationality."[30] Marcuse argues that the ostensible "value-freedom" of science that Weber insisted upon ends up valorizing the status quo by recapitulating its (capital-industrial) rationality as the form of rationality in general. Thus, even as Weber bemoans the reification and alienation of such rationalization, he takes it nonetheless to be the destiny of the West.

Weber's skepticism towards socialism amounted to a critique of the vision of overly planned and intensively bureaucratized society. But his argument is premised on the prevailing conception of rationality that is, Marcuse argues, altogether capitalist. Thus it in fact constitutes an exhaustive critique of *capitalist* reason, if it can still be called "reason." As Marcuse, reading Weber, emphasizes, the progress of industrial capitalism is in fact accompanied by an ever-greater extension of bureaucratic control. And if bureaucracy is "formally the most rational kind of control thanks to its 'precision, constancy, discipline, tautness and reliability, i.e. calculability both for the masters and all those interested,'" then Marcuse is by no means the first to conclude that it is not the expert but in fact "the apparatus [that is] in control here."[31] Bureaucratization is, accordingly, reification: "The expert administration of the apparatus as the most formally rational type of domination: this is the reification of Reason—the apotheosis of reification."[32]

Precisely because of the reification bureaucracy entails, it solicits, or proves ultimately to defer to, an instance that is not entirely bureaucratic and may well be completely irrational. And, Marcuse argues, Weber in fact has a name for this lacuna in his own account of rationalization – charisma. Even as charisma, which revolutionizes or overturns traditional modes of authority, tends to establish itself in the long term in modes of bureaucratic rule, so bureaucratic authority always ultimately relies on a moment that, in the absence of more traditional authorities, must be charismatic. As he concludes, "Weber's concept of reason ends up in irrational charisma."[33] The intensification of bureaucracy as rational control is complemented by the bureaucratic potentialization of the irrational instance to which the bureaucracy is ultimately beholden.

What seems at first to be Marcuse's attempt to condemn Weber's account of rationalization ends with a reflection on his own time and place, the United

---

[30] Marcuse, "Industrialization and Capitalism," 7.
[31] Marcuse, "Industrialization and Capitalism," 12.
[32] Marcuse, "Industrialization and Capitalism," 12.
[33] Marcuse, "Industrialization and Capitalism," 13.

States of the mid-1960s, in what amounts to an ambivalent concession to Weber: "It is hard, in the darkening houses of bondage, to see reason of any kind."[34] Perhaps it was an insight of this sort that Weber was already attempting to express in Central Europe before and following the First World War, an insight, namely, into the irrationality that underpins, but undermines, everything that is called "rationalization." As Marcuse concludes: "Or was *irony* hidden in Max Weber's concept of reason, the irony which understands and disavows? Perhaps he is saying: and is that what you call reason?"[35]

In approaching the problem of the rationality of rationalization in its various forms, Weber's writings problematize their own ostensibly scientific mode of exposition and argumentation by driving the analysis to the point of incredulity. In Marcuse's reading, Weber's work turns into a Kierkegaardian kind of text, which undermines the legitimacy and authority of the rationality to which it pretends to appeal, and so exhibits the structure of a pseudonymous argument in which the reader is drawn into an interrogation, at once textual and existential, that cannot be coherently concluded on the level of statement. Just as Kant, addressing the dialectic of reason around the problem of the *Schwärmer*, was driven to produce a "witty" little work, Weber is obliged, unwittingly perhaps, to engage the paradoxical irrationality of rationalization in a studious and altogether earnest form of irony. On account of the problems of presentation into which their investigations lead them, each becomes, to quote Kierkegaard writing of himself, "a peculiar kind of poet or thinker."[36] The study of the fanaticism of rationalization and of rational fanaticism thus seems to solicit a peculiar mode of exposition that operates in an indirect, ironic, if not altogether literary manner.

## Abstraction (Hegel, Marx)

In *Fanaticism: The Uses of an Idea*, Alberto Toscano departs from a line in Hegel: fanaticism is "enthusiasm for the abstract."[37] In exploring the deployment of the

---

34 Marcuse, "Industrialization and Capitalism," 17.
35 Marcuse, "Industrialization and Capitalism," 17.
36 Søren Kierkegaard, Preface to "Two Discourses at the Communion on Fridays," in *Without Authority*, ed. and trans. Edna Hong and Howard Hong (Princeton: Princeton University Press, 1997), 165; translation modified.
37 Alberto Toscano, *Fanaticism: On the Uses of an Idea* (London: Verso, 2010), xi. "*Fanatismus*, d.i. eine Begeisterung für ein Abstraktes," in Georg Wilhelm Friedrich Hegel, *Vorlesungen über die Philosophie der Geschichte*, in *Werke in zwanzig Bänden* (Frankfurt am Main: Suhrkamp, 1979), 12:431.

idea of the "fanatic," Toscano aims to articulate the attraction, but also the philosophical traction, of what he calls a "politics of abstraction."[38] In his account, abstraction is related to fanaticism because it always stands for a kind of transcendence. It presents the possibility of a radical or revolutionary departure from the established order, a departure that is freed from the conservative constraints of a political thought and practice that delimits itself according to the critical criteria of the feasible and the legitimate. These are the precincts of the "politics of finitude," propounded by conservative and post-structuralist theorists alike, of which Toscano is wary. Behind the depoliticizing and anti-theoretical term "fanatic," Toscano uncovers a field of theoretical and even practical political activity that the proponents of the "politics of finitude" polemically discredit rather than earnestly engage.[39]

Abstraction, in Toscano's account, presents the promise of a politics that is "fanatical" insofar as it refuses to tolerate the historical constraints of the concrete order of things as well as the pragmatic and disciplinary constraints of the "common sense" logic of a down-to-earth political praxis. But, as Toscano also acknowledges in the conclusion of his study, this kind of abstraction is called for because of the proliferation of concrete structures of abstraction in everyday life: "An assertion of the rights of the abstract in politics must nonetheless be accompanied by an effort to account for the emergence, autonomization, and power of real abstractions."[40] Indeed, the moment in which fanaticism could plausibly be defined in terms of abstraction was also one in which a particular kind of abstraction would be realized – by means of bureaucratization. For the time of Hegel, or the time announced in Hegel, was one in which bureaucracy emerged as a machinery equipped with both the justification and the means for putting abstraction into reality. This transformative process was precisely what Weber would later call "rationalization."

It is not by chance that in *Elements of the Philosophy of Right* (1820), Hegel would produce the first significant theory of bureaucracy in his discussion of the third and culminating moment of ethical life, the State, at a time when philosophy was becoming, or in any case was suspected of becoming, an increasingly professionalized if not bureaucratic activity. As Schopenhauer would observe in his caustic reflections on university philosophy, it was Hegelianism that facilitated a state-sponsored professionalization of philosophy, which henceforth op-

---

**38** Toscano, *Fanaticism*, xii.
**39** Toscano, *Fanaticism*, 141.
**40** Toscano, *Fanaticism*, 251.

erated "on behalf of the government" (*im Auftrage der Regierung*), betraying its true vocation, namely, the free investigation of truth:

> It was the way that university philosophy served the purposes of the state that earned Hegelry such unprecedented ministerial favor. For it, the state was "the absolutely perfected ethical organism," and the whole purpose of human existence went into the state. Could there be a better preparation for future legal clerks and upcoming civil servants than this [...] ?[41]

Abandoning its formerly lofty aspirations, university philosophy dedicated itself to forming the human in the image of the petty bureaucrat devoted to serving the state as the ultimate good. The bureaucratization of philosophy produced a philosophy that venerated bureaucratization.

Schopenhauer's account which, given his own history in Berlin, is not without *ressentiment*, is true to this extent: the apotheosis of the (Prussian) state in Hegel's *Philosophy of Right* does indeed include a curiously uncritical advocacy of bureaucracy. If the state was "the march of God in the world," the implementation of this divine plan was the occupation of an ostensibly dispassionate, upright, and mild category of professionals, the bureaucracy, which assumed the status of the "universal estate" (*allgemeiner Stand*).[42] The executive activity of these representatives, along with the judiciary and the police, of executive power or governmentality (*Regierungsgewalt*), consisted in administering relations between the universal interest of the state and particular corporate interests that make up civil society. Hegel referred to this executive activity as *subsumption* – "the subsumption of *particular* spheres and individual cases under the universal."[43]

Despite the divine character of their worldly calling, Hegel's bureaucrats are, to be sure, the very opposite of fanatics, at least as Hegel defines the latter in the remark to § 270 devoted to the relation of religion to the state. There fanaticism (*Fanatismus*) emerges as "a *polemical* kind of piety," which is the misplaced, because purely subjective, conviction that one can operate as an individual or community without reference to the state, or, rather, with reference to a totality that transcends the state.[44] The fanatic is the figure who turns this polemical piety

---

[41] Arthur Schopenhauer, "Über die Universitäts-Philosophie" (1851), in *Parerga und Paralipomena, Sämtliche Werke*, ed. Wolfgang Löhneysen (Frankfurt am Main: Suhrkamp, 1989), 4:182. Translation mine.
[42] Georg Wilhelm Friedrich Hegel, *Elements of the Philosophy of Right* (1820), ed. Allen Wood, trans. H. B. Nisbet (Cambridge: Cambridge University Press, 1991), § 258; § 303.
[43] Hegel, *Philosophy of Right*, § 273.
[44] Hegel, *Philosophy of Right*, § 270.

into a political program. Fanaticism is, accordingly, the attempt to transform the state by bringing every particular aspect of the ethical life it organizes and articulates into an immediate relation to the subjectively conceived totality in what proves to be a completely destructive project: "it would wish to find the whole in every particular, and could accomplish this only by destroying the particular."[45]

Ultimately fanaticism's insistence on wholeness, if not holiness, betrays itself in its particularity. The attempt to abstractly assert itself as universal would destroy everything particular including itself, "for fanaticism is simply the refusal to admit particular differences."[46] In the *Philosophy of Right*, fanaticism is addressed in the discussion of the constitution of the state because the fanatic is not only the inner enemy of the state, but presents the antithesis of the concept of the state. For fanaticism is the vision of a totalitarian order of things that has no patience for the task of subsuming particularities, which properly characterizes the state.

In contrast, the bureaucracy is, according to Hegel, the organ of the state responsible for subsumption. As such, the bureaucracy is the antidote to fanaticism and the bureaucrat is the anti-fanatic insofar as his task is to ensure that there is no particularity that cannot be subsumed concretely under the rules and ends of the state. The bureaucracy is responsible for implementing measures that admit particularity, articulating society in accordance with the universality of the state, such that it becomes what, in Hegel's account, it is – the concrete universal.

The term subsumption, taken from logic, recalls in particular Kant's *Critique of Pure Reason* (1781) where it describes the operation of the transcendental schema to subsume the multiplicity given in intuition under the categories of understanding. A younger Hegel, a Hegel who had not yet been called by the state to Berlin, had, in preparatory work for the *Phenomenology of Spirit* (1807), been one of the most sophisticated critics of the Kantian notion of subsumption – on account precisely of the abstraction he took it to involve.[47] Yet in *Philosophy of Right* he seems to accept subsumption very much in the sense he otherwise criticizes. If bureaucracy as executive power is tasked "with the concrete knowledge and oversight of the whole,"[48] then the question of the abstract violence involved in such governance by concrete knowledge is raised. While the abstraction of fa-

---

[45] Hegel, *Philosophy of Right*, § 270.
[46] Hegel, *Philosophy of Right*, § 270.
[47] See, for example, Georg Wilhelm Friedrich Hegel, *The Science of Logic* (1812), ed. and trans. George di Giovanni (Cambridge: Cambridge University Press, 2015), 515–525.
[48] Hegel, *Philosophy of Right*, § 300.

naticism is characterized by a complete refusal of particular distinction, "for fanaticism wills only what is abstract, not what is articulated,"[49] the bureaucrat imposes an abstract articulation on social reality, one that does violence to the particularities which are distorted, rather than preserved, by the concrete process of administrative subsumption.

The problem of subsumption was central to Marx's "Contribution to the Critique of Hegel's Philosophy of Right" (1843–1844), in which he reproaches Hegel for identifying the executive power with the image or self-image of Prussian bureaucracy. Rather than reflecting in a manner that would be properly philosophical on what subsumption might involve (for there are many kinds of subsumption, "Applied mathematics is also subsumption, etc."), Marx observes that Hegel uncritically takes the politically and philosophically questionable model of subsumption characteristic of bureaucracy in the modern state as the true expression of subsumption: "Then he takes any one of the empirical forms of existence of the Prussian or modern state (just as it is), anything which actualizes this category among others, even though this category does not express its specific character."[50] While the question that needed to be asked was: "Is this the rational, the adequate mode of subsumption?"[51] Hegel instead presumed that an existing organization would correspond unproblematically to the category in his thought. In so doing, Marx concludes, "Hegel gives *a political body to his logic:* he does not give the *logic of the body politic.*"[52]

The problem with the organ of subsumption that Hegel settles upon – the Prussian bureaucracy – is that it is by no means universal, neither a "universal estate" nor in the service of the universal. The subsumption in the *Philosophy of Right* is, therefore, subsumption under particular interests that are not ultimately those of the state or, rather, not those of the genuinely universal mode of political life that would transcend and ultimately do away with the state. Marx proceeds to identify a series of tensions as a result of which, far from being a universal class, the bureaucracy proves to be a "contradictory identity" that, like the institution of the state it represents, presents a spurious, purely abstract, sublation (*Aufhebung*) in the unfolding of ethical life.[53]

---

[49] Hegel, *Philosophy of Right*, § 5.
[50] Karl Marx, "Contribution to the Critique of Hegel's Philosophy of Law," in *Marx & Engels Collected Works* (New York: International Publishers, 1975), 3:48.
[51] Marx, "Critique of Hegel's Philosophy of Law," 48.
[52] Marx, "Critique of Hegel's Philosophy of Law," 48.
[53] Marx, "Critique of Hegel's Philosophy of Law," 48.

Because it plays a formal role in the articulation of the state, bureaucracy is "state formalism" (*Staatsformalismus*).⁵⁴ On account of this formalism, it tends to render the state a formal state, a "state as formalism," in which the real ends of the state are confused with the formal ones of the bureaucracy.⁵⁵ Thus Marx argues that the bureaucratization of the state is its reification. Rather than appearing as it is or ought to be, a substantial ethical totality in which each member takes part, it appears as a formal bureaucratic structure over and against society at large. The bureaucracy is thus generative of "practical illusions," here the "illusion of the state," that the left Hegelians, and most famously Marx, associated with religion. Bureaucrats are the theologians of the state: "The bureaucracy is la république prêtre."⁵⁶

The bureaucracy brings about a doubling of the state: "Each thing has therefore a double meaning, a real and a bureaucratic meaning, just as knowledge (and also the will) is both real and bureaucratic."⁵⁷ This bureaucratic vision of the world not only obscures reality but provides the criteria for real bureaucratic activity: "The really existing, however, is treated in the light of its bureaucratic nature."⁵⁸ And while the bureaucratic attitude pretends to accord with objective standards of scientific knowledge, it is in fact subordinated to a secretive knowledge, the principle of which is the authority of the closed hierarchy of the bureaucracy itself. Thus the out-of-touch "spiritual" attitude of bureaucracy towards the external domain of its intervention turns out to be informed by a "crass materialism" within the institution itself, which instills a passive and mechanical attitude of obedience.⁵⁹ Far from subordinating his private interests to those of the state, or as Hegel has it, finding his substantial satisfaction in the ends of the state, the bureaucracy becomes the locus of the bureaucrat's private striving – "a *chasing after higher posts*, the *making of a career*" – to which he does indeed devote his life.⁶⁰

While bureaucracy fosters this rapacious "materialist" struggle to climb within the bureaucracy, outwardly it is constantly compelled to justify its existence to itself and to the outside world by voraciously expanding the sphere of its application: "Whilst the bureaucracy is on the one hand this crass materialism, it manifests its crass spiritualism in the fact that it wants to *do every-*

---

54 Marx, "Critique of Hegel's Philosophy of Law," 45.
55 Marx, "Critique of Hegel's Philosophy of Law," 46.
56 Marx, "Critique of Hegel's Philosophy of Law," 46.
57 Marx, "Critique of Hegel's Philosophy of Law," 47.
58 Marx, "Critique of Hegel's Philosophy of Law," 47.
59 Marx, "Critique of Hegel's Philosophy of Law," 47.
60 Marx, "Critique of Hegel's Philosophy of Law," 47.

*thing.*"⁶¹ The exigency of bureaucratization emerges in Marx's account as a compensation for the bureaucracy's own anxieties about its reality. Since "it is purely an *active* form of existence and receives its content from without," it can only prove its existence by means of the bureaucratic colonization and transformation of reality.⁶² As a result, for its own validation, the bureaucratic attitude treats the world as if it were only there to be bureaucratically determined and controlled: "For the bureaucrat the world is a mere object to be manipulated by him."⁶³

That Hegel himself is aware of these tensions is indicated by the less philosophical aspects of his discussion, which consider the importance of bureaucratic salaries and examinations and, crucially, of checks and balances inside and outside of the bureaucracy itself to prevent bureaucratic abuses. Of these Marx draws out the most striking, those ostensibly operative in the bureaucrat himself:

> *In the civil servant himself* – and this is supposed to humanise him and make "behaviour marked by dispassionateness, uprightness and kindness" "customary" – "direct moral and intellectual education" is supposed to provide the "spiritual counterpoise" to the *mechanical character* of his knowledge and of his "actual work." As if the "mechanical character" of his "bureaucratic" knowledge and of his "actual work" did not provide the "counterpoise" to his "moral and intellectual education"!⁶⁴

The bureaucrat, that representative of the universal estate, proves in Hegel's own account, and for Marx this is decisive, to be cut off from his humanity: "The human within the official is supposed to secure the official against himself."⁶⁵ Like the world in which he operates, the bureaucrat emerges as a "dualistic category," in which the humane and the bureaucratic stand over and against one another.⁶⁶ In Marx's view, the realized concept of subsumption would break down the walls that separate the bureaucracy from humanity, and the bureaucrat from his humanity, to give rise to a genuinely universal class in which every particular would be capable of, and interested in, taking part. In the meantime, the critique of bureaucracy consists in bringing to light the faulty, perhaps fraudulent, forms of subsumption that contribute to inhuman and potentially fanatical excesses of administration.

---

61 Marx, "Critique of Hegel's Philosophy of Law," 47.
62 Marx, "Critique of Hegel's Philosophy of Law," 47.
63 Marx, "Critique of Hegel's Philosophy of Law," 48.
64 Marx, "Critique of Hegel's Philosophy of Law," 53.
65 Marx, "Critique of Hegel's Philosophy of Law," 53; translation modified.
66 Marx, "Critique of Hegel's Philosophy of Law," 53; translation modified.

## The Confusion of the Present Age (Kierkegaard I)

Marx's critique of the bureaucratic model championed by Hegel proceeds immanently and so maintains the ideal of subsumption in a purer or more perfect form. Writing some years later, Kierkegaard would approach the matter from the opposite direction: What is the possibility for registering truly exceptional, that is to say unsubsumable, phenomena under conditions of bureaucratization that he also associated with "Hegelianism"? The problem, of which both bureaucracy and Hegelianism were symptomatic, is not the faultiness of contemporary modes of subsumption but the prevailing principle of the subsumability of all things.

Between 1846 and 1847, at a moment that marked a threshold in his authorship, Kierkegaard worked on a book that he would not ultimately publish in book-form, known as *The Book on Adler*.[67] The occasion of the book was the case of Adolf Peter Adler, who claimed to have had a "revelation" in 1842. In a way that resonates with Valéry's question, *How is a Swedenborg possible?*, Kierkegaard is fascinated by this bureaucrat, a minister of the State Church, who, while claiming to be nothing less than an apostle, seems to have thought that everything could go on as if nothing out of the ordinary had happened, expecting even to hold onto his office – and his state pension.[68] Just as, a century earlier, the case of Swedenborg had provided Kant with the opportunity to ironically demonstrate the metaphysical confusion of his contemporaries, Kierkegaard sought to use the "phenomenon" of Adler to illustrate what he called "the religious confusion of the present age."

The confusion as illustrated by Adler comes down to this: an inability to understand the meaning of the words that define the apostle's singular calling, "authority" and "revelation." Operating *e concessio* – without judging the veracity of Adler's claims – *The Book on Adler* proceeds as a "kind of literary review" that addresses itself to his discourse.[69] Analyzing Adler's published writings as well as his depositions to church authorities – who proceeded on the assumption

---

**67** Søren Kierkegaard, *The Book on Adler*, ed. and trans. Edna Hong and Howard Hong (Princeton: Princeton University Press, 1998). Although Kierkegaard accorded great importance to them, revising and returning to them over many years, he would ultimately publish only a small part of the writings around Adler as a "minor ethical-religious essay" with the title "The Difference between a Genius and an Apostle," under the pseudonym, H. H., published in Søren Kierkegaard, *Without Authority*, ed. and trans. Edna Hong and Howard Hong (Princeton: Princeton University Press, 1997).
**68** Kierkegaard, *Adler*, 148.
**69** Kierkegaard, *Adler*, 320.

that he had lost his mind – the "ministering critic" (Kierkegaard was undecided about whether to use a pseudonym in this case) shows that Adler has no sense of what it means to claim to be an apostle – and so fails to be one.

On account of the revelation that he communicates and the authority that it implies, an apostle ought by definition to present an irreconcilable "offense" to the established order that he claims to transcend and thus threatens to transform:

> And when the single individual continues along this road and goes so far that he does not as the ordinary individual *reproductively renew the life of the established order* **within himself** by willing, under eternal responsibility, to order himself within it but wants to renew *the life of the established order by bringing a new point of departure* for it, *a new point of departure in relation to the basic presupposition of the established order*, when he by submitting directly to God must relate himself transformingly to the established order – then he is the extraordinary.[70]

An apostle is extraordinary – if not altogether fanatical. Adler, in contrast, is remarkable for being the most un-remarkable of would-be apostles. Far from presenting a new point of departure, much less an "offensive" one, he sought to comprehend the exceptionality of the situation he claims for himself in the most ordinary and amenable of ways. Although he had burnt his Hegelian papers upon receiving the "revelation," the critic argues that Adler reverts to "Hegelianism" – indeed becomes a "satire on Hegelianism"[71] – insofar as his writings seek, so to speak, to subsume the revelation, ultimately straying so far as to interpret the event in the aesthetic terms of genius. By attempting to translate the revelation into ethical-universal terms (in his capacity as a Hegelian and minister of the State Church) and finally into aesthetic ones (as a poet), Adler shows his confusion regarding his claim to be an apostle.

Far from being exceptional even in this regard, Adler presents an "epigram" of the age. For like Adler, the age has no conception of the extraordinary, such that even if something extraordinary were to happen, a revelation, for example, it would be taken – just like the overused *word* "revelation" – to mean something altogether trivial and banal. That in the present age a bureaucrat can take himself to be an apostle without abandoning his office, let alone overturning the world order, is indicative of an age that cannot tell the difference between a revelation and a mere "miscellaneous announcement."[72]

---

[70] Kierkegaard, *Adler*, 150.
[71] Kierkegaard, *Adler*, 91.
[72] Kierkegaard, *Adler*, 18.

All this is to say: if Adler really means what he says, he would have found another means of saying it. The question that the case of Adler poses is the question of how, in the present age, something as extraordinary as a revelation might be communicated or, alternatively, in what guise the extraordinary might express itself. *How is an Adler possible?* thus becomes, *How would an Adler be possible, if only he could communicate what he claims he has to say?* Thus *The Book on Adler*, insofar as it is concerned first of all with diagnosing the present age, is really about authorship rather than apostledom. At a moment when Kierkegaard was ostensibly abandoning pseudonymous authorship, the study of Adler served as the occasion for reflection on, and as a guide to the problem of, authorship with which his distinctive writings had been engaged.[73]

*The Book on Adler* was to have opened with an introduction on the difference between "premise" and "essential" authors. And in the course of the subsequent review of his work, Adler would emerge as a prototypical premise author. For he premises his writings on the authority and revelation they purport to convey, yet prove utterly to lack. The confusion is betrayed not simply in the fact that his writings fail to convey the revelation they claim to communicate but in presuming to be in possession of an appropriately authoritative and revelatory discourse in the first place. An essential author (*en væsentlig Forfatter*), in contrast, would not make authority and revelation the premise of his discourse but would work towards the conclusion that these are missing: "In order to find the conclusion, it is first and foremost necessary to perceive very vividly that it is lacking and thereby in turn very vividly to miss it."[74] Bringing into focus the absence of authority and the evacuation of the revelatory character of language constitutes the paradoxical conclusion to which, in the present age, essential authorship must be devoted.

---

[73] That *The Book on Adler* was to serve as a guide of sorts to the complexities of Kierkegaard's authorship is indicated in the "Editor's Preface" to the version of the book that was to be published under the pseudonym Petrus Minor. With regard to the *Two Ethical-Religious Essays* by H. H., in which Kierkegaard finally published part of the reflections occasioned by Adler, he would observe: "The significance of this little book (which does not stand *in* the authorship as much as it relates totally *to* the authorship and for that reason also was anonymous, in order to be kept outside entirely) is not so easy to explain without going into the whole matter. It is like a navigation mark *by which* one steers but, please note, in such a way that the pilot understands precisely that *he is to keep a certain distance from it.* It defines the boundary of the authorship," *On My Work as an Author*, in Søren Kierkegaard, *The Point of View*, ed. and trans. Edna Hong and Howard Hong (Princeton: Princeton University Press, 1998), 6.

[74] Kierkegaard, *Adler*, 8.

Insofar as the concept of authority has been "completely forgotten in our confused age,"[75] every author is, as Kierkegaard would repeat with regard to his own writings, "without authority." Only by means of an essential authorship that thematizes the irony of its lack of authority might a perspicuous presentation of the confusion of the present age be achieved. If there is a transcendent "religious" sphere of the sort about which Adler thought he was speaking, it cannot be authoritatively revealed, but only, and indefinitely, disclosed by means of an indirect form of communication that abandons the premise of being properly authoritative or revelatory. In other words, essential communications will be literary ones. Rather than announcing themselves as apostles, exceptional figures will henceforth have to resort to the kinds of communication analyzed in this book – from such peculiar expressions as Bartleby's "I would prefer not to," or Stevie's "Poor! Poor!" in *The Secret Agent* or finally Gregor's chirping, hissing and snapping, to the perplexities betrayed in Kohlhaas's vain talk of forgiveness, Kurtz's expiring exhalation, "The horror! The horror!" or the officer's painful insistence on the dubious principle that "guilt is always beyond doubt" in *In the Penal Colony*.

## 1848 – More Confusion (Kierkegaard II)

In 1848, following "the events that have now changed the shape of Europe," the conclusion about the inevitable confusion of extraordinary phenomena in the present age seemed no longer to be the last word on the matter; they solicit further writing. This is indicated in the "postscript for one who has read the essay" that Kierkegaard drafted later that year.[76] For it seemed that in the sphere of the political something extraordinary, perhaps even conclusive, could come to pass; that in politics "a new departure in relation to the established order" could emerge.[77] A postscript was needed to address the fact that the extraordinary seemed indeed to reveal itself with authority – just in the field of the political.

In the post-1848 postscript, the "religious confusion" is reconfigured in political terms, or rather the "confusion of the present age" is construed as the confusion of the religious and the political. While for the "ethical-religious" investigation "*the point of departure is from above, from God, and the formula is this paradox, that an individual is used,*" in the political case "*the point of departure*

---

75 Kierkegaard, *Adler*, 4.
76 Kierkegaard, *Adler*, 315.
77 Kierkegaard, *Adler*, 317.

*is* from below, *from that which is lower than the established order,* since even the most mediocre 'established order' is still preferable and superior to the flabbiest of everything flabby – the crowd."[78] It is apparent, however, that the two kinds of "new departure" – the religious and the political, revelation and revolution[79] – are difficult to distinguish. Indeed the distinction seems ultimately to be only one of intensity: the political "goes more easily, less paradoxically, more directly."[80]

While the extraordinary but amorphous power in modern mass movements appears to be something quite inhuman, even superhuman, it is not for that reason to be mistaken for something divine, or for what had once – in another age – been called the divine. But in addressing the confusion of the religious and the political Kierkegaard himself seems to get into some confusion. For rather than bringing the difference into focus, he instead presents modern politics in a confused figure. Drawing in extravagant style on contemporary reactionary discourse, he describes "a prodigious monstrosity with many heads or, more correctly and accurately, a thousand-, according to the circumstances, a hundred-thousand-legged monstrosity, the crowd, an irrational enormity, or an enormous irrationality," adding among other things that it is an enormous "abstraction" with power that should be measured, like that of a machine, in "horsepower."[81]

Although, presumably to avoid confusion, he avoids the more religiously coded terms – fanatic, enthusiast – with his description of this overdetermined biopolitical agent, Kierkegaard recites the distinctly Protestant tradition critical of enthusiasm emblematized in Luther's coinage of the term *Schwärmerei*. Abandoning the impression of genuine transcendence implied in the term *Enthusiasmus* as it was conveyed from its ancient Greek sources, the German word *Schwärmerei* – literally, *swarming* – would be deployed by Luther to disparage

---

78 Kierkegaard, *Adler*, 317.
79 It would be a mistake to assume that Kierkegaard was opposed to revolution in principle. In 1846, Kierkegaard published a "Literary Review" of an anonymous novel entitled *Two Ages* (1845) that set the present age in relief against the revolutionary one. In Kierkegaard's reading, the present age proves devoid of the passion that lends genuine form and significance to words and deeds, whereas: "The age of revolution is essentially passionate; therefore it is essentially *revelation*, revelation by a manifestation of energy," Søren Kierkegaard, *Two Ages: The Age of Revolution and the Present Age: A Literary Review*, ed. and trans. Edna Hong and Howard Hong (Princeton: Princeton University Press, 1978), 66. In 1848, just as Marx was observing that historical events repeat themselves first as tragedy and then as farce, Kierkegaard seems, in a similar fashion, to consider the confusion of contemporary events as evidence of the loss of the possibility of revolution in the revelatory sense.
80 Kierkegaard, *Adler*, 317.
81 Kierkegaard, *Adler*, 317–318.

enthusiasm by writing it off as creaturely delusion. As I will discuss in more detail in the following chapter, Luther applied the term primarily in polemical attacks on those "Protestants" – among them, Thomas Müntzer and his followers – who protested too much. The swarm evoked images of the amorphous, frenzied, contagious mass violence associated with the behavior of insects, birds, and vermin. Like Kierkegaard's headless millipede, it was supposed to emphasize the impersonal, debased, animalistic character of certain kinds of excessive or eccentric and above all mass behavior, as well as disparaging the abject conditions that might have given rise to such outbursts.

The danger for Luther, as for Kierkegaard, is the presumption of being able to communicate a confusion without succumbing to it. Whether addressing the case of a *Schwärmer*, like Adler, or a swarm, like the "crowd," as soon as one starts denouncing *Schwärmerei*, one risks being swept up in the confusion. The infectious confusion of the *Schwärmer* can spread to all those inclined, without irony, to use the word "Schwärmer" as if they know what they mean to say. For *Schwärmerei* is the confused expression of an authority that it lacks. No one then is more *schwärmerisch* than the person who, as if passing a judgment from on high, claims to have identified among those "exceptional" phenomena that do not seem to conform to the limits of the human or of human explanation a definite case of *Schwärmerei*. If the term *Schwärmerei* is used to denote the confusion of the religious and the political, it does so only by presenting, rather than clarifying, the confusion in which it is caught up.

No less *schwärmerisch*, however, is the position that dismisses the "religious" in the name of the political on the assumption, recited ironically by Kierkegaard, that "everything these days is politics."[82] In further notes related to the Adler materials written in reaction to 1848, the confusion of the age consists in the attempt to resolve religious questions politically: "temporality wants to explain in a temporal [deleted: and worldly] way that which in temporality must be an enigma and which only eternity can and will explain."[83] Exemplary in this regard is the project Kierkegaard attributes to modern politics, the politics of the crowd but also of the bureaucracy of government administrations, namely, "to solve the problem of *equality* of all people in the medium of worldliness."[84] Since worldliness is constituted by difference, the world being nothing other than such differences, modern politics finds itself attempting to institute equality "in the medium of which the essence is difference."[85]

---

82 Kierkegaard, *Adler*, 320.
83 Kierkegaard, *Adler*, 230, Kierkegaard's emendation.
84 Kierkegaard, *Adler*, 229.
85 Kierkegaard, *Adler*, 229.

Here the Church minister, Adler, who, in confusing himself with an apostle, expressed the leveling of the difference between the temporal and the eternal, finds his political counterpart in the government minister who, here appearing more like what Hegel called a fanatic, purports to be authorized to level temporal differences once and for all. Kierkegaard presents the image of Europe in a paradoxical state of catastrophic stagnation in which an endless succession of ministers representing one administration after the other attempts, in the name of "progress," to implement equality and is driven to distraction by the violence to life and mind of that apocalyptic project:

> Even if all travel in Europe ceased because one would have to wade in blood, and all cabinet ministers became sleepless in order to ponder, and even if ten [*changed from:* a hundred] cabinet ministers lost their minds every day, and every next day ten [*changed from:* a hundred] new ones began where the others left off and also lost their minds – there still would not be essentially one step of progress.[86]

The swamp-like image of this stagnant bloody stream of insane bureaucrats emerges as the epigram of the age of fanatical bureaucratization.

## Another Kind of Despair (Kierkegaard III)

In 1849 Kierkegaard published *The Sickness unto Death* under the pseudonym, Anti-Climacus.[87] The text can be read as an analysis of the confusion of the present age undertaken with attention to its fundamental mood – despair. For, although in the text despair is presented as an essential disposition of the human being as such, the book exhibits in its very mode of analysis the contemporary crisis of faith. If in *The Book on Adler* the problem of authority emerges under conditions in which the meaning of authority has been "completely forgotten," in *The Sickness unto Death*, the problem of despair corresponds to – and overcompensates for – the felt loss of the transparency and security that are supposed to be offered by "faith."

In the opening to the main body of the text, Anti-Climacus develops a definition of the self that subtly displaces the tautological Fichtean model of the I=I that he seems otherwise to be reciting. For he argues that the self is not

---

[86] Kierkegaard, *Adler*, 229–230.
[87] Søren Kierkegaard, *The Sickness unto Death: A Christian Psychological Exposition for Upbuilding and Awakening by Anti-Climacus* (1849), ed. and trans. Edna Hong and Howard Hong (Princeton: Princeton University Press, 2011).

only the synthesis of the relation in which the self reflects on itself to achieve a relation of identity, but one that structurally relates in turn to another. In relating to itself as a synthesis, the self always also relates to a third that exceeds the synthesizing operation and is its condition: "The human self is such a derived, established relation, a relation that relates itself to itself and in relating itself to itself relates itself to another."[88] Far from being, in principle, self-identical and self-determining, the human self exhibits a structural dependency on, or vulnerability to, an alterity that cannot be recovered, synthesized, or appropriated by subjective activity. This feeling of disintegration and debilitation that goes beyond and is more fundamental than the duplicity disclosed in "doubt" (*tvivl*) is what he calls "despair" (*fortvivlelse*).

Thus, in contrast to Fichte, the self is not absolutely self-positing – although it may, in despair, claim that it is. For this reason, for those who are self-aware – and as a structural function of this awareness – there are two possible kinds of despair: "If a human self had itself established itself, then there could be only one form: not to will to be oneself, to will to do away with oneself, but there could not be the form: in despair to will to be oneself."[89] Despair, in short, can either take the form of a desire to lose or annihilate the self or of a sovereign, but ultimately spurious, potentialization of the self. For the dogmatic "Christian," Anti-Climacus, faith is the only antidote to despair. For in faith the self becomes "transparent" to and securely "rests in" the alterity felt in despair, which it recognizes as the very source of the self, the "establishing power" from which all authority and revelation emanates.[90]

It is far from clear, however, whether the kind of subordination required of faith is even a possibility in the present age. In the "Editor's Preface" that he had prepared for *The Book on Adler*, Kierkegaard wrote:

> The calamity of our age in politics, as in religion and as in everything, is disobedience, not being willing to obey. One only deceives oneself and others by wanting to make us think that it is doubt that is to blame for the calamity and the cause of the calamity – no, it is insubordination – it is not doubt about the truth of the religious but insubordination to the authority of the religious.[91]

---

**88** Kierkegaard, *Sickness*, 13–14.
**89** Kierkegaard, *Sickness*, 14. There is a third unconscious form: "the despair that is ignorant of being despair," 42.
**90** Kierkegaard, *Sickness*, 14.
**91** Kierkegaard, *Adler*, 5.

In *The Sickness unto Death,* insubordination returns in the concept of defiance. While "no despair is entirely free of defiance,"[92] defiance in its emphatic sense refers to an intensification of the second form of despair. Such defiance is not simply the refusal to submit to the alterity one feels, and thus to acknowledge the passivity betrayed in passion. Rather, it passionately mobilizes the very power that it refuses to recognize in a hyperbolic gesture of self-assertion. Accordingly, the defiant self paradoxically gains its extraordinary force from the fundamental impotence it feels: "Then comes defiance, which is really despair through the aid of the eternal, the despairing misuse of the eternal within the self to will in despair to be oneself."[93]

Bureaucratic fanatics are defiant figures of despair. As such they are exemplary of the "calamity" of the present age – if that calamity is not itself the phenomenon of fanatical bureaucratization. Their despair, however, is of another kind that does not correspond to the dialectic analyzed by Anti-Climacus – in despair, to will/not to will to be oneself. Instead the following schema suggests itself: in despair, to will to be a bureaucrat; or in despair, not to will to be a bureaucrat. Fanatical bureaucrats like Michael Kohlhaas, Kurtz in *Heart of Darkness* or the officer in *In the Penal Colony,* defiantly attempt to establish themselves as if they were the self-determining, sovereign subjects that they feel they are not. They do so, however, by drawing on the authorization of bureaucracy and "selflessly" investing themselves in, indeed sacrificing themselves to, its rationalizing project. Thus bureaucracy paradoxically takes the place of the "establishing power" that it at the same time evacuates of all authority.

Meanwhile the peculiar fanatics, Bartleby, Stevie in *The Secret Agent* and Gregor in *The Metamorphosis,* are not, despite their apparent weakness or passivity, to be mistaken for exhibiting despair of the first form – in despair, not to will to be oneself. The intensity of their passion betrays rather a mode of defiance but one that, insofar as it does not seek power or assume sovereign postures, cannot be assimilated to the second form of despair either – in despair, to will to be oneself. Insisting, so to speak, on the futility of their curious comportments, they "prefer not to," that is, they prefer not to be oneself and prefer other things than to will. Lacking the transparency and security afforded by faith, these figures are nonetheless devoted to the excavation of their passion and to the exposition of the alterity that obscurely moves them. Without seeking to identify with or consciously to control it, they cultivate a taste for their "own" peculiarity.

---

[92] Kierkegaard, *Sickness,* 49.
[93] Kierkegaard, *Sickness,* 67.

# 2 A New Kind of *Schwärmer*: Kleist's *Michael Kohlhaas*

Schiller's *The Criminal out of Lost Honor* (1786/1793) and Kleist's *Michael Kohlhaas* (1808/10) articulate a threshold in the history of the modern entanglement of politics and life. Demonstrating how the inhuman treatment of the criminal produces brutalized and abandoned outlaws consigned to a world outside the law, Schiller's novella advocates reintegration and rehabilitation implemented by a more humane and indeed humanizing administration. Schiller's literary intervention thus calls for a governmental program of institutional transformation of the sort that was, in fact, already under way, and that would, for better or worse over the centuries to follow, involve an intensifying bureaucratization of life.

Written just two decades later, Kleist's *Michael Kohlhaas* presciently diagnoses the perversion of Schiller's bureaucratic-humanitarian project. While Schiller's bureaucracy novella presents the passionate case for bureaucracy, Kleist's treats bureaucracy as a kind of pathology. It investigates the ambivalence of the "righteous feeling" (*Rechtgefühl*) that emerges under conditions in which bureaucratization has so colonized life as to leave no concrete position, no coherent outlet for expression, and not even the possibility to conceive of justice outside of bureaucracy. As a function or functionary of this ambivalence, which betrays an incommensurability between life and law, between nature and bureaucracy, a new figure emerges, the bureaucratic *Schwärmer*. Assuming a paradoxical position with regard to the law – at once righteous and outrageous – Kohlhaas attempts, with nothing less than an apocalyptic fervor, to bureaucratically bring about an end to all bureaucracy. Ultimately he despairs of this project, for the justice he seeks, from the point of view of the law, is not only negligible but futile. Only once he has given himself up to the authorities does he begin to sense that his righteous passion calls for peculiar, un- and anti-bureaucratic measures.

## Wolves and Mad Dogs

Drawing on the historical case of the criminal Friedrich Schwann, Schiller's *The Criminal out of Lost Honor* sets out to provide, as the subtitle states, "a true history" of the infamous figure named Christian Wolf in the text.¹ The literary study purports to analyze aspects of the case that escape the purview of the law and thereby provide the anthropological rationale for a program of legal reform. While Christian Wolf is executed for his crimes, "the autopsy of his vice will perhaps instruct humanity and – it is possible, even justice."²

The criticism of the law in Schiller's story is that it lacks discrimination: it operates on a blunt distinction between human and wolf. In this respect, it carries on an absolutist legal tradition that distinguishes the Christian and his inner life from the criminal, or Wolf, who is defined on the basis of outward, and so perceptible, actions. As Hobbes, who marks the beginning of this tradition, states: the concern of the law is not *sin*, the larger category of all possible deviation, but exclusively *crime*, defined as perceptible transgression given that *crimen* derives from *cerno*, to perceive.³ *The Criminal out of Lost Honor* analyses the conditions for such legal discrimination, for crime can be the discernible deviation from the law only once the field of distinctions that facilitates such discernibility has been produced.⁴ These proto-criminal discriminations define what is called "honor." Honor (or its loss) is entirely a function of the system of discriminations that generates social perceptions.

Schiller's account of Christian Wolf's unfortunate case slips at a certain point into the first person so that the reader is made party to the criminal's perceptions and brought to see how, in fact, the criminal is a product of the discriminations of the law. During his time in prison, Wolf observes, "I saw myself as a martyr of natural right, and as a victim of the law."⁵ He sees himself as a victim of the determinations of nature – "Nature had neglected his body"⁶ – as well as of those of society – he is poor and the inn, known as "The Sun," that he runs with his mother, is failing. Indeed, the underlying criticism of the narrative is that the indignities of nature are amplified, rather than alleviated, by those im-

---

1 Friedrich Schiller, *Der Verbrecher aus verlorener Ehre. Eine wahre Geschichte*, reprinted in *Schillers Werke. Nationalausgabe*, vol. 16, *Erzählungen*, ed. Julius Petersen and Hermann Schneider (Weimar: Böhlau, 1954). The text was initially published under the title *Verbrecher aus Infamie* (1786) and subsequently as *Der Verbrecher aus verlorener Ehre* (1792). Translations mine.
2 Schiller, *Verbrecher*, 9.
3 Thomas Hobbes, *Leviathan* (1651), ed. Richard Tuck (Cambridge: Cambridge University Press, 1996), 202.
4 See Joseph Vogl and Ethel Matala de Mazza, "Bürger und Wölfe. Versuch über politische Zoologie," in *Vom Sinn der Feindschaft*, ed. Christian Geulen, Anne von der Heiden, and Burkhard Liebsch (Berlin: Akademie, 2002).
5 Schiller, *Verbrecher*, 12.
6 Schiller, *Verbrecher*, 10.

posed by the law and, as a result, he is destined to be a wolf and not free to be a man. Wolf becomes a criminal – a poacher (*Wilddieb*) – because he is already regarded as one and, by the time he is sent to prison, feels utterly abandoned by humanity, coming to hate "everything resembling the human" at which point, at least to judge from an encounter with a child after his release, he has also "ceased to appear human."[7] Driven to murder by the wolfishness of his social condition, which overwhelms his Christian conscience, he embraces his infamy to become a famous outlaw, gaining the tenuous "honor" of leading a gang of bandits.

The program for legal reform proposed by the narrator in the light of the case of Christian Wolf thus announces the historical transition that Foucault describes as a turn from sovereignty to governmentality.[8] Shifting away from the absolutist logic of exclusion – to recall the classical sense of infamy as exile or civic death – an enlightened approach is proposed that would internalize criminality, and with it all hitherto infamous forms of life, by elaborating a more sophisticated, but also more invasive, system of distinctions by which deviants may be categorized, contained, and ultimately re-formed for human society. Whence the significance of the closing line of the story when, rather than awaiting identification by the authorities, the former outlaw freely *incriminates* himself declaring: "I am the innkeeper of The Sun" and thus shows himself worthy of such an enlightened project.[9] By demonstrating that the human/wolf distinction is a legal construct and an inhumane one, the implicit anthropology of Schiller's narrative calls for a more inclusive and more discriminating model for legal-political recognition in what amounts to an immense bureaucratic project of humanization.

The story of the lost honor of the unhappy Wolf might be set against that of an unfortunate but ultimately "honorable" dog that appeared in Kleist's *Berliner Abendblätter* of October 9, 1810 under the title "Daily Police Notices."[10] It too

---

**7** Schiller, *Verbrecher*, 13.
**8** Particularly significant here are Foucault's lectures on the "abnormal," to which Vogl and Matala de Mazza also refer, Michel Foucault, *Abnormal: Lectures at the Collège de France, 1974–1975*, ed. Valerio Marchetti and Antonella Salomoni, trans. Graham Burchell (New York: Picador, 2003). See also Achim Geisenhanslüke, *Die Sprache der Infamie: Literatur und Ehrlosigkeit* (Paderborn: Fink, 2014); Michael Ott, *Das ungeschriebene Gesetz: Ehre und Geschlechterdifferenz in der deutschen Literatur um 1800* (Freiburg: Rombach, 2001).
**9** Schiller, *Verbrecher*, 29.
**10** All references to Kleist from Heinrich von Kleist, *Sämtliche Werke, Brandenburger Kleist-Ausgabe* (BKA), ed. Roland Reuß and Peter Staengle (Basel: Stroemfeld/Roter Stern, 1988–2010). References to BKA II/1 *Michael Kohlhaas* (1808 and 1810) will be by page number only; citations

concerns the political-zoological problem of accurate discrimination on the part of the law.[11] The account is based on a police report drawn up by the president of the police at the request of the king no less, regarding cases of rabid dogs near his residence in Charlottenburg. In his capacity as editor (*Redacteur*), Kleist made some slight alterations to the report, alterations touching not so much on its substance as its composition. Slight though they were, they changed the tone significantly. It is no longer, as the title suggested, a police notification but an ironic commentary on an excessive bureaucratic response to a perceived but, by its own subsequent admission, unfounded threat to public health and safety. The dog (with a leash still around its neck) ends up in the courtyard of a certain privy councilor, whose healthy dogs set upon it, and cause the otherwise clueless (*rathlos*) authorities to assume it to be rabid (*so hielt man ihn für toll*). The dog and the other dogs it had bitten are promptly shot and buried, Kleist adds, *honorably* (*ehrlich*).[12] This addition shows that the dogs had in the end been loyal and law-abiding domestic animals untouched by the wild rage the authorities mistakenly supposed had infected them. Meanwhile the excessive police measures, Kleist adds, gave rise to the rumor – the very rumor that had presumably solicited the king's request for a report – that a rabid dog had been ravaging the city biting domestic animals and humans alike. Kleist's subtle redacting demonstrates that a kind of contagious and indiscriminate violence, a certain rabidity in short, is here both the supposition and the effect of bureaucratic practice, the ostensible function of which is to ensure that all such dangerous, infectious, and otherwise mad and maddening behavior is excluded from the polity.

Until well after Louis Pasteur's development of a vaccine in the late nineteenth century, rabies presented, of course, a significant biopolitical problem. A zoonotic disease that can be transmitted between animals and humans, it cultivated, as the police report shows, potentially destabilizing anxieties that can-

---

of other Kleist texts will be in-text under BKA with volume and page number, letters by date. Translations mine.

**11** The phrase proposed by Vogl and Matala de Mazza to refer to the use of human-animal distinctions in political discourse in "Bürger und Wölfe," see also *Politische Zoologie*, ed. Anne von der Heiden and Joseph Vogl (Zurich: Diaphanes, 2007).

**12** Reprinted in BKA II/7: 45. For the original report, see Arno Barnet in collaboration with Roland Reuß and Peter Staengle, "Polizei-Theater-Zensur. Quellen zu Heinrich von Kleists 'Berliner Abendblättern,'" *Brandenbürger Kleist-Blätter* 11 (1997): 70; also Roland Reuß's introductory remarks regarding authorship in the *Abendblätter* with regard to similar cases of *Redaction*, "Geflügelte Worte." For a brief commentary on this paragraph and the context of its publication, see Reinhold Steig, *Heinrich von Kleist's Berliner Kämpfe* (Berlin, Stuttgart: Spemann, 1901), 365– 367.

not be separated from an intrinsic horror at the proximity between human and animal. It not only unsettled the tenuous distinction between the animal and the human but also exhibited the thoroughgoing dependence of humans, their political and domestic life if not indeed their "humanity," on domesticated animals that could, however, suddenly turn to unfathomable violence, *Wut*, in German, or *rage*, as the disease is known in French.[13]

But the real problem, as it appears in the *Berliner Abendblätter*, is that of discrimination. For, in cases in which perceptible criteria for a decision that distinguishes between a loyal dog and a raging one are lacking, the distinction rests on an arbitrary decision of a perplexed bureaucracy that thus risks exhibiting its own lack of discrimination. The oversight assigned to the bureaucracy, which articulates lawful distinctions and regulates public perceptions, betrays a constitutive tendency to oversight in the sense of misapprehension which in turn leads to inflationary measures. It is no doubt for these reasons that the authorities took exception to Kleist's notice. He was obliged to publish a retraction in the subsequent issue in which he attributed the title "Daily Police Notices" to an "oversight" (*Versehen*) on the part of the *Redaction* (BKA II/7:50).[14]

While *The Criminal out of Lost Honor* expresses its own historical moment by soliciting the governmental machinery of the modern administrative state, Kleist's work anticipates and critically reflects on the biopolitical consequences of Schiller's program. The question of bureaucracy and its "humanitarian" responsibility for the political-zoological administration of the polity, and the inverse, if not opposing, question of the inhumanity of bureaucracy and the raging madness to which it can give rise, are central to Kleist's novella, *Michael Kohlhaas*. Indeed, it is altogether indicative of the distinction between Schiller and Kleist that, rather than a wolf, a dog appears at the decisive moment when Kohlhaas, "one of the most upright [*rechtschaffensten*] and at the same time outrageous [*entsetzlichsten*] men of his time" (63), questions whether he should stay within the legal system and submit to the injustice with which it is complicit,

---

**13** Under the Grimm's entry, "Tollwut," a compound that apparently entered circulation around 1810, Kleist's epigram to *Robert Guiscard* (1808) is cited, which refers to his recent *Penthesilea* (1808): "Nein, das nenn' ich zu arg! Kaum weicht mit der Tollwuth die Eine [*Penthesilea*] / Weg vom Gerüst, so erscheinet der gar mit Beulen der Pest," *Phöbus* 4–5 (April/May 1808): 70.
**14** The article, according to the correction, should have appeared above the title and so without a title of its own. Remarkably, the usually punctilious editors of the *BKA* have in fact "corrected" the "oversight" and moved the title below the Charlottenburg episode with a note referring to the "typographical error" in the subsequent edition. On the significance of *Versehen* in Kleist's literary work, see Walter Müller-Seidel, *Versehen und Erkennen: Eine Studie über Heinrich v. Kleist* (Cologne: Böhlau, 1967).

or abandon the law and become an outlaw.[15] When his rightful claims have been turned down once again by the authorities, Kohlhaas exclaims to his wife: "I'd rather be a dog if I'm to be trodden underfoot, than a human!" (107). The phrase, as with almost all utterances attributed to Kohlhaas, is ambiguous: it could suggest that in order to *remain a man* and so not be treated like a dog, Kohlhaas must become an outlaw. But perhaps the distraught Kohlhaas is literally expressing the mad thought that it is, under such conditions, preferable *to be a dog*. If he is to be downtrodden by the law or its representatives and so treated inhumanly, he must renounce his human claims and indeed his humanity. By assuming, rather than suffering, the posture of the dog, without, however, abandoning the polity to become a wolf, he discloses a new – non-legal, inhuman, para-political – sphere of action. Only with difficulty can one distinguish a raging man from the inscrutable violence of a rabid dog.

There is a long history of rabies analogies in political discourse. One famous and possibly pivotal instance is Luther's intervention during the peasant wars, "Against the Robbing and Murdering Hordes of Peasants" (1525), in which the rebellious man is likened, in his infectious and lawless violence, to a rabid dog who ought to be treated accordingly:

> Therefore let everyone who can, smite, throttle and stab, secretly or in public, remembering that nothing can be more poisonous, harmful or devilish than a rebel. It is just like having to kill a rabid dog; if you do not strike him, he will strike you, and a whole land with you.[16]

The text emerged in print only after the Battle of Frankenhausen that proved a definitive defeat and turning point in the peasant uprisings (as well as marking the capture, torture, and execution of the evangelical revolutionary, Thomas Müntzer) and so the document was circulating while the peasants – revolting

---

**15** On the relation between *The Criminal out of Lost Honor* and *Michael Kohlhaas* with particular attention to the relation between literature and law and their respective elaboration of subjectivity, see Geisenhanslüke, *Die Sprache der Infamie*, 116–157; Susanne Lüdemann, "Literarische Fallgeschichten. Schillers 'Verbrecher aus verlorener Ehre' und Kleists 'Michael Kohlhaas,'" in *Das Beispiel. Epistemologie des Exemplarischen*, ed. Jens Ruchatz, Stefan Willer, and Nicolas Pethes, *Literatur Forschung* 4 (Berlin: Kadmos, 2007); and Bernd Hamacher, "Geschichte und Psychologie der Moderne um 1800 (Schiller, Kleist, Goethe). 'Gegensätzische' Überlegungen zum 'Verbrecher aus Infamie' und zu 'Michael Kohlhaas,'" *Kleist-Jahrbuch* (2006): 60–74.
**16** Martin Luther, "Against the Robbing and Murdering Hordes of Peasants" (1525), reprinted in *D. Martin Luthers Werke* (Weimar: Böhlau, 1883–2009), 18:358. Hereafter cited as WA (*Weimarer Ausgabe*) with volume and page number.

or not – were in fact being systematically slaughtered, disciplined, and contained by the authorities.[17]

It is not unlikely that Kleist had in mind Luther's text against the robbing and murdering hordes of peasants and, more generally, Luther's political-theological dispute with Müntzer[18] when he wrote the story, set in the mid-sixteenth-century, about a horse-dealer whose "righteous feeling" (*Rechtgefühl*) made him "a robber and a murderer" (64).[19] In turning to the time of Luther, Kleist, however, also engages contemporary debates around 1800 in which the biopolitical and political-theological determinations of the human with regard to nature were at stake. The focus is no longer, as it was for Schiller, on criminality, but rather on a righteous extremism that confounds discrimination for which Luther had coined the political-zoological term: *Schwärmerei*.[20]

## *Schwärmerei*

Luther applied the category of *Schwärmerei* to the radical Reformation in general, but most consistently to discredit those figures, and the followers who

---

**17** On Luther's pamphlet and its reception, see Roland Herbert Bainton, *Here I Stand: A Life of Martin Luther* (New York: Abingdon-Cokesbury Press, 1950), 280–4. For a sociological assessment of the significance and outcome of the Peasant Wars, see Peter Blickle, *Die Revolution von 1525*, 4th expanded edition (Munich: Oldenbourg, 2004), esp. 245–278.
**18** An early attempt, to which Kleist could have had access, to provide a comprehensive account of Thomas Müntzer that was inspired by the French Revolution was published by Georg Theodor Strobel, *Leben, Schriften und Lehren Thomä Müntzers, des Urhebers des Bauernaufruhrs in Thüringen* (Nürnberg: Monath und Kussler, 1795). The pro-Luther account of Müntzer's role in the peasant insurrection published by Melanchton, "Historie Thome Müntzers, des anfangers der Döringischen vffrur" (1525), was accessible in a number of editions of Luther's works.
**19** *Michael Kohlhaas* takes up the story "from an old chronicle" as the subtitle announces, of the historical trader, Hans Kohlhase, who in the 1530s, having been abused by a Saxon aristocrat whose people stole a pair of his horses, and without, in his eyes, receiving just retribution for damages, started a feud against the Saxon polity, leading a brigand existence along the border with Brandenburg, at one point corresponding with Luther and in some accounts even meeting him, until his eventual capture and execution. On Kleist's sources, see Roland Reuß, "Nachrichten von Hans Kohlhase," *Berliner Kleist-Blätter* 3 (1990): 44–54.
**20** For a survey on the *Schwärmer* in literature around 1800, see Manfred Engel, "Das 'Wahre', das 'Gute' und die 'Zauberlaterne der Begeisterten Phantasie.' Legitimationsprobleme der Vernunft in der spätaufklärerischen Schwärmerdebatte," *German Life & Letters* 62.1 (2009): 53–66; also Victor Lange, "Zur Gestalt des Schwärmers im deutschen Roman des 18. Jahrhunderts," in *Festschrift für Richard Alewyn*, ed. Herbert Singer and Benno von Wiese (Cologne: Böhlau, 1967); Jörg Paulus, *Der Enthusiast und sein Schatten: literarische Schwärmer- und Philisterkritik im Roman um 1800* (Berlin: Gruyter, 1998).

seemed inevitably to swarm around them, who claimed to enjoy an immediate communication with the divine.[21] Contesting such inspirationalism, he turned to the natural world, adapting a zoological term associated with the behavior of certain vermin in order to insinuate that the transcendence of the *Schwärmer* was a function not of a voice from on high but of an all-too-earthly animal swarming from below. In its most radical form, the *Schwärmer* was a figure, who, like Müntzer, was convinced not only of the divine order investing his soul but that the implementation of that order, the Kingdom of Heaven on earth, was the task of all true believers.[22] Thus the *Schwärmer* in general mistakes his inner voice, his dreams and visions, for communications from god when they are merely symptoms of his own sin-imbued creaturely existence. And in the cases in which *Schwärmerei* escalates into a political imperative to change the world as an instrument of god's work, it inevitably expresses itself, as Luther saw things, in the sort of contagious, amorphous and merely destructive violence associated with the image of the swarm.

In the late eighteenth century, as discussed in Chapter One, the theological, political, and zoological concerns concentrated in the term *Schwärmerei* would return, especially in the wake of the French Revolution and the revolution in philosophy brought about by Kant.[23] The question of the human, of the freedom or transcendence proper to the human, as well as the related concern about the religious, ideological, delusional, pathological, and in any case dangerous excesses of inhumanity to which humans appeared prone, was the heated subject of

---

[21] For a thorough examination of the wide-ranging uses of the term and other polemical language see Günter Mühlpfordt, "Luther und die 'Linken.' Eine Untersuchung seiner Schwärmerterminologie," in *Martin Luther: Leben, Werk, Wirkung*, 2nd edition, ed. Günter Vogler (Berlin: Akademie, 1986), 325–346; for a survey of Luther's position with regard to politics and economics in the light of his theology and the unfolding social and political upheaval of the period, see chapter III, "Prophets, Enthusiasts, Iconoclasts, Fanatics and the Peasant War," in Martin Brecht, *Martin Luther: Shaping and Defining the Reformation 1521–1532* (Minneapolis: Fortress Press, 1990).

[22] See Hans-Jürgen Goertz, *Thomas Müntzer: Mystiker, Apokalyptiker, Revolutionär* (Munich: Beck, 1989) and his earlier *Innere und äussere Ordnung in der Theologie Thomas Müntzers* (Leiden: E.J. Brill, 1967). Writing explicitly in response to the exigencies of the early twentieth century, Ernst Bloch insists on a stricter equivalence between politics and theology in his *Thomas Münzer als Theologe der Revolution* (Munich: Wolff, 1921).

[23] On the trajectory of *Schwärmerei* or "enthusiasm" in the wake of the Reformation through the Enlightenment, see Lawrence Eliot Klein and Anthony La Vopa, eds., *Enthusiasm and Enlightenment in Europe, 1650–1850*, The Huntington Library Quarterly 60.1 (1998); Michael Heyd, *Be Sober and Reasonable: The Critique of Enthusiasm in the Seventeenth and Early Eighteenth Centuries* (New York: Brill, 1995); and Lothar Kreimendahl and Norbert Hinske, eds., *Die Aufklärung und die Schwärmer*, (Hamburg: Meiner, 1988).

the so-called *Schwärmer*-debates in the German public sphere.[24] If, throughout the Enlightenment, *Schwärmerei* had been deployed polemically to refer, in a vague and indeterminate sense, to all that was opposed to Enlightenment reason (superstition, prejudice, irrationality, etc.),[25] by the end of the eighteenth century such an opposition was no longer tenable. On the one hand, anti-Enlightenment thinkers like Herder criticized the relentless, cold, deadening, and architectonic character of rationality,[26] while its defenders found themselves driven to distinguish between a healthy, productive, *Enthusiasmus* that was indispensable to the progress of reason and an unhealthy and destructive *Schwärmerei*.

Both parties, however grudgingly or gleefully, acknowledged the dangerous excesses to which reason itself proved vulnerable. While Kant's entire critical project can be seen as responding to this problem, the watershed is perhaps most clearly indicated in the general definition of *Schwärmerei* he conditionally proposes in the middle of the second *Critique*: "If *Schwärmerei* in its most general meaning is a principled overstepping of the limits of human reason [...]" (*Wenn Schwärmerei in der allergemeinsten Bedeutung eine nach Grundsätzen unternommene Überschreitung der Grenzen der menschlichen Vernunft ist...*).[27] *Schwärmerei* emerges as a propensity for over-principled action, an excess of a certain "rationality" that transgresses the limits of *human* reason and so the limits of humanity by reason.[28]

In 1805–1806, about the time Kleist would have been thinking about adapting the story of the historical Hans Kohlhase,[29] Fichte and Schelling engaged in one of the last of the *Schwärmer*-debates. At stake was the cosmo-political question, which today would be called an eco-political one, concerning the limits and

---

**24** For a survey of the *Schwärmer*-debates at the end of the eighteenth century in Germany, see Anthony La Vopa, "The Philosopher and the *Schwärmer*: On the Career of a German Epithet from Luther to Kant" in *Enthusiasm and Enlightenment in Europe*.
**25** See Norbert Hinske, "Die Aufklärung und die Schwärmer – Sinn und Funktion einer Kampfidee," in *Die Aufklärung und die Schwärmer*, 6.
**26** Johann Gottfried Herder, "Philosophie und Schwärmerei, zwo Schwestern" (1776), in *Herders sämtliche Werke*, ed. Bernhard Suphan (Berlin: Weidmann, 1877–1913), 9:497–98.
**27** Immanuel Kant, *Critique of Practical Reason*, trans. and ed. Mary Gregor (Cambridge: Cambridge University Press, 2015), 73. German references to *Akademie-Ausgabe* (AA) with volume and page number, Immanuel Kant, *Gesammelte Schriften*, ed. Preussische [later Deutsche] Akademie der Wissenschaften (Berlin: Reimer, later de Gruyter, 1900 ff). AA, 5:85.
**28** For a discussion of this conditional definition and the allure of "moral *Schwärmerei*," see Peter Fenves, "The Scale of Enthusiasm," in *Enthusiasm and Enlightenment in Europe*, 122–135.
**29** It was in 1805, while working at the Berlin *Finanzdepartement*, that Kleist's intention to write a story based on that of Hans Kolhase was first mooted. He was supposedly introduced to the story early that year by his friend Ernst von Pfuel, see Roland Reuß, "Nachrichten von Hans Kohlhase," *Berliner Kleist-Blätter* 3 (1990): 44.

scope for the exploitation of nature by reason. Fichte, in his *Wissenschaftslehre*, was representative of the rational domination of the earth that licensed the liquidation of nature as part of the thoroughgoing technical-political take-over and development of the world for the sake of the human. Schelling, attending to what he called *Naturphilosophie*, attempted to articulate an altogether different relation to nature that was not centered on the human subject and that acknowledged nature's fundamental difference from all subjective representations. In this debate the Fichtean subject emerges as the archetype of the bureaucrat treating nature as though it were in need of a comprehensive "rationalization," while Schelling considers nature to be self-organizing in ways that cannot be subsumed by subjective-bureaucratic determinations. If Kleist's *Michael Kohlhaas* can be read as a late contribution to the *Schwärmer*-debates, it is because it restages this quarrel in order to expose an antinomy within bureaucratization that is the common source of the *Schwärmerei* that Fichte and Schelling reproach each other for and that I have called bureaucratic fanaticism.

Whatever Kleist might have known of their debate, which was carried out in a very public and polemical fashion in popular lectures and publications, the problematic follows as a logical progression from the "Kant crisis" described in his 1801 letters to Wilhelmine von Zenge. Indeed, Ernst Cassirer has argued that this epistemological crisis, which was in fact occasioned by what Kleist referred to as the "recent so-called Kantian philosophy," was brought about by a reading of Fichte's *The Determination of the Human* (1800), an essay intended as a popular introduction to his system of thought.[30] The second section of Fichte's text, entitled "Knowledge" (*Wissen*), concludes, very much in the spirit of Kleist's distraught letters to von Zenge,[31] with an exposition of the radical contingency and ultimate instability, if not completely illusory character, of the known world: "There is nothing lasting, neither in me nor outside of me, but only constant change. I know nothing about being, not even about my own. There is no being. – *I myself* know nothing at all, and do not exist. *Images* are: they are all that there is [...]. All reality is transformed into a wondrous

---

**30** Ernst Cassirer, *Heinrich von Kleist und die Kantische Philosophie* (Berlin: Reuther & Reichard, 1919). Ludwig Muth questions the validity of Cassirer's assumption and argues instead that Kleist's crisis emerged from a reading of the second part of Kant's *Critique of the Power of Judgment* on teleological judgment; nonetheless he admits that Cassirer's characterization of the nature of the crisis is exact, *Kleist und Kant; Versuch einer neuen Interpretation* (Cologne: Kölner Universitäts-Verlag, 1954). Muth's suggestion will be returned to below.
**31** Esp. Kleist's letter dated March 22, 1801, Berlin. BKA IV/1.

dream [...]."³² Such passages on the shifting grounds of human knowledge would no doubt have been enough to nauseate an existential reader and aspiring natural scientist such as Kleist.³³

Cassirer is of the opinion that "Faith" (*Glaube*), the closing section of Fichte's essay, would have been too obscure and profoundly unsatisfying for the likes of Kleist.³⁴ Crucial in the light of the subsequent *Schwärmer*-debate, however, is Fichte's acknowledgement in the transition from "Knowledge" to "Faith" that a philosophy that departs from the subject cannot sustain itself without recourse to something beyond the grasp of human knowledge and action. An instance of transcendence – that has simply to be believed – is needed to first motivate and guarantee the self-determining activity of the subject. Fichte calls this pre-determining or predestining instance the voice of conscience (*Stimme des Gewissens*). It expresses itself in the pure form of the command: Act! "Not mere knowing, but rather *acting* according to one's knowledge, is one's determination: [...] you exist in order to act; your action and your action alone determines your worth."³⁵ The voice of conscience, speaking for reason as such, demands nothing less than the rational-scientific conquest of nature as the measure of the determination of the human.³⁶

In his Schellingesque *Differenzschrift*, Hegel had some years before characterized the difference between Fichte and Schelling around the question of difference itself. Fichte, he argues, refuses to recognize a "real opposition" (*reale Entgegensetzung*).³⁷ That is to say, there is no real alterity for Fichte, certainly not of the non-subjective, "natural" sort attested to in Schelling's *Naturphilosophie*. It is true that in *The Determination of the Human*, as elsewhere in Fichte, nature is always only there for the human subject's self-realization (as *not-I*) and so ultimately, one could say, nature is not anything other than "human nature": "The nature in which I have to act is not something foreign, called into existence without reference to me, into which I could never penetrate. It is formed by my own laws of thought, and must be in harmony with them."³⁸ It is this refusal to ac-

---

**32** Johann Gottlieb Fichte, *Die Bestimmung des Menschen* (1800), reprinted in *Fichtes sämmtliche Werke*, ed. I. H. Fichte (Berlin: Veit & Comp., 1845–1846), 2:244. Translations mine.
**33** See Kleist's letter to Wilhelmine von Zenge, Berlin, March 23, 1801. BKA IV/1.
**34** Cassirer, *Heinrich von Kleist und die Kantische Philosophie*, 48.
**35** Fichte, *Bestimmung*, 249.
**36** Fichte, *Bestimmung*, 298.
**37** Georg Wilhelm Friedrich Hegel, *Differenz des Fichteschen und Schellingschen Systems der Philosophie* (1801), in *Werke* (Frankfurt am Main: Suhrkamp, 1979), 2:99. The publication of Hegel's *Differenzschrift*, and the subsequent collaboration of Hegel and Schelling editing the *Kritisches Journal der Philosophie*, marked the definitive break between Schelling and Fichte.
**38** Fichte, *Bestimmung*, 258.

knowledge the real difference of nature by means of an anxious potentialization of subjectivity that leads Schelling later to call Fichte a *Schwärmer*.

In *Laying out the True Relation of Naturphilosophie to the Improved Fichtean Doctrine* (1806), Schelling asks:

> If a stubborn striving to force through and insist on the universality of his subjectivity by means of his subjectivity, to annihilate all nature wherever possible, and in contrast to apply the unnatural as a principle, and inflexibly to insist on the validity of a one-sided system of knowledge in its most glaringly truncated form as scientific truth – if such striving may be called swarming [*Schwärmen*], then who in all this time has more aggressively, loudly and in the most proper sense swarmed than Herr Fichte himself?[39]

Drawing further on the time and terminology of Luther and his critique of iconoclasm (*Bildstürmerei*), Schelling even refers to Fichte's "storming of nature" (*Naturstürmerei*). For a subjectivity that responds to a "voice" demanding relentless activity and that refuses to recognize the reality of anything other than what is determined by such activity, is not only itself unnatural and alienated but threatens to annihilate the nature upon which it depends. Fichte's *Wissenschaftslehre* is *Schwärmerei*.

Schelling was reacting to Fichte's claim in the lectures he gave in Berlin in 1805, *Fundamental Features of the Present Age* that "all *Schwärmerei* is and will necessarily be *Naturphilosophie*."[40] As Fichte sees things, Schelling's *Naturphilosophie* is symptomatic of the desolation of the contemporary epoch, the fundamental feature of which is the need for metaphysical assurance to ground, orient, and enliven experience. *Naturphilosophie* claims to find such assurance in the hidden non-empirical ground of nature, rather than recognizing it, as Fichte would have it, as the proper calling (*Bestimmung*) of man. *Naturphilosophie* is *Schwärmerei* because it has no conception of human action in nature but only, like the swarming creatures evoked in the term, of the human's captivation by it. The *Naturphilosoph* threatens to lose himself in nature. The term "Schwärmerei" in this late *Schwärmer*-debate thus describes polar attitudes regarding the

---

**39** Friedrich Wilhelm Joseph von Schelling, *Darlegung des wahren Verhältnisses der Naturphilosophie zu der verbesserten Fichte'schen Lehre* (1806), reprinted in *Schellings sämmtliche Werke*, ed. K.F.A. Schelling (Stuttgart: Cotta, 1856–1861), I.7:47.
**40** Johann Gottlieb Fichte, *Die Grundzüge des gegenwärtigen Zeitalters* (1806), reprinted in *Gesamtausgabe der Bayerischen Akademie der Wissenschaften*, ed. Erich Fuchs, Reinhard Lauth, Hans Jacobs, and Hans Gliwitzky (Stuttgart: Frommann-Holzboog, 1964–2012), 1/8:118. Lectures given at the *Akademie der Wissenschaften* in Berlin 1804–1805 and first published in 1806. Individual lectures also appeared in *Eunomia* (January 1805), 1:18–35; and *Geschichte und Politik* (1805), 1:1–23.

contested determination of the human with respect to nature in the first decade of the nineteenth century.

## Bear Nature and Bureaucracy

In their respectively *schwärmerisch* ways, both Fichte, departing from the side of the subject, and Schelling, departing from the side of nature, sought to resolve the epistemological perplexity that had brought on Kleist's crisis in his reading of "so-called Kantian philosophy" in 1801. Ludwig Muth has argued that this crisis was first of all a career crisis insofar as Kleist planned to become a scientist. It stemmed therefore (contra Cassirer) most likely from a reading of the second part of Kant's *Critique of the Power of Judgment*, the critique of teleological judgments.⁴¹ The crisis would then hinge on Kant's argument that the study of nature as a living and comprehensive totality could only be a critical or regulative supposition for natural science. It can only operate on the supposition of a "technique of nature" (*Technik der Natur*) that remains, however, a human technique for the study of nature and cannot be taken as the organizing structure of nature itself.⁴² For Kleist this means that the status of scientific knowledge, as he despairingly writes to von Zenge, is radically undecidable: "Everything that you say to me in objection, *may* be true, without, however, the doubt being erased."⁴³ In the following years, he would abandon his scientific ambitions and become, among other things, a writer.

If in his literary writings Kleist found the means to "aesthetically" work through his crisis, one place that takes up the problem of the "technique of nature" is to be found in "On the Marionette Theater" (1810). The text presents a dialogue between a narrator and the renowned "first dancer of the opera," who appear to be involved in a purely aesthetic discussion concerning the "natural grace of man." (BKA II/7:325). To illustrate a point, the dancer tells a story about a visit to a Russian estate. The sons of the nobleman Hr. v G… were keen fencers and the elder challenged the dancer who, however, out-fenced him with ease. The boy admits "that he had found his master," but adds, "everything in nature finds its own," and proposes to lead the dancer to his master – only this master-fencer turns out to be a bear (BKA II/7:329). What follows is a description of the dispositive of man vis-à-vis nature around 1800: "The bear stood, as I

---

**41** Muth, *Kleist und Kant*, esp. 58–75.
**42** See Immanuel Kant, *Critique of the Power of Judgment*, ed. Paul Guyer, trans. Paul Guyer and Eric Matthews (Cambridge: Cambridge University Press, 2001), 263–284. AA, 5:§72–78.
**43** Letter to Wilhelmine von Zenge, Berlin, March 28, 1801. BKA IV/1.

in astonishment came up before him, on his hind legs, with his back against the post to which he was chained, his right paw raised ready to strike, looking me in the eye: this was his fencing posture" (BKA II/7:329). As if in parody of Fichte's voice of conscience, which by inciting one to act lends reality to one's otherwise dream-like knowledge, the bemused dancer finds himself instructed to attack the animal – "Thrust! Thrust!" (*Stoß! Stoß!*) – but it is as if his every movement is anticipated, the bear parries the attacks and ignores the feints (BKA II/7:329– 330).

If the dancer's anecdote about fencing hovers ambiguously between aesthetics and violence, between a dance and a struggle unto death, it also recalls the contours of the Fichte-Schelling dispute. On the one hand, the bear appears to respond according to the rules of fencing, rules, what is more, that it has been trained to follow by Hr. v. G.... On the other hand, it exceeds those rules and so fences like no other fencer in the world, as though it reads the soul of its antagonist (BKA II/7:330). If, in the first case, nature (the bear) has been trained (by the Herr) to correspond to the engagements of the fencer, in the second, nature is the determining ground of the fencer's action, a ground that remains unfathomable to the fencer himself. In both cases, however, there is a correspondence between the postures, positions, and posing of man and those of nature.

Most striking, however, is the fact that the thrusts should be met with a parry at all – and not, for example, with a bite. The wild possibility of a radical, or animal, alterity is closed and locked up in advance. The bear may outwit the fencer but the fencer need never abandon the certainty that the bear will play by the rules. The game, in that sense, is of course rigged – the symbol for which is the post to which the bear is chained. This then would be the "technique of nature," that Kant intended as a heuristic device and that Kleist here presents as the concrete apparatus that had to be in place before the training regime of the *Herr* could even begin. Drawing attention to the artificiality of this set-up, Kleist's image discloses what is supposed to be foreclosed, namely, the possibility of an encounter with *bear* nature.

The dancer's anecdote thus indicates a fundamental duplicity in "nature" that might be traced throughout Kleist's work and roughly corresponds to the extremes presented by Fichte and Schelling. The fencing bear, like the domestic dog, stands for nature that is there *for* man. It appears according to more or less human rules so as to be there just for his engagement, enjoyment, and edification. But there is also another side of nature, the bear or rabid side, that has nothing to do with human enterprises and conscious aspirations, that perplexes the laws of human perception, and disrupts the expectations of teleological judgment – and thus altogether flouts the (human) laws of nature.

Kleist's intervention into the epistemology of nature is to shift the debate from philosophy to a history of techniques for the apprehension of nature. Around 1800 the technical-political apparatus that set out to arrange and preserve the rational representation of the world was already called "bureaucracy." The "fragile setup" (*gebrechliche Einrichtung*) of this bureaucratically instituted world (79), as well as the sometimes excessive and arbitrary violence by which it is maintained, is exposed in such cases of mistaken oversight as the one involving the supposedly rabid dogs in Charlottenburg reported in the *Berlin Abendblätter*. The problem of *Schwärmerei*, which in the Fichte-Schelling dispute revolves around the relation between human and nature, in Kleist gets caught up in the problem of bureaucratization.

In *Michael Kohlhaas*, Kleist captures the historical moment at the turn of the nineteenth century by presenting the convergence, on the one hand, of the global claims for the rational organization of the world in philosophy and, on the other, the global deployment of rationality by means of bureaucracy for economic and political ends that sought to restrain nature and convert it into "natural resources" suitable for human purposes.[44] When Fichte writes in *The Determination of the Human:* "Nature must gradually be brought to a condition in which her regular dynamic can be securely reckoned and counted upon, and in which her force invariably demonstrates a fixed relation to the power that is determined to rule her – that of the human,"[45] Kleist draws the consequence that this is not a philosophical but a bureaucratic problem of implementation. Nature has to be processed in order to become what it is in the philosophy of Fichte and, indeed, what it would be for the industrial and colonial expansion of the nineteenth century.

Setting *Michael Kohlhaas* in the sixteenth century, Kleist presents a threshold in this political history of nature in which the universal aspirations of bureaucratic organization are in their infancy. For it was with the advent of the age of expansion and the reformations that it solicited, first of all of space – even in those territories (like the German ones) least disposed to take part in the first wave of colonial exploration – that the contours of the nineteenth-

---

[44] Among his many attempted career paths, Kleist had been himself a bureaucrat: he studied Kameralia and Jurisprudenz in Frankfurt am Oder in 1799, prepared for the Prussian civil service exam in Berlin in 1800, and in 1805 worked in the Berliner Finanzdepartement and later as Diätar in the Domänenamt in Königsberg. Indeed, Rupert Gaderer refers to *Michael Kohlhaas* as Kleist's "bureaucracy novella" not only because it is about bureaucracy but equally because it emerged "doubtless out of bureaucracy itself," *Querulanz: Skizze eines exzessiven Rechtsgefühls* (Hamburg: Textem-Verlag, 2012), 42–43.

[45] Fichte, *Bestimmung*, 268.

century conception of the relation between human and nature might be discerned. And of course Michael Kohlhaas, in the very opening of the story, is subject to this territorial re-organization (and the costs of its implementation) when he is brought to a halt in the pouring rain by a felled tree across the road and instructed to pay a toll. The tree, which now contributes to the legal articulation of space, is indicative of the interpenetrating relation between nature, law and its implementation that goes to the very heart of the story. As he prepares to proceed, Kohlhaas remarks to the toll collector, who, cursing the bad weather, irritably hurries him to move on: "'Yes, old man, if that tree had stayed in the wood, it would have been better for me and for you'" (65).

## Humanity?

*Michael Kohlhaas*, the first part at least, is a story about a horse dealer who, through the abuse of power of a series of corrupt bureaucrats, is wrongfully deprived of his two fine black horses, *Rappen*, which are subsequently exploited beyond recognition. Kohlhaas then devotes himself tirelessly to a futile legal battle with the corrupt authorities and ultimately resorts to a full-scale partisan insurrection for the *restitutio in integrum* of the horses. It is his apparently exclusive insistence on his own private right that led Ernst Bloch famously to call Kohlhaas a "legal-pedant (literally, a paragraph-rider) out of righteous feeling" (*Paragraphenreiter aus Rechtsgefühl*).[46] And on these grounds, Carl Schmitt, in his study of the partisan, dismisses the novella as apolitical, "because he [Michael Kohlhaas] fought exclusively for his own injured private rights, not against a foreign conqueror and not for a revolutionary cause."[47] Crucial to Bloch's reading, however, is an opposite tendency: Kohlhaas's apparently obsessive positive-legal preoccupation is actually a symptom of a more fundamental *political* concern that he is, however, unable to express – except perhaps in the unspeakable and unexpected outburst of para-military violence to which he resorts.[48]

---

[46] Ernst Bloch, *Naturrecht und menschliche Würde* (1961), reprinted in *Gesamtausgabe* (Frankfurt am Main: Suhrkamp, 1985), 6:93.
[47] Carl Schmitt, *Theorie des Partisanen. Zwischenbemerkung zum Begriff des Politischen* (Berlin: Duncker & Humblot, 1963), 91. Translation mine.
[48] Andreas Gailus argues that "the political" in Kleist emerges from "the double foundation of anarchic passion and its structuring in the symbolic machinery of the state," *Passions of the Sign: Revolution and Language in Kant, Goethe, and Kleist* (Baltimore: Johns Hopkins University Press, 2006), 128. In these terms, Kohlhaas's excessive and ultimately incoherent adherence to

Bloch opens his brief but illuminating reading of the novella in *Natural Law and Human Dignity* (1961) with this reflection:

> Something just forgotten can suddenly appear to be inordinately important. All that is missing concentrates itself in the disappeared memory. Thus there is an empty passion for positive right as soon as it, at a not particularly important point, happens not to be maintained. The passion would thus be worthy of a better cause, or rather: it stands for a better cause [*eine bessere Sache*].[49]

The excessive affect to which even a petty abuse by the juridical-bureaucratic system gives rise can express itself in a passion for positive right – but only because, under the prevailing regime, it is not possible to find the terms with which to formulate the just, or "more worthy," object of its feeling. At the extreme, the insistence on a single case of violated right becomes the rallying point against the broader, structural injustice of the legal system as a whole. The more tenaciously and rigorously one insists on positive right and the semblance of justice it promises, the more one ultimately denies the law and exhibits its structural injustice, to the point that one may even present a revolutionary or chiliastic threat to the legal order as such.

Bloch in this way accounts for the inflationary excess of Kohlhaas's actions with regard to his seemingly petty legal claim. Although he does not use the term, Bloch's reading of Kleist thus provides an essential contribution to the "theory of the partisan." Contra Schmitt, for whom Kohlhaas is just an outlaw or criminal, one can say following Bloch: Kohlhaas *is* a partisan, a partisan without a cause.[50] For, in what is certainly a characteristic, if not definitive, condition of partisanship, the terms are lacking in which he might express his injured righteous feeling. Bloch's reading further suggests that what is at stake in the figure of Michael Kohlhaas and his famous *Rechtgefühl* is nothing less than "natural law and human dignity," to cite the title of Bloch's book.[51]

---

the symbolic order betrays an "affective surplus" that is parapolitical or partisan in its expression.

**49** Bloch, *Naturrecht*, 93. Translation mine.

**50** For a study explicitly responding to Schmitt that elaborates the ways in which *Michael Kohlhaas* anticipates the Prussian partisan theories developed subsequently by the likes of Clausewitz and Gneisenau, see Wolf Kittler, *Die Geburt des Partisanen aus dem Geist der Poesie: Heinrich von Kleist und die Strategie der Befreiungskriege* (Freiburg: Rombach, 1987), 291–324.

**51** Kohlhaas's "Rechtgefühl" is often, as is the case in Bloch, transliterated into the more conventional German word *Rechtsgefühl*. I will maintain Kleist's strange term, untranslated, when specifically discussing Kohlhaas.

If this is the case, the horse dealer's one use of the term "human" (*menschlich*) is altogether revealing, for it indicates a significant displacement of the question of natural right and human dignity in the early nineteenth century context to which Bloch assigns Kleist's particular contribution in his chronologically ordered study.[52] It has to do directly with the unfortunate horses – which present, after all, the occasion for the entire course of events – and the abject condition in which Kohlhaas finds them when he returns to Tronkenburg:

> How great was his astonishment when, instead of his sleek, well-fed black horses, he saw a haggard pair of skinny mares, with bones on which one might have hung things as on hooks, their manes and coats entangled from want of grooming and care; the true image of misery in the animal kingdom! Kohlhaas, to whom the horses neighed with a feeble movement, was beside himself with indignation, and asked what had befallen his horses? (74)

Kohlhaas who, when he established that the demand for an official pass had no legal basis, "smiled at the lean Junker's joke," and returned to Tronkenburg, "without any particular bitter feeling beyond that of the general privation of the world" (73), only begins to get really agitated upon the sight of this true image of misery in the animal kingdom. Confronting the castellan, who happens to be passing, the distressed Kohlhaas demands to know who permitted Junker von Tronka to use his horses for field work, adding: "is this really human?" (75). Here Kohlhaas is, to be sure, asking about the legality of the case: what *right* has von Tronka to use his horses in this way? But the question: "is this really human?" (*ob das wohl menschlich wäre?*) appears to indicate a concern that goes beyond the question of right, certainly of positive right. The question of the human emerges here ambiguously between a legal claim and the suffering animals; at the very point of application of the bureaucracy's – possibly illegal and questionably humane – penetration of the animal kingdom. This hesitant question, posed by Kohlhaas as if involuntarily in the midst of a discourse on Right as he, at the same time, attempts to get the horses to show some sign of life, already presents in tentative form the structural ambivalence of his *Rechtgefühl*.

Perplexing in Kohlhaas – first of all to himself – is that he feels at once driven to make his claims in a seemingly rigorist legal-bureaucratic discourse of positive right, and incited to engage in some incommensurable course of action that

---

[52] It is striking that the author of *Thomas Münzer* (1921) makes no reference to the sixteenth-century religious movements that form the background of Kohlhaas's (as well as the historical Kohlhase's) agitation.

seems instead to indicate, as Bloch argues, an obscure claim of "natural right" or perhaps, rather, a claim on behalf of a "nature" that can have no rights at all. At stake would be something like a human comportment to nature or a natural human comportment, one that does not dehumanize itself in an exploitation of nature of the sort exhibited in the image of the abused horses.

In Fichte, as we have seen, the determination of humanity comes from a voice of conscience that announces itself as the very form of reason with the sheer command: Act! There is, beyond the decisive penetration of this imperative, no positive passive, or passionate, dimension to his conception of humanity. Even recognition (*Anerkennung*), Fichte's innovation in the field of natural right, which involves an acknowledgment of the self-determining activity (that is, the humanity) of the other, consists in a purely inter-active encounter in which each mutually self-limits their own sphere of action.[53] This led the young Hegel to criticize Fichte's notion of humanity for being too bureaucratic and lacking any recognizably human feeling – trust, desire, love – that must involve a *real* moment of passivity in its receptivity of the other.[54]

In a similar way, reason announces itself in Kant in the form of the categorical imperative that discloses the "humanity in our person" (*Menschheit in unserer Person*). But in Kant the imperative is accompanied – and this is definitive of the *humanity* of this reasonable being (a being with human sensibility) – by an ambivalent feeling that is at once painful and pleasurable, humiliating and elating that he calls *Achtung*.[55] *Achtung* is the feeling of one's own humanity and the "acknowledgment" of the humanity enshrined in any other person. As such it is the paradoxical feeling of belonging to a rational community that transcends the sensible realm, a sense therefore of distinction from the sensible world in which one is nonetheless called upon to perform rational-moral activity.[56]

---

[53] See the conclusion to Fichte's deduction of the legal relation (*Rechtsverhältnis*) in *Foundations of Natural Right in accordance with the Principles of the Wissenschaftslehre* (1796), reprinted in *Fichtes sämmtliche Werke*, 3:44.

[54] Hegel writes in 1801, "the law must prevail, even if as a result trust, joy, and love, all the potencies of an authentic ethical identity, were to be, as one says, destroyed root and branch," *Differenzschrift*, 87.

[55] On *Achtung*, see Immanuel Kant, *Groundwork of the Metaphysic of Morals* (1785), AA, 4:401; "On the Incentives of Pure Practical Reason," in *Critique of Practical Judgment* (1788), AA, 5:71–88; and the "Analytic of the Sublime" in *Critique of the Power of Judgment* (1790), AA, 5:244–278.

[56] For a discussion of *Michael Kohlhaas* in the context of late eighteenth-century debates on "moral feeling," see Joachim Rückert, "'...der Welt in der Pflicht verfallen....' Kleists 'Kohlhaas's als moral- und rechtsphilosophische Stellungnahme," *Kleist-Jahrbuch* (1988/89): 384–390.

In Kleist's work, especially in *Michael Kohlhaas* – and in this respect it is profoundly an expression of the nineteenth century – there is nothing like a transcendent voice of reason addressed to the human. Rather, there is an impersonal imperative of "rationalization" in which, in the name of "humanity," or in any case, "humanitarianism," different spheres of rationality are bureaucratically implemented in response to differing and changing political, economic, and technical circumstances. Under such conditions, a new, and thoroughly ambivalent, feeling emerges – *Rechtgefühl* – that is intimately related to bureaucratization: it is at once a bureaucratic feeling, an impulse to translate everything into rational-bureaucratic terms, and a feeling that reacts against the relentless colonization of bureaucracy and its brand of rationalism. *Rechtgefühl* is, as it were, torn between a sense of awe for bureaucracy and a thorough disdain for it.

In *Michael Kohlhaas*, this feeling is related to humanity, if only in the form of a question. It announces itself not in a categorical voice of reason, nor even in a human voice or human figure or face, but in the inarticulate animal address of his neighing horses. The question of humanity does not, therefore, pose itself as distinct from, but rather expressly in relation to, the natural world. Humanity appears as a relational category, one not defined by, or exclusive to, relations between human beings, but relating to a receptivity of "nature," beyond the distinctions that qualify nature for the purposes of its rational-technical exploitation and even aesthetic contemplation.

It is immediately following the question of humanity that Kohlhaas first exhibits a disposition towards extreme "righteous" action: "The horse dealer's heart beat against his chest, he felt strongly inclined to fling the good-for-nothing fatso into the mud, and put his foot on his brazen face. Yet his *Rechtgefühl*, which resembled a gold scale, still wavered..." (76). His pivoting *Rechtgefühl* checks itself, which is to say: his righteous *feeling* in the face of his miserable horses and the impudent bureaucrat is checked by his feeling of *Right* that would officially establish the facts of the case and the applicable legal situation. The rest of the account traces his oscillation between these two competing responses to this thoroughly ambivalent affect – expressed at once in an articulate but incoherent legalism and an inarticulate excitement and violent agitation against the legal system as a whole.

Kohlhaas's confusion, born of the contrariety inherent in his *Rechtgefühl*, escalates to the point that, maddened by his impotence and the recalcitrance of the castellan, Kohlhaas refuses to even recognize his mistreated *Rappen*. The castellan mockingly tells von Tronka, "he refuses to recognize the horses are his" (78), to which the indignant Kohlhaas responds: "These *are* not my horses, righteous lord! These are not the *horses* that were worth thirty gold coins! I want my well-groomed and healthy horses back!" (78). He declares furthermore that he would rather hand the pair over to the knacker (*Abdecker*), than accept that such treatment can pass for lawful behavior: "He left the nags standing in the courtyard without troubling himself further about them, and, insisting that he knew how to get justice (*Recht*) himself, swung onto his bay horse and rode off" (78–79). Abandoning the horses in the name of the law, Kohlhaas appears to show no further concern for the horses themselves. It would seem, finally, that the matter is indeed for him a purely legal issue. But even here an ambivalence is discernable, for he abandons them out of the conviction that the law ought not to tolerate such abuse – abuse of his rights and of his property to be sure, but also, perhaps, of the horses themselves.

When subsequently, receiving a further letter refusing to consider his case, Kohlhaas, in a rage, likens horses to dogs – "Kohlhaas, for whom it was not to do with the horses – he would have suffered the same had it concerned a pair of dogs – Kohlhaas frothed at the mouth in rage, upon receiving this letter" (100) – the remark is fully consistent since Kohlhaas's *Rechtgefühl* is a function not of economic worth but of legal recognition.[57] All that remains ambiguous is whether the recognition in question is that of Kohlhaas, the legal subject and human being, including the recognition of his property, or whether a more obscure recognition that might relate to a horse or even a dog may be at stake. That the latter may indeed, under the prevailing dispensation, not constitute a "legal" form of recognition at all and so imply a "natural-legal" critique of the law, one that furthermore would be a displacement of what was (and is) known as "natural law," explains the problematic nature of Kohlhaas's extremism. At the very least, it implies that the two kinds of recognition – of the human and of the horse/dog – may not in fact be as far apart as the discriminations produced and preserved by the law, even the tradition of natural law. This, in any case, is suggested by Kohlhaas's own use of the political-zoological analogy with the dog as he, enraged – foaming at the mouth – considers going rogue (107).

In short, it remains ambivalent (including for Kohlhaas himself) whether his strong feelings and extreme action emerge out of the question of the legal recognition of *his* rights or, rather, on behalf of that stratum of existence that does not have, and cannot be recognized as having, rights at all, a recognition then of the rightful, and so human, relation to and in nature. Kohlhaas, the horse dealer,

---

[57] Tim Mehigan, relying on a justification Kohlhaas gives Luther of his uprising, discerns in *Michael Kohlhaas* the introduction of a Rousseauian concept of "legal recognition" that would supplant a Hobbesian model of absolutist legitimation, *Heinrich von Kleist: Writing after Kant* (Rochester: Camden House, 2011), 79.

does not merely trade horses for human use, he is concerned that they are *humanely* handled. In a phrase that captures a prevailing legal-political reading of *Michael Kohlhaas,* Joachim Bohnert writes: "Kohlhaas does not want two horses back, he wants *the right* to get two horses back."[58] Bohnert is right, but only if at the same time the opposite thesis is also maintained: Kohlhaas does not want his rights, he wants the *horses* back as they were.

## Nags – and other negligible things

It is remarkable that every reader knows, but seems nevertheless inclined to neglect the fact, that at the heart of *Michael Kohlhaas* is an unfortunate pair of *Rappen.*[59] This has much to do with the legalistic character of Kohlhaas's claims and no doubt also with the narration which seems, for the sake of credibility, hard put to provide a plausible presentation that accords with the prevailing perceptions of reality and habits of reading. This bureaucratic tendency, which might be called the structure of oversight inherent even to the techniques of reading and writing literature, will be explicitly thematized with the arrival of the gypsy woman (*Zigeunerin*) in the second part of the novella. But not to be distracted from the horses: literally in the middle of the 1810 version the otherwise elusive or overlooked animals reappear, abandoned once again, in the middle of the market place in Dresden:

> As the crowd started to disperse, the knacker from Döbbeln, whose business was done and not wanting to hang around any longer, tied the horses to a lamppost, where they stood all day, without anyone taking any care of them, subject to the sport of street urchins and layabouts. As a result, owing to their lack of care and protection, the police had to take the matter in hand, and at nightfall called the knacker of Dresden to remove the horses to the knacker's yard on the outskirts of the city until further notice. (194–195)

The *Rappen* have been reduced to nags (*Schindmähren*). They have no value and are nobody's property, or rather are not worth enough to count as property, and so, from the point of view of the law, as Graf Wrede remarks to the *Elector*, are dead: "my lord, they *are* dead: they are in the legal sense of the word dead, be-

---

[58] Joachim Bohnert, "Positivität des Rechts und Konflikt bei Kleist," *Kleist-Jahrbuch* (1985): 49.
[59] Francisco Larubia-Prado discusses this neglect of horses in the critical reception of *Michael Kohlhaas* in, "Horses at the Frontier in Kleist's Michael Kohlhaas," *Seminar: A Journal of Germanic Studies* 46.4 (2010): 330–50. His reading, however, risks slipping into the tendency he criticizes by seeking to establish what horses in general stand for – in German literary history as well as in Kleist's text.

cause they have no value" (198). No longer of any value as living beings, these animals, in the interests of public health and safety, have to be disposed of. This is the work of the knacker (*Abdecker*, literally: un-coverer; or *Schinder*, flayer), whose office it is to extract the last remnant of economically useful life from the covering and protection of the law and so fulfill in fact the death that the law had already ascribed to them. The knacker belongs to the para-legal bureaucratic machinery that disposes of all that the law does not recognize and, in so doing, serves to produce and preserve the precincts in which the legal system imposes itself. Thus the knacker was assigned to the contours of the polity and of the living body more generally, often in addition being responsible for carrying out the torture in interrogations and even doubling for (or competing to be) the principality's executioner or hangman. These peculiar representatives of the law were themselves segregated within the legal system as well as within the broader context of social life – theirs was a "dishonorable trade" (*unehrliche Beruf*). Their work of purification and decontamination was itself considered unclean and infectious even if a primary positive product of *Abdeckerei* was soap. In German, the dishonorable office of the knacker became, by association with the carrion he treated, a derogatory synonym, like similarly disparaging designations derived from carrion – *Schelm, Luder,* or *Aas* – for the dangerous, infectious, vermin-like underclass of society that lives a tenuous existence on the edge of the law.[60]

It was, of course, just this political-zoological category of the population that was drawn to Kohlhaas during the course of his partisan uprising. At the very moment that he described himself as "a deputy of Michael, the Archangel" (140), a veritable army of such vermin-like figures – "carrion rogues" – were swarming around him. And, in response to his media as much as his military campaign, even that inscrutable animal, "public opinion," seems to have been ready, as Luther warns the Elector, to voice support (160). The extraordinary force of Kohlhaas's uprising is attributed by the narrator, first of all to purely economic factors and coincidental historical circumstances – the "rabble" (*Gesindel*) of unemployed former-soldiers returning from the recent war in Poland (129) – but it is clear that economic considerations cannot explain the extent and extremism of his following. Instead, his *Schwärmerei* is functionally related by the narrator to the "peculiar position that he assumed in the world" (138). Kohlhaas's appeal to rogues and citizens alike seems to be a function of his posture of transcendence – he refuses fundamentally to take part in a world that

---

[60] See Jutta Nowosadtko, *Scharfrichter und Abdecker: der Alltag zweier "unehrlicher Berufe" in der frühen Neuzeit* (Paderborn: Ferdinand Schöningh, 1994).

does not measure up to his *Rechtgefühl*. But this "peculiar position" (*sonderbare Stellung*) or, as Luther puts it, "contrary position" (*trotzige Stellung*) reflects the bi-polarity of his *Rechtgefühl* (153). Despite his grand political-theological declarations, it is never clear whether the righteous passion that inspires the uprising comes from on high or from down below, where Kohlhaas finds such teeming enthusiasm for his obscure cause.

It is only when Kohlhaas abandons the para-military aspect of his undertaking and returns, under amnesty, with the guarantee that his legal case will be pursued in Dresden, that the people turn against him on account of the anarchy that the recognition of his legal claim, which proves not to be a claim of positive law in the ordinary sense, brings about. In brief, when the bureaucracy attempts "in good faith" to fulfill the terms of Kohlhaas's apparently legal claim – the restitution of his *Rappen* – the whole legal system and political infrastructure come under threat in a riot that marks the turning point in the first part of the story.

There is first of all a scene of recognition. Kohlhaas is called upon by Chamberlain Kunz, one of the more influential of von Tronka's relatives, to inspect the two horses brought into town by the knacker from Döbbeln on the basis that they are thought to be those rescued from the burning stables in Tronkenburg. Von Tronka himself, it is reported, had been unable to recognize the horses given their abysmal state. Arriving at the marketplace, Kohlhaas approaches the knacker's cart and "from a distance of twelve feet where he stopped, glanced at the animals, which stood on trembling legs, their heads bent down to the ground, not eating the hay that the knacker had laid before them..." (189–190). Then, turning to the chamberlain, he states, "the knacker is quite right; the horses tied to the cart belong to me!" and departs the scene (190).

Without so much as laying a hand on them, Kohlhaas "recognizes" the *Rappen*, although in all likelihood they are not his. Thus his gesture of recognition, which seems to operate on the level of established rights within the law, instead insists on the perceptibility of those aspects of nature rendered indiscernible by the discriminations of the bureaucratic system. Furthermore, Kohlhaas's statement, "the knacker is *quite right*" (*der Abdecker hat* ganz recht) is not an expression of agreement with the knacker, who himself admits knowing nothing about where the horses came from. It is rather a polemical legal claim. He places right on the side of the knacker and so on, or along, the edges of the legal-bureaucratic dispositive for the disposing of useless, used-up or unseemly life – the very site out of which earlier the question of humanity had emerged.

If the bureaucracy attempts to take Kohlhaas's apparently legal claim seriously and so to restitute the *Rappen* and recognize the right of the knacker, it will no longer be executing its habitual legal functions, but will enter a peripheral terrain of political-natural indetermination that threatens to thoroughly undermine it. Most immediately, it risks dissolving a fundamental, if never pristine, distinction between human society and its outside, a distinction that in the sixteenth century and still to a degree around 1800 (to judge from Schiller's *The Criminal out of Lost Honor*) revolved around the perceptions and discriminations concentrated in the concept of "honor." The baffled bureaucrats do not appear to be fully aware of the consequences of Kohlhaas's claim but these are not lost on the good citizens of Dresden crowding around the scene.

The occasion of the riot that follows is an apparently insignificant, but as it turns out thoroughly over-determined, incidence in which the discriminations of social, political and what is called "human" life seem to be concentrated. While negligible in itself, it generates a revolting air around it. The whole turn of events is prompted by a pile of shit (*Mistpfütze*) from the hapless horses who stand ignored and neglected at the center of the scene. After Kohlhaas's curt recognition of the horses and departure from the square, the Chamberlain promptly pays the knacker and instructs one of his men to take charge of the horses: "The boy (*Knecht*), who upon being called by his lord had left a circle of friends and relatives he had among the people (*Volk*), indeed stepped, as it happened, a little red in the face, over a big pile of shit that had formed at their feet, towards the horses" (190–191). The blush that marks this unpleasant but otherwise innocuous transgressing of the boundaries of propriety outrages the boy's self-respecting relatives who are looking on. A cousin, Meister Himboldt, pulls him back shouting: "Don't touch those nags!" (191) and, himself stepping gingerly over the dung-heap, tells the Chamberlain to find a knacker's boy for such work. The boy, emboldened by the solidarity of the people (*Bürger*), adds that "the horses would first have to be made honorable (*ehrlich*)" (192–193) before he would perform the service. The Chamberlain, flying into a rage, rips the hat that bears the emblem of his house off the boy's head and expels him from his service. And so the riot begins.

The crowd of "citizens" (*Bürger*) turns into a raging mob that strips the bureaucrat of every sign of his legal person and protections (his insignia and sword are torn from him), even threatening to kill him. In a structural inversion of Kohlhaas's uprising, the *Rechtgefühl* of the law-abiding citizens of Dresden is provoked by an apparently negligible infraction that paradoxically turns them into a riotous rabble, indeed into a veritable swarm, for the sake of their rights as honorable citizens.

The riot is a turning point in the novella as it marks the breaking of Kohlhaas's will and the turn in public sentiment against him:

> This incident, however little the horse dealer was in fact to blame for it, nevertheless aroused, even among the most moderate and better members of the public, a mood in

the land that put the outcome of his dispute in serious jeopardy. His comportment to the state was considered altogether unacceptable, and in private homes and public spaces, the opinion arose that it would be better to publicly do him an injustice and strike down the whole case once again, than to give him justice, obtained by violent defiance, for such a negligible cause, merely to satisfy his raving obstinacy. (195)

Treated, illegally, like a common criminal, Kohlhaas is sentenced to a dishonorable execution. He is to suffer an ignominious death, treated not unlike his *Rappen* at the hands of the very same knackers (*Schinderknechten*): "so he was condemned to be drawn and quartered by knackers with red-hot pincers, and his body burnt between the wheel and the gallows" (227–228).

While leading an ad hoc army and declaring himself "a sovereign, alone under God" (128), Kohlhaas could enjoy broad public support by assuming a posture of transcendence with regard to the discriminations that articulate the world and the polity. In so doing, Kohlhaas presents the characteristic indiscriminacy of the *Schwärmer* in the alluring aspect of one who refuses all discriminations but his own. When, however, he returns to the precincts of the law to press his claim, "for such a negligible cause" as the restitution of his *Rappen*, he becomes an unbearable irritation and anxiety to the citizens who would rather be done with him. Indeed, as evidenced in the riot, his apparently trifling legal claim could prove to be more of a threat to the social order than the armed uprising. Here, his suddenly repellent *Schwärmerei* poses the threat of radical indiscriminacy by undermining the discriminations of the law without, however, asserting a higher instance of distinction.

The occasion of Kohlhaas's *Rechtgefühl* is indeed, from the official point of view, a "negligible cause." But as the riot shows, it is precisely such apparently negligible instances that can stoke a person's, or the people's, *Rechtgefühl* and incite an uprising. It would seem, therefore, that the object of *Rechtgefühl* is always a negligible cause, its intensity a function of its nullity. The more negligible the cause, the more impassioned is the righteous passion provoked, precisely because it relates to an exigency that is not recognized and cannot be formulated within a given dispensation. In a manner that recalls the properly political "better cause" (*bessere Sache*) of which Bloch claimed Kohlhaas's *Rechtgefühl* was a confused expression, the "negligible cause" (*nichtige Sache*), on account of its apparent triviality, eludes bureaucratic oversight and throws the whole infrastructure of political life into question. Such negligible but enraging issues not only reveal the arbitrariness and the violence of the discriminations that organize and grade the perceptible world, but further expose another apparently indiscriminate realm in which the relation between nature and politics, human and animal, the seemly and the unsightly might be articulated differently.

Kohlhaas is rescued from his dishonorable punishment by the bureaucratic intervention of Brandenburg orchestrated by Kohlhaas's acquaintance, the former City Captain and current Arch-Chancellor, an intervention motivated not first of all by concerns for justice or the principle of protecting its citizens, but for an extraneous reason: to irritate, without however openly opposing, Saxony in order to appease its ally, Poland, which at the time was in dispute with the House of Saxony. Henceforth, Kohlhaas becomes entangled in an increasingly complex set of overlapping and interlinking bureaucracies.

## Bureaucrat against Bureaucracy

Turning away from the nags, as one is inclined to do, and focusing instead on the terms in which Kohlhaas expresses his *Rechtgefühl*, one can recapitulate his extremism in the following way: it presents a bureaucratic attempt to do away with bureaucracy. Kohlhaas emerges, paradoxically, as a paradigmatic exponent of bureaucracy. He is a bureaucratic fanatic. For the self-understanding of his hyperbolic agitation – expressed in his declarations and justifications – reveals the political-theological eschatology that underwrites bureaucratization itself. At stake, in short, is the project of realizing a perfect organization of the world – a just order – in which bureaucracy would no longer be needed or would, rather, disappear in the seamless operation of order itself.

The perplexity of Kohlhaas is that he is altogether convinced by, and speaks eloquently in the terms of, the order he so vigorously opposes. This is because, as we have seen, his Recht*gefühl* is not just a *feeling* of injustice, but one that feels compelled to express itself and attain recognition in legal-bureaucratic terms. While it is provoked by a frustrated encounter with bureaucracy – with its violence and also its recalcitrance – it responds in the form of a dogged legal battle that culminates in his being called, "a good-for-nothing querulant" (100). Querulance emerged, as Rupert Gaderer has shown, as a characteristic affective reaction – itself chronically bureaucratic – to the period of intensive bureaucratization signalled by the implementation of the Prussian *Allgemeines Landrecht* (1794).[61]

Kohlhaas, according to his own account, goes from being a law-abiding citizen to a querulant and finally becomes an outlaw or *Rechtsfanatiker* because not

---

[61] See Gaderer, *Querulanz*. Kohlhaas's encounter with bureaucracy is indeed the point of departure for Gaderer's reflections on the emergence of an excessive "Rechtsgefühl" around 1800 betrayed in the criminalization and subsequent pathologization of querulance, which he treats as a side effect of (Prussian) bureaucratization.

the *law*, but the *legal system* supposed to implement the law, fails to acknowledge his case. Thus, although he will ultimately declare a "provisional world government" (141), he can remain persuaded that his partisan rebellion is not aimed at revolution but at a restitution of the law, specifically, at rescuing the instance of the law from the clutches of a corrupt bureaucracy. Kohlhaas does not see himself as an outlaw at all; on the contrary, at the height of the rebellion he positions himself as the only true representative and partisan of the law.[62]

Throughout the first part of the novella, Kohlhaas operates on the principle that if he could only approach the law itself, justice would be done: "The Lord (*Herr*) himself, I know, is just…" (108). Though it ostensibly refers to an earthly authority (the lord of the land, *Landesherr*), this statement is what it sounds like: a theological supposition. If only the law could be accessed immediately, purified of its corrupt worldly mediations, it would be just. Kohlhaas's political-theology is thus premised on the assumption that the law is strictly distinguishable from the short-comings and the failures of the offices and institutions in which it is enshrined. But the law – as the whole novella is there to show – is not distinct from, and cannot be extricated from, the network, however corrupt, inept, and nepotistic, that surrounds it in an impenetrable, thoroughly mediated, mediatized, and ultimately also militarized manifestation. The law is but an effect of the bureaucracy through which it functions.

Kohlhaas's crisis here reiterates Kleist's Kant crisis: just as the certain and eternal character of knowledge of nature was undermined by the finite conditions that facilitate one's access to it, the conviction that the law is just is undermined by the very mediations by which it is made manifest. The critical moment emerges here: if the law is not to be separated from bureaucracy, then there is no just order and all apparent order, as represented by prevailing legal institutions, is ultimately nothing but a sedimentation of a history of arbitrary violence and thus fundamentally unjust. Not until the end of the novella, however, will Kohl-

---

[62] From this perspective, it is possible to arrive, as Joachim Bohnert has done, proceeding in the manner of a jurist on the basis of articulate utterances alone, at precisely the opposite reading to Bloch: Kleist in general and Kohlhaas in particular is concerned *exclusively* with positive law. If, as Bohnert is persuaded, the law is always and unproblematically presupposed in Kleist, then his literary texts are not concerned with the question: *What is right?* but with the incommensurability between law and facticity, specifically between the law as it ought to be and the concrete forms in which it takes effect in reality; Bohnert, "Positivität," 46. See also Theodore Ziolkowski, who argues that Kleist's youthful interest in natural right gradually gives way to a preoccupation with its codification and implementation, the decisive instance of which, upon which Kleist ostensibly hangs his faith, is the *Allgemeines Landrecht*. Positive law, Ziolkowski implausibly claims, is consistently affirmed in Kleist's literary work, "Kleists Werke im Lichte der zeitgenössischen Rechtskontroverse," *Kleist-Jahrbuch* (1987): 28–51, esp. 48.

haas acknowledge this situation which compels him to an equally fanatical attempt to abandon the precincts of the law altogether.

In the first part, Kohlhaas's efforts to attain access to ever higher authorities exhibit a particular structure, an asymptotic scale of querulance: the more openly and insistently one seeks to penetrate the bureaucracy and appeal to the very instance of the law, the more one is repulsed by it. This culminates in the fatal attempt of Kohlhaas's wife, Lisbeth, to approach the *Landesherr* in person to submit their supplication: "It seems she had pressed forward too boldly towards the person of the sovereign, and, without any fault on his part, owing to the sheer zeal of one of the guards who surrounded him, she had received a blow on the breast from the shaft of a lance" (113). The intensifying drive towards the law only pushes one further from it into criminal querulance and ultimately into lawless banditry. It is along these lines also, on account of being "expelled from the community of the state," that Kohlhaas will later seek to justify his actions to Luther (151).

So when Kohlhaas proves unable to arrive at a pure instance of the law in the world, he turns outlaw but justifies this action on the basis of a turn inward – he finds the missing instance of the law in himself, in the very immediacy of his *Rechtgefühl*. Already, before his wife's death, when Kohlhaas finds out that, despite the legal proceedings, his horses are still being used for fieldwork in Tronkenburg, "in the midst of the pain of seeing the world in such disorder, quivered a sense of inner contentment, to see matters in his own breast once again in order" (101). Not long afterwards, convinced that the *Herr* expresses himself in his own person, Kohlhaas will call himself "a sovereign, alone under God, free from empire and world" (128). The turn inward licenses him, much in the manner of that paradigmatic *Schwärmer*, Thomas Müntzer, to administer the law as if doing "god's work" in the world – a phrase Kohlhaas's wife actually uses (91). At the height of the rebellion, Kohlhaas sees himself as the incorruptible legal representative of a transcendent law, "a deputy of Michael, the Archangel" (140), in combat with corrupt earthly powers or rather with the inherent corruption of earthly powers. Kohlhaas here assumes the paradigmatic posture of the fanatical bureaucrat.

## Doubt, Forgiveness, and the Givenness of *das Vergebliche*

The inner state of disarray into which he is thrown by the letter signed by Luther, "the contents of which accused him of injustice," is betrayed when "his face flushed a deep red" (148). Max Kommerell observes that Luther's letter, "divides (*entzweit*) Kohlhaas against himself" and, indeed, Luther's intervention marks the beginning of the breakdown of Kohlhaas's righteous conviction by introducing doubt (*Zwei-fel*) into the legal-bureaucratic interpretation of his feeling.[63] Henceforth Kohlhaas no longer feels convinced of the Right of his *Rechtgefühl*, which now betrays itself for what it always already was: a form despair.

The critical turning point is marked during the interview with Luther, by an approach to the window. Luther asks: "But hadn't you done better, all things considered, to have forgiven the Junker for your Redeemer's sake, taken the horses, meager and worn-out as they were, and ridden them home to fatten them up in your stables at Kohlhaasenbrück? – Kohlhaas answers: may be! as he went over to the window: may be not!" (155). Reflected in the window, Kohlhaas finds himself divided and cast into doubt. If, throughout, he has taken his *Rechtgefühl* to be an obligation to do the right thing, where "right" was understood to be according to the letter of the law, whether the law of the land or the law that expressed itself within himself, then the thought – "may be! [...] may be not!" (*kann seyn!* [...] *kann seyn, auch nicht!*) – shatters the univocality he had ascribed to the law.

The desperate thought that things could have been otherwise, that in retrospect it might have been right to do the "wrong" thing, or, above all, that what appeared right may turn out to have been *all for nothing*, is particularly painful for Kohlhaas with regard to his wife. It is no coincidence, then, that he expresses precisely this anxiety with reference to Lisbeth, who otherwise had not been mentioned since her death, attempting to suggest that his entire rebellion had been to give her death some validity: "Kohlhaas responded, as the tears rolled down his cheeks: worthy lord! It has cost me my wife; Kohlhaas wants to show the world, that she did not die in some illegal dealing [*ungerechten Handel*]" (154). Recalling his wife, Kohlhaas is obliged to reflect on their relationship, which brings into focus the unjust violence that follows from the righteousness of his rigorous interpretation of his *Rechtgefühl*. Simply put: his bureaucratic attitude, which instrumentalized his marriage, expressed itself in the abuse of his wife. Kohlhaas had mistaken his marriage for a merely legal arrangement. The most revealing scene of this misapprehension is the moment that he announces to his neighbor, without so much as consulting his wife, his intention to sell all his property: "Lisbeth, his wife, went pale at these words. She turned, picked up her youngest, who was playing on the floor behind her, and cast a look, in which death was painted, past the red cheeks of the boy, who played with the ribbons on her neck, towards the horse dealer and a sheet of paper that he held in his

---

[63] Max Kommerell, *Geist und Buchstabe der Dichtung: Goethe, Schiller, Kleist, Hölderlin* (Frankfurt am Main: Vittorio Klostermann, 1991), 291.

hand" (101–102). Lisbeth falls victim to the bureaucrat in her husband's person, the violence of which is the mark of death in her gaze as she looks at her husband with his papers even before she, for the sake of their marriage, approaches the *Landesherr* herself and is killed in the attempt.

When, therefore, Lisbeth on her death-bed sits up and, taking the Luther Bible from which Kohlhaas was reading to her, pages through until she voicelessly indicates a verse: "Forgive your enemies; do good to those that hate you," there are good grounds to suppose this message is a dying sign of forgiveness for her husband: " – she pressed his hand with a profoundly soulful look, and died" (115).[64] The gesture may also of course implore Kohlhaas to forgive von Tronka. Whatever its intention, the communication of forgiveness, coinciding with the moment of her death, produces a convulsion in Kohlhaas that is expressed in the 1808 version of the novella by four dashes: " – Kohlhaas thought: – – – – ; kissed her, his tears flowing freely, closed her eyes, and dismissed the priest" (50). Forgiveness is, for Kohlhaas, at the critical moment of the loss of his "dearest Lisbeth," if not unthinkable then inconceivable – it can be thought only under erasure typographically conveyed by what are in German referred to as *Gedankenstriche*, struck-out-thoughts. In the 1810 version, in contrast, the dashes are replaced by an articulate but no more coherent thought, which indeed seems to be a kind of parapraxis: "so may God never forgive me, as I forgive the Junker!" (115). Given that it marks the very point at which Kohlhaas becomes an outlaw, this unhappy and incoherent thought, in its confusing syntax, ambiguous mood, and seemingly catachrestic use of *vergeben* (to forgive), may be taken as a precise formulation of his fanatical state of mind. In the interview with Luther, where both forgiveness and his wife come under discussion, Kohlhaas's thought finds an arguably clearer iteration, "Kohlhaas wants to show the world, that she did not die in some illegal dealing" (154): he had embarked on his desperate partisan expedition in order that the death of his wife would not have been for nothing – *vergebens* (in vain).

For Kohlhaas, in his bureaucratic mode, the opposition between something and nothing is the difference between a *Rechtssache* (legal case) and a *bloße Sache* (mere thing) or, in the words of the citizens of Dresden, a *nichtige Sache* (negligible cause). This is why, early on, he treats his marriage as no more than a legal arrangement and approaches the case of the horses in exclusively legal terms. But the ambiguity of his *Rechtgefühl* also betrays a sense that things

---

[64] The citation from the Luther Bible is incorrect – or is, in any case, a paraphrase. Matthew 5:44: "Ich aber sage euch: Liebet eure Feinde; segnet, die euch fluchen; tut wohl denen, die euch hassen; bittet für die, so euch beleidigen und verfolgen."

are less straight-forward: that a marriage is not simply a legal relation, a horse not simply expendable property. The *Rechtgefühl*, in other words, discloses the fact that there *are* things that are not recognized by the law and so are not recognizable at all – except as *vergeblich* (of no avail). Under a bureaucratic regime, the *vergeblich* is *given* but only as a *feeling*, one that demands recognition it cannot, under prevailing conditions, receive. In this regard, *Michael Kohlhaas* presents a conundrum familiar to, if not peculiar to, modern political movements and militancy that seem to be driven by affect at the cost of articulacy. For the provocation of *Rechtgefühl* will always be a "negligible thing," and the desperate and apparently disproportionate agitation that it solicits will always face the reproach of lacking a coherent political project and so of being "all for nothing."

At the end of the meeting, Kohlhaas asks for communion to which Luther responds: "The Lord, whose body you desire, forgave his enemy," and asks whether Kohlhaas is not, for his part, ready to likewise forgive the *Junker* ("gleichfalls [zu] vergeben," 157). The traumatic word *vergeben* once again lacerates his *Rechtgefühl*, exhibited in a hesitation marked in the text and the reddening of his face, as he refuses to forgive and so give up as *vergeblich* the death of his wife and the fate of his horses (157–158). He is also, however, and perhaps first of all, refusing to abandon the sense of *Recht-* in his *Rechtgefühl* and so concede that it is *just* a feeling, or perhaps, recalling his initial response to his wife's death, it is a – – – – *gefühl*. Kohlhaas is still too attached to his sovereign sense of self as a righteous and upright subject to acknowledge what his doubts are already intimating: that his feeling may be a rightless feeling or a feeling for the rightless, and so a demand for an inarticulate "right" that attests to an other-than-bureaucratic attitude to the world.

Denied the communion he desires, Kohlhaas leaves Luther's office "with a pained expression" (159). For Kohlhaas, communion is not first of all, as Luther here presents it, a sacrament of forgiveness but – in what is also consistent with Lutheranism – a sacrament of the "real presence" of the *Herr*. Or rather: given that, since his wife's death, "forgiveness" only means whatever is expressed in his incoherent thought responding to her dying message, if he seeks communion for "forgiveness," it is precisely in order to be given the right meaning of the word and so resolve his confusion. He seeks, in other words, immediate communication with the *Herr* that would stand as the instance of Right he longs for in order to orient, order, and set to right his words and actions. Communion, were it ever to be granted, would finally and determinately provide the *communiqué* – the official document or directive – appropriate to his *Rechtgefühl*.

Thus although he does not take Luther's advice, Kohlhaas is after all "Lutheran," above all because he believes in the healing power of the word – in grace by scripture. But for him this means – in contrast to Luther, for whom the word of grace arrives unfathomably and is strictly distinguished from the law – that the highest instance will express itself in an official, signed, law-binding communication. As Carol Jacobs argues, until the very end of the novella, Kohlhaas is driven by a blinkered belief in the transparency and efficacy of "text and law underwritten by the name of authority."[65] In the absence of such an instance, Kohlhaas, signing in the name of the archangel Michael, had even taken to drafting such official communications himself. Kohlhaas's faith consists in the conviction – one that may have its roots in the media revolution of the Reformation and certainly anticipates the preponderance and authority assigned to paperwork in the nineteenth century[66] – that Right can only be expressed in writing. His fanatical bureaucratic faith in script alone is exhibited in its relentless one-sidedness throughout the first half of the novella, which reaches its crisis in Luther's office. The rest of the text presents the shattering of this fixation with the official word that solicits, however, the embrace of a no less *schwärmerisch* communion of a different kind. Turning from a frustrated drive towards a transcendent instance of the law, Kohlhaas will henceforth direct himself towards a transcendence in nature beyond the law. To return to the terms of the debate between Fichte and Schelling: having assumed the posture of a Fichtean bureaucrat and become indeed a kind of *Schwärmer*, Kohlhaas will now be attracted by the obscure appeal of *Naturphilosophie*.

## The Gypsy Soothsayer

At the center of the last part of the novella, in the midst of all the coming and going of bureaucrats and bureaucratic communications between Vienna, Dresden, and Berlin regarding the case of Kohlhaas, is a note (*Zettel*). The note allegedly contains a prophecy scrawled in charcoal by a "gypsy soothsayer," sealed with her lead ring and handed to the surprised Kohlhaas in a lead capsule with the cryptic remark, "'an amulet, Kohlhaas, the horse dealer, look after it well, it will one day save your life!'" (242). Essential, despite the facetious officiousness of the ceremony performed by the gypsy (*Zigeunerin*) in front of the

---

[65] Carol Jacobs, *Uncontainable Romanticism: Shelley, Brontë, Kleist* (Baltimore: Johns Hopkins University Press, 1989), 140.
[66] See Cornelia Vismann, *Files: Law and Media Technology*, trans. Geoffrey Winthrop-Young (Stanford: Stanford University Press, 2008). On the central role of print media in the Reformation, see Mark Edwards, *Printing, Propaganda, and Martin Luther* (Berkeley: University of California Press, 1994) and Johannes Burkhardt, *Das Reformationsjahrhundert: deutsche Geschichte zwischen Medienrevolution und Institutionenbildung 1517–1617* (Stuttgart: Kohlhammer, 2002).

Elector of Saxony (264–265), is that the note is in no way an official document. Instead its unauthorized force generates an agitation in the second part of the novella that seems structurally and functionally to parallel Kohlhaas's partisan uprising in the first. For a time, Kohlhaas, on account of the note, finds himself in a peculiar position of passive resistance within the very bowels of the legal system, a curious freedom under conditions of arrest. Ultimately, however, swinging to the opposite extreme, Kohlhaas will seek an extra-legal and distinctly unbureaucratic satisfaction from the mysterious message – in the end, he will swallow it.

The transfer of the note, it emerges, had taken place shortly after his wife's death and so on the very eve of his uprising. It is recounted from two perspectives – that of Kohlhaas and that of the Elector of Saxony – and presents in almost theatrical fashion the encounter of legal-political authority with a peculiar "order" of an altogether different kind. Representative of the latter is the *Zigeunerin*, who belongs to that "free" but rightless group of people, who, since at least the sixteenth century, were treated de facto and for the most part also de jure as "vogelfrei" (literally: free as a bird), which is to say vulnerable to the arbitrary violence of "law-abiding" people in the German-speaking lands as well as to the systematic discrimination of their various legal systems.[67] The scene takes place in the marketplace of Jüterbock, just on the Saxon side of the border

---

[67] Apart from a brief period of protection afforded them by King Sigismund's *Freibrief* (1493) and a period of relative recognition when they were able to enlist as soldiers during the Thirty Years War, the Roma, who first arrived in central Europe at the beginning of the fifteenth century, have consistently been assigned to a state of political and economic rightlessness, beginning with the Augsburger Reichstag declaring them "vogelfrei" in 1498. In the eighteenth century, alongside more general efforts to strengthen territorial integrity along borders and regulate the movement of "foreigners," there emerged extremely discriminatory legislation and police measures against *Zigeuner* – including corporal punishment and execution for entering particular territories – along with some more or less violent projects of assimilation. Around 1800 an ambivalence emerges that is related to the one elaborated in Schiller's *The Criminal out of Lost Honor*. For the most part authorities continued to be repressive on the "traditional" model but there were growing pressures – especially from the emergent human sciences – to consider more conciliatory or rather disciplinary policies. On the history of the Roma and Sinti in Germany, see Klaus-Michael Bogdal, *Europa erfindet die Zigeuner: Eine Geschichte von Faszination und Verachtung* (Berlin: Suhrkamp Verlag, 2014) and Joachim Hohmann, *Geschichte der Zigeunerverfolgung in Deutschland*, revised edition (Frankfurt am Main: Campus Verlag, 1988), esp. 13–51. Of the historical period presented in *Michael Kohlhaas*, Hohmann writes: "The freedom of the gypsies at the time of the Reformation, the Peasant Wars and the renewed threat of the Turks – in 1529 the Turks besieged Vienna – was largely a *Vogelfreiheit*. It remained the case that any 'law-abiding man' ('*rechtschaffene Mann*') could kill a gypsy encountered on his property and be exempt from punishment," 17, translation mine.

with Brandenburg, where she appears in her capacity as soothsayer (*Wahrsagerin*, literally, truth-sayer) – a vocation that Kohlhaas, in his account, more than once refers to as her "science" (*Wissenschaft*) – and attracts a large audience from the populace among whom she enjoys some repute. It is doubtless this popularity that prompts the *Landesherren* of Brandenburg and Saxony, who happen to be passing by during a recess from an official meeting, "to destroy the reputation of this peculiar (*abenteuerliche*) woman by playing a joke on her before the eyes of the people" (260).

Brandenburg demands a sign from the *Zigeunerin* by which he might that very day test the truth of her sayings. The sign, she responds, would be their encounter, in the very marketplace where they were standing, with the roebuck being reared in an enclosure in the Elector's park some distance away. Improbable though it was that the prediction would come to pass, Brandenburg orders that the buck be killed immediately. As a result of this excessive measure, the prediction comes true: the deer is killed but is mistaken for prey by a hunting hound, which, as it is trained to do, promptly brings the dead animal to the marketplace and drops it obediently at his master's feet.

Brandenburg, who presumes, in a manner exemplary of the bureaucratic attitude, to have nature at his command, is here humiliated by the sheer fact of his implication in a natural world that exceeds the scope of his oversight. For her part, this implication in nature is precisely what the gypsy soothsayer sees; it constitutes the very field of her "science." The electors are forced to acknowledge that the tamed and domesticated realm which is subject to the rationality of human command is itself but a precarious configuration embedded in a broader scheme of nature. The electors and the entire organization they represent are exposed in their passivity to, and within, a natural order, the distinctions of which – and above all the distinction *from* which – they understood themselves to be in a privileged position to administer. This threat of violence to the bureaucratic order, a threat that cannot in turn be bureaucratically accounted for or contained, is concentrated in the rest of the novella in the *Zigeunerin*'s note, which contains, she says, a prophecy about the demise of the house of the Elector of Saxony.

As the example of the roebuck shows, the workings of the unofficial but all-enveloping scheme of nature appear, by the standards of the *Landesherren* – and also those of the reader – thoroughly improbable. Their reality or effectivity would remain for the most part unobserved and unacknowledged, mere aberrations of chance, were it not for the attention drawn to them by the *Zigeunerin* whose very person and "true sayings" are occasion for considerable incredulity. According to a romantic trope, the *Zigeunerin* is here a medium of nature, but a nature that in its abjection and apparent violence and disorder is anything but what is usually called "romantic."[68] Over and against the *Landesherren*, she stands for a different attitude to nature, with which, by their own discriminations, she is identified. Indeed, recalling again the dispute between Fichte and Schelling, the *Zigeunerin* here appears as a peculiar kind of *Naturphilosoph*.

It is often remarked that the *Zigeunerin* appears in the text as a "supernatural" intervention, an irritation in the otherwise realistic-seeming presentation of the narrative.[69] Her appearance therefore solicits, on the level of the narrative, the same sort of anxieties about the arbitrariness of official discriminations that had ended in a riot in Dresden. Indeed, the *Zigeunerin*, insofar as she figures in the novella in defiance of the rules and norms of plausible narration, seems to offend "the republican freedom of the reading public," to cite the narrator of Schiller's *The Criminal out of Lost Honor*,[70] and has accordingly generated a veritable outcry in the history of the novella's reception.[71] It is as if the *Zigeunerin*'s appearance were an offense to serious literature or a ridiculous concession to the ignorant or superstitious exigencies of the "popular" readership. The vilified position of the *Zigeunerin* – within the novella as well as by readers of it – makes her the unmistakable exponent of *das Vergebliche*, the *given* but *negligible* stratum of existence (the hapless horses, for example) which is present throughout *Michael Kohlhaas* but tends to be overlooked by readers and characters alike. Concentrated in the figure of the *Zigeunerin*, it – frustratingly – cannot be so easily overseen. The disconcerting effectivity of her apparently supernatural intervention is thus really on the side of nature.

In the falling out between Fichte and Schelling that culminated in each calling the other a *Schwärmer*, the *Naturphilosoph* is defined, as is suggested by Hegel in his *Differenzschrift*, by an essential trait: the recognition of a "real opposition" (*reale Entgegensetzung*).[72] Such a real opposition has obscurely imposed itself in much of *Michael Kohlhaas* as what I have called the givenness

---

[68] On "romantic" associations of the gypsy with nature and naturalness, as well as with a keen sense of perception or observation, see Hohmann, "Zigeunerverfolgung," 60–63; also Katie Trumpener, "The Time of the Gypsies: A 'People without History' in the Narratives of the West," *Critical Inquiry* 18.4 (1992): 865.

[69] Reuß has argued that some of this anxiety has to do with the way that the *Zigeunerin* functions in the apparently realistic narrative to upset the assumed relation between language and reality, "'Michael Kohlhaas's und 'Michael Kohlhaas,'" *Berliner Kleist-Blätter* 3 (1990): 36–38.

[70] Schiller, *Verbrecher*, 8.

[71] For a survey of the critical reception of the *Zigeunerin*, see Claudia Breger, "Wider die Ordnung der Dinge? Die 'Zigeunerin' in Kleists *Michael Kohlhaas* und das Textbegehren ihrer Interpretation," in *Ortlosigkeit des Fremden: "Zigeunerinnen" und "Zigeuner" in der deutschsprachigen Literatur um 1800* (Cologne: Böhlau, 1998).

[72] Hegel, *Differenzschrift*, 99.

of *das Vergebliche*. In Kleist's story, the *Zigeunerin* presents a perspective that recognizes this real alterity in nature over and against the attitude of the *Landesherren* who, like Fichtean bureaucrats, imagine that every discrimination or determination is, in the last instance, of their own making or at least of the same sort.

The "opposition" is complicated in Kleist by the same displacement exhibited in the dispositive restraining the bear in "Marionette Theater." For the confrontation with *real* bear nature that would show itself if the apparently graceful fencing bear were released from its post would be characterized by a double determination: a negative exclusion from the lawful side of the law and a positive expression of the (apparent) lawlessness of nature. In *Michael Kohlhaas*, the *Zigeunerin* appears as the mediator in the porous space between the order of appearances as maintained by the bureaucracy and unbound nature that the bureaucracy excludes – as *vergeblich*. Indeed, the political-zoological category historically associated with the precarious existence of the *Zigeunerin* – *vogelfrei* – expresses the ambivalent place of nature in the novella: it indicates an abject side produced, or rather processed, by bureaucracy and mercilessly subject to the law and the machinery of its discriminations but also, and at the same time, an unbridled side of nature, free from the law, because inaccessible to, or incommensurate with, bureaucratic determinations. The *Zigeunerin*, herself victim of this legal exclusion, is in the "free" position that offers insight into the workings of bureaucracy *and* nature, and specifically into the interpenetration of one by the other.

It is then a real opposition, although one he fails to recognize as such, which at the beginning of the story is communicated to Kohlhaas in the neighing address of the *Rappen* that first gave rise to his *Rechtgefühl*. As we have seen, Kohlhaas sought throughout the first part of the novella to interpret this feeling as a legal-political injury when in fact it attests to a more radical passivity: a *real*, albeit legally *vergeblich*, encounter with nature. The *Zigeunerin* arrives in Kohlhaas's prison – as a secret agent – ostensibly sent by Chamberlain Kunz to persuade him to hand over the note, but, on her own agenda, to mediate between Kohlhaas, the legal subject, and the altogether different nature that he *feels* but has persistently, and with escalating consequences, failed to account for. To recall the expression to which his *Rechtgefühl* first gave rise, she arrives to mediate the question of his humanity, which now appears to be the question of this felt relation of real difference.

Unlike the authorities who are constitutively incapable of recognizing the *Zigeunerin* (for the same reason that it is impossible to recognize the abused *Rap-*

*pen* ),[73] Kohlhaas, his bureaucratic attitude atrophying in prison, not only recognizes her but discerns traits about her that recall his lost wife:

> The horse dealer noticed a curious similarity between her and his dead wife, Lisbeth, such that he might have asked whether she was her grandmother. For it was not only that the traits of her face, her hands, still beautiful even in their bony structure, and especially the use she made of them as she spoke, reminded him vividly of her: also a mark, with which the neck of his wife was identified, he noticed on hers. (275)

Kohlhaas here begins to be able to read the "natural" grammar of gestures freed from the exigencies of lawful-bureaucratic articulation. These are associated with his loving and faithful wife and, as if for the first time, he begins to recognize the "true" aspects of that relationship which had been obscured and done violence to by his legalistic rigorism. These gestures of the hands recall also, of course, the unconscious activity of Kohlhaas's own hands in Tronkenburg as he, despairingly asking about the humanity of their treatment, handles his horses. The *Zigeunerin* serves to bring to light such unacknowledged extra-legal but perhaps truly *human* mediations.

Kohlhaas, however, has little patience for such subtleties. Where in Tronkenburg he ultimately abandons his horses to pursue a legal claim, at this point he abandons legal discourse and would be done with language altogether in favor of an inarticulate expression of unmitigated violence. No longer seeking to underwrite his use of force with reference to the authority of the law, he now embraces the raw terms of an unlawful, suicidal violence. To one messenger sent by Saxony to procure the note, he responds: "you can bring me to the scaffold, but I can do you harm, and I want to!" (250). And when the *Zigeunerin* tells him the significance of the note he carries, suggesting he might use it to save his life, "Kohlhaas, rejoicing over the power that was given him to mortally wound his enemy in the heel at the very moment that it trampled him in the dust, answered: Not for the world, little mother, not for the world!" (277). Even after her words of caution, which she offers as she holds up his son – "not for the world, Kohlhaas, the horse dealer, but for this beautiful little blonde boy!" –

---

[73] The Elector of Saxony observes, "the government had searched all over the land for the woman in vain" (268). The representatives of the bureaucracy can only misrecognize or mistake her for someone else. When Chamberlain Kunz goes to Berlin with the assignment of recovering the note from Kohlhaas, he randomly picks a "rag woman" (*Trödelweib*) from among the "riffraff" on the streets of Berlin to act the part of the *Zigeunerin* as a secret agent for Saxony, at which point the narrator is obliged to acknowledge the improbability of the tale, for "the chamberlain had made the most monstrous mistake," and had inadvertently chosen the *Zigeunerin* herself (274).

he persists in justifying the extreme course of action, to which the gypsy meditatively replies, "he may be right in certain respects" (278).

As Kohlhaas presses her with further questions, the interview is interrupted. To avoid the arriving police, she slips away, saying: "When we meet again, you will not lack understanding (*Kenntnis*) about all this!" (279). Kohlhaas, however, is sentenced and so the second meeting and the promised knowledge never comes. If, however, the understanding in question relates to her observation, "that he *may* be right in *certain* respects" – and it is conceivable that this would have been the whole of the *Zigeunerin*'s lesson – it is unlikely that Kohlhaas would have proved amenable to such knowledge. For throughout he has responded to the immediate urgency of his *Rechtgefühl* with an attitude of all or nothing. To be right is to be wholly right, rightness *in some respects* is no right at all.

Under the influence of the *Zigeunerin* (and her note), Kohlhaas has undergone a conversion: he no longer mistakes his *Rechtgefühl* for a legal issue, but sees it rather as arising from the violence of his legal separation from nature. He responds, however, in a typically extreme manner: by desiring to eliminate the difference it discloses. Failing to become a *Naturphilosoph* in the style of the *Zigeunerin*, he thus remains a *Schwärmer*. Rather than recognizing the distinction from nature as one that calls for a "human" articulation other than the rational-technical discriminations imposed by the bureaucracy, it is the distinction itself that Kohlhaas cannot tolerate. Indeed, he is, in the end, as staged in the closing scene at the place of execution, ready to suffer at the hands of the law rather than suffer the distinction from nature.

When, at the place of execution, he finally – if altogether improbably – receives retribution according to the letter of the law, the moment is thus an anticlimax, an occasion for a few formal arrangements and nothing more. As Kohlhaas is led to the scaffold, to satisfy the sentence of the imperial law, "for breaching the peace of the land" (289), he approaches the Elector of Saxony who stands disguised in the crowd. Taking out the note, he reads it to himself and swallows it, at the sight of which, the Elector falls "unconscious, into convulsions," while Kohlhaas turns to the scaffold, "where his head fell beneath the executioner's axe" (290). With this defiant gesture Kohlhaas incorporates the indeterminate para-legal force presented by the note, striking a blow against the man, but really against the representative of the bureaucracy that he had devoted his life – and so much more – to contesting.

If this is revenge, however, it is not his – it is the nemesis of nature. For the action is by no means that of a self-possessed subjectivity, nor can it be construed as a human action – except insofar as the altogether inhuman behavior of a *Schwärmer* is all-too-human. He swallows the note to annihilate the self that *feels* its separation from nature. The blade only accomplishes the disintegration of his person that is already underway. Kohlhaas does not forgive, but he gives himself up and not for a higher instance, nor ultimately for the sake of a *vergeblich* one, but simply for nothing. He thus fails to do justice to his *Rechtgefühl*, he fails to attest to and articulate the right relation to an apparently *nichtige Sache* as well as to the *vergeblich* aspect of nature more broadly.

In describing the oscillation of Kohlhaas's extremism, Kleist's novella presents the contours of a nineteenth-century problematic: the question of the human, posed in terms of the just relation to, and mediation of, the difference between nature and humanity. Throughout the century this question, as adumbrated in *Michael Kohlhaas*, will find, for the most part, only bureaucratic or nihilistic – in short, *schwärmerisch* – responses. The chapters that follow will trace the oscillation of these two extremes in the work of Melville, Conrad, and Kafka, analyzing their distinctive literary attempts to explore the scope of political existence and excavate peculiar forms of life at the limits of an increasingly administered social world and ever more exhaustively exploited global landscape.

# 3 Architecture of the Office: Melville's *Bartleby*

*Bartleby* (1853/56) is not simply a story set in an office, but a story about bureaucratization.[1] The term, "bureaucracy," was just entering the English language from Europe at the time Melville was writing and the novella addresses this moment with a reflection on the historical transformation of the sense and function of "the office," a transformation, namely, in the relation between space and obligation that defines any official post or position. In this regard, the process that would come to be called "bureaucratization" referred to a reorganization of the office as a secure space, walled off from the cares of life, so as, in principle, to more effectively – not to say recklessly or ultimately fanatically – take care of the demands of "humanity."

Melville's bureaucracy novella articulates this emergent office architecture and provides a veritable glossary of terms to characterize the bureaucratic type who would operate in it, conscientiously fulfilling his official duties. The snug office, in which the lawyer can easily engage in his various avocations, is upset by the entrance of Bartleby. Insofar as he neither does his duty nor makes demands on the offices of others, insofar as he thus undermines the very principle of the office, Bartleby throws the office into confusion, and the lawyer into despair.

## Of Offices

A bust of Cicero is mounted in the lawyer's office. Every self-respecting office needs one. Although, one suspects, the author of *The Offices* (*De officiis*, 44 BC) may well have had reservations about his plaster-of-paris bust decorating a lawyer's office as so much office furniture.[2] Later in the text, Bartleby will

---

[1] Herman Melville, *Bartleby, the Scrivener*, in *Piazza Tales and Other Prose Pieces, 1839–1860, Scholarly Edition*, vol. 9, ed. Harrison Hayford, Alma A. MacDougall, and G. Thomas Tanselle and others (Evanston: Northwestern University Press: 1987). Citations in-text.

[2] In 1849 Melville purchased a collection of Cicero published by Harper's Classical Library including two volumes of *The Orations* translated by William Duncan and a third volume containing Thomas Cockman's translation of *The Offices*, followed by William Melmoth's translation of "Cato: or, An Essay on Old Age" and "Laelius: or, An Essay on Friendship," *Cicero: The Orations Translated by Duncan, the Offices by Cockman, and the Cato and Lælius by Melmoth* (New York: J. & J. Harper, 1833). Citations from Cicero from this edition.

keep "his glance fixed upon" the effigy, which sits just above the lawyer's head when he sits at his desk, while preferring not to answer the lawyer's "friendly" questions (30).

It is often remarked, in a condemnatory manner, that this pale imitation of Cicero perfectly accords with the fallen, counterfeit condition exemplified by the Wall Street lawyer and his office. Be that as it may, the historical displacements suggested by the appearance of this very modern vestige of classical culture in the office – the transfer of Cicero from the rostrum, where he fulfilled his public offices, or even from his retirement in Tusculum, where he wrote *De officiis*, into a dingy down-town city office of nineteenth-century New York, and equally the transformation of the soundness and solidity of the marble sculpture of classical antiquity into the plaster mass reproductions of an office ornament[3] – present, in abbreviated form, a long and convoluted history of the office that converges in *Bartleby*. Indeed, the bust of Cicero indicates the singular contribution of Melville's text to the "archeology of the office."

Such an archeology has recently been undertaken by Agamben in *Opus Dei: An Archeology of Duty*, devoted to excavating the political theology of the office.[4] The decisive moment in Agamben's account is when, in the course of the institutionalization of the early Church, the concept of liturgy (*leitourgia*), which applied to the performance of the sacraments, the "ministry of the mystery," became bound to that of the office (*officium*). The efficacy of the mystery – that, on each occasion, it could be treated as real and not simply a performance – came to be defined as a function of the *office* of the person who performed the service and vice versa. What this meant was the transformation of a priest's calling into a priest's office, which then in itself (regardless therefore of the priest's personal moral and spiritual stature) guaranteed the efficacy of the liturgy. By marrying the notion of efficacy to that of the office,[5] a formidable formula

---

[3] In the lecture he gave on tour in 1857 and 1858, recalling his 1849 visit to Italy, a reconstruction of which is available under the title "Statues in Rome," Melville approached his conclusion saying: "Here, in statuary, was the Utopia of the ancients expressed. The Vatican itself is the index of the ancient world, just as the Washington Patent Office is of the modern. But how is it possible to compare the one with the other, when things that are so totally unlike cannot be brought together? What comparison could be instituted between a locomotive and the Apollo? Is it as grand an object as the Laocoön?" in *Piazza Tales*, 408.

[4] Giorgio Agamben, *Opus Dei. Archeologia dell'ufficio* (Torino: Bollati Boringhieri, 2012); references to the English translation, *Opus Dei: An Archaeology of Duty*, trans. Adam Kotsko (Stanford: Stanford University Press, 2013).

[5] Agamben notes: "While modern scholars derive the etymology of *officium* from a hypothetical *\*opficium*, 'the fact of actualizing a work" or "the work effectuated by an *opifex* (artisan) in his

had been found for institutionalizing the relation to transcendence while also facilitating the geo-political propagation of the Church – and ultimately not only the Church.

In the early modern period, especially in the work of Samuel Pufendorf, this notion of the efficacy of the office was generalized, according to Agamben, into a moral concept that defined the very agency of human beings. The classic notion of virtue in morality was displaced by the discourse of duty to a transcendent law or command and thus marked a distinctly modern shift towards the privileging of what Agamben calls a second "ontology" of obligation. This, he argues, culminates in the primacy of the practical in Kantian philosophy, insofar as the world of freedom that coincides with the absolute obligation of the categorical imperative is taken to be, in contrast to the world of appearances, the thing-in-itself. Drawing on a similar distinction made by Heidegger in his *Introduction to Metaphysics* (1935) and elsewhere, Agamben claims that this "ontology" of efficacy and obligation, of "what ought to be," has long operated in parallel with the classical ontology of "what is," and has come to be the defining criteria of reality: "being and acting today have for us no representation other than effectiveness."[6] Thus Agamben's archeology provides not simply an account of the emergence of an ontology of the office, but also the claim that there has been a veritable colonization of ontology by the office.

One intriguing outcome of Agamben's study is that, in contrast to the eminently place-bound significance of both liturgy and office in their classical and Christian contexts, the spatial dimension seems to have disappeared altogether from the contemporary modality of the office. Insofar as it *ought to be*, the modern office *is not there*. Even as efficacy is increasingly the measure of all that "is," the office itself cannot be said to exist. Agamben writes, "In office or duty, being and praxis, what a human does and what a human is, enter into a zone of indistinction, in which being dissolves into its practical effects and, with a perfect circularity, it is what it has to be and has to be what it is."[7] The unusual phrase, at least in the English translation, "in office or duty," indicates the difficulty. Since the office is pure operativity, "in" can only have a temporal sense, as in the more correct but even more awkward phrases "in officing" or "in duty-

---

*officina*' [...] it is significant that the Latins instead traced it back to the verb *efficere*," *Opus Dei*, 146–147.

6 Agamben, *Opus Dei*, 14. This distinction is associated with the imperative as opposed to the indicative mood; see Agamben, *Qu'est-ce que le commandement?* trans. Joël Gayraud (Paris: Payot & Rivages, 2013).

7 Agamben, *Opus Dei*, 14.

ing," which better inflect the way that the operations disappear in the process of bringing forth their effects. In short, the office obfuscates itself in effectivity.

The implication that, with the rise of its effectiveness and also its pervasiveness, the office disappears, indeed that it is all the more effective and pervasive to the degree that it disappears, is remarkable given the fact that, above all in the period contemporary with this development – from the end of the eighteenth century onwards – a growing proportion of humanity would be spending a substantial part of their waking lives *in* an increasingly ubiquitous, more or less concrete place generically called, in English, "the office." Agamben's archeology of the office needs to be supplemented by a reflection on this more concrete architecture of the office. Putting the escalating imperatives of duty and efficacy into relation with the proliferation of modern office structures brings both under the rubric of "bureaucratization" – a term Agamben does not mention. And yet it is precisely such an analysis that might contribute to the understanding of the contemporary preponderance of the ontology of effectivity, as well as of biopolitics, with which his work is concerned.

The bust of Cicero that adorns the nineteenth-century office in Melville's text is the trace of this other, architectural, dimension of the history of the office. It was Cicero in *De officiis* who, as Agamben emphasizes, first brought life, the conduct and governance of life, under the auspices of the office. Whereas the Stoic term that Cicero translates, *kathēkon*, refers to the appropriate action on particular occasions where law or virtue are silent or undecided, for Cicero all such situations, which after all make up a significant part of everyday life, are to be assimilated under the broader category of the office: "for they [offices] take in every part of our lives, so that whatever we go about, whether of public or private affairs, whether at home or abroad, whether considered barely by ourselves, or as we stand in relation to other people, we lie constantly under the obligation to some duties."[8] Thus, his treatise is devoted to the systematization of occasional offices into an official way to conduct life as a whole: "certain directions and precepts, according to which on all occasions it is our duty to govern our lives and actions."[9]

In the nineteenth century, a very different governance of life by the institution of the office was well underway. That Cicero would be a model, if only a plaster one, for the emergent bureaucratic class is by no means surprising. But whereas for the republican Cicero, the office related to the conduct of life above all as a public and engaged citizen, the modern office, as a function of

---

8 Cicero, *The Orations*, 3:9.
9 Cicero, *The Orations*, 3:11.

the mass societies in which it emerged, is characterized by an increasing separation of the office from the "life" that it in different ways administers. With the increasingly thoroughgoing officiating of life, but the commensurate alienation of the office, Cicero's project is both consummated and evacuated, as indicated by the plaster reproduction of his bust that provides, so Melville's text suggests, generic decoration for the countless nondescript offices proliferating in the mid-nineteenth century. The significance of "Cicero" is exhausted in the official function of the bust, which is, quite simply, to bestow on the workplace the one-time austere, but like the sculpture itself, long-since thoroughly banal, name of "office."

## In the Office

The lawyer's introduction is, to readers today at least, typical of a bureaucrat. Indeed, if not in so many words, he promptly admits to being more of a bureaucrat than a lawyer. If lawyers are those who typically seek out publicity – confrontations and debates, renown and recognition – Melville's lawyer in contrast retreats from the public realm into the obscure space of the office, a space that is neither public, certainly not in a sense that Cicero, for instance, would have recognized, nor private or domestic. It presents instead a curiously asymmetrical threshold between public service, private enterprise, and personal undertaking, which addresses the law and the economy but in a highly mediated fashion, emblematized in the figure sitting behind a desk. In this obscure space the lawyer applies himself with characteristic "prudence" and "method" to what he calls his "avocations" (14).

As a description of office work, the term "avocation" is illuminating. It originally meant distraction from one's calling, from the real work to be done, but by the nineteenth century a perverted sense had emerged, the one the lawyer uses, meaning work in general or the business at hand. He is suited to his career as a bureaucrat precisely because it is not a vocation, nor a calling, but a manner to evade the very question of a calling, insofar as he devotes himself instead to the "business" of the office. Office work is avocation – work that is more of a distraction than a vocation and that is carried on behind a desk and ideally behind closed doors rather than in the public eye – even when holding a public office.

Indeed, Melville's lawyer was, in the period he recounts, holder of the "good old office" (14), as he refers to it, of Master in Chancery. As such, he had to do with so-called equity cases that were not covered by the norms and regulations of common law. As Master in Chancery, the lawyer's office treated cases that fell between the habitual distinctions of private, familial, and public life on the one hand (divorces, property disputes, foreclosures, protection of persons in need of juridical protection), and private enterprise and public interest on the other (corporate and church property, bankruptcy, fraud, accidents).[10] The story takes place on the eve of the new 1846 New York constitution which, the lawyer rues, robs him of his secure and remunerative post, but at the same time signaled a rationalization of the New York legal system by removing the antiquated office and generalizing the application of the law under a simplified hierarchy of courts. The interstitial legal space that was subject to the discretion of the Chancery would be inculcated and regularized by a more efficient bureaucratic system. A sense of this "rationalizing" momentum is conveyed in *Bartleby*.

Master in Chancery also means the master of, but more specifically perhaps *in*, an enclosure, or an enclosed space.[11] It thus provides a generic name for the retreat from public space into the walled chambers of the office block in which the lawyer can "do snug business" (there is more than one suggestion that his interests, if not his avocations, may not have been altogether "equitable") in "the cool tranquility of a snug retreat" (14). His "original business" – "that of a conveyancer and title hunter, and drawer-up of recondite documents of all sorts" – was in real estate, whence his questionable relation with the notorious New York property baron, John Jacob Astor (19). It was not, however, with the "real" of real-estate that he had to do but rather the "recondite" paperwork that conditions the efficacy of modern realty.[12] One of the functions of his office

---

[10] Melville's brothers Gansevoort and Allan Melville had both worked in the offices of the New York's Court of Chancery. For a summary of the scope of the equity jurisdiction of the New York Chancery before it was abolished in 1847, see James Folts, *"Duely & constantly kept": A History of the New York Supreme Court, 1691–1847 and an Inventory of its Records (Albany, Utica, and Geneva Offices), 1797–1847* (Albany: New York State Court of Appeals and New York State Archives, 1991), 19.

[11] John Carlos Rowe points out that *cancelli*, from which chancery (along with chancellor) is derived, referred to the "lattice, railing, grating," behind which the "keeper of the barrier," the "secretary," was stationed, "Ecliptic Voyaging: Orbits of the Sign in Melville's 'Bartleby the Scrivener," in *Through the Custom-House* (Baltimore: Johns Hopkins University Press, 1982), 122. The resonance of *cancelli* is also discussed in Cornelia Vismann's media-theoretical reading of *Bartleby*, *Files: Law and Media Technology*, trans. Geoffrey Winthrop-Young (Stanford: Stanford University Press, 2008), 29–38.

[12] John Jacob Astor, known as the "Landlord of New York," was, Stephen Zelnick writes, "probably the most hated man in New York when he died in 1848," in "Melville's 'Bartleby': History, Ideology, and Literature," *Marxist Perspectives* 2 (Winter 1979–1980): 75; regarding Astor's relation to Trinity Church and the foreclosures with which a Master in Chancery would have been associated, see Thomas Dilworth, "Narrator of 'Bartleby': The Christian-Humanist Acquaintance of John Jacob Astor," *Papers on Language and Literature* 38 (2002): 49–75.

retreat, both in his original business and as Master in Chancery, therefore, is as a pristine separation from the reality or the effects of his professional avocations.

This separation of the force and effect of his office work is reflected in his language and comportment in the office. While Agamben emphasizes the functional relation between office and the speech act of the command, it is striking that *inside* the office, commands are rarely given. Although it operates on the basis of an implied or imputed obligation, the lawyer's language is rarely in the imperative. Indeed, it hardly ever even risks the status of a full statement. His mode rather is understatement, his presentation self-consciously self-deprecating, and he orders his office by hints, leading questions, suggestions, accommodations, compromises and only unwillingly rises to the confrontation of a direct statement or command. At all costs, he avoids any act or speech act (for example, firing or dismissing an employee) that would coincide with its execution. In short, the lawyer does not lay down the law in the office; the office rather is built on official assumptions on which the lawyer's language and behavior rely.

This method of governance by assumption, which evades the travails of decision and the risks entailed in entering the sphere of real action, provides occasion for one of the more amusing episodes in the story. The lawyer, who later acknowledges an incapacity to act more forthrightly – "Turn the man out by an actual thrusting I could not" (35) – makes his first attempt to dispense with Bartleby by paying (indeed, overpaying) him and expressing the assumption that he will be gone the following day. En route home he reflects with some satisfaction on his "masterly management in getting rid of Bartleby" (33) and remarks on the elegance of his procedure: "Without loudly bidding Bartleby depart – as an inferior genius might have done – I assumed the ground that depart he must; and upon that assumption built all I had to say" (33). When it then turns out that Bartleby has not in fact left the office, the dumbfounded lawyer even considers pressing on with his assumption:

> Yes, as before I had prospectively assumed that Bartleby would depart, so now I might retrospectively assume that departed he was. In the legitimate carrying out of this assumption, I might enter my office in a great hurry, and pretending not to see Bartleby at all, walk straight against him as if he were air. [...] It was hardly possible that Bartleby could withstand such an application of the doctrine of assumptions. (35)

This mad conceit of carrying on regardless, which would amount to abnegating existence for the sake of his assumption, is in fact thoroughly consistent with the attitude of the bureaucrat even if here it marks the beginning of the increasingly frantic breakdown in his method and self-assurance. The office does not simply put theory into practice, it behaves as if theory were already practice, as if it already presented practical reality.

Indeed, the whole walled-in structure of the office is set up to facilitate precisely this attitude – the suspension of the given according to the doctrine of assumptions. On the one hand, the office shuts itself away from the world to present a site of obscurity with regard to the public and private, the economic and political aspects of its activities; on the other, it shuts out the world the better to be able to carry on its business. The two office windows that the lawyer early on describes do not look out onto the world but rather onto walls.[13] The first looks onto a white wall, the interior of a skylight shaft running down the building, affording a view "deficient," as the narrator wryly remarks, "in what the landscape painters call 'life'" (14). The second offers by contrast "an unobstructed view of a lofty brick wall, black by age and everlasting shade" (14), standing within ten feet of the window. Some have suggested that the structure of the office is reminiscent of a (Platonic) cave.[14] But the snug bureaucrat's attitude is not the comfortable ignorance of those chained before shadowy projections they take for reality. Rather the shaded enclosure of the office enables him to operate with projections of what reality ought to be.

It is not simply that bureaucratic institutions are notorious for their lack of transparency. A certain opacity is, rather, constitutive of the office such that the facticity of what is can be walled-off and shut out so that the imperative of efficacy may be fulfilled. If "what the landscape painters call 'life'" (14), names the domain of human interaction among humans and between humans and nature, the office is structurally "deficient" – alienated, blinkered, insulated from life – in order to facilitate its efficacy on life. The office itself is, indeed, a deontological or a de-ontological space. To recall Agamben's distinction, it presents an enclosed site of suspension where *what is* is placed, as it were, in abeyance or walled off, while *what ought to be* is assumed, according to the doctrine of assumptions.

The characteristic "office work" consists therefore, paradigmatically in *Bartleby*, in copying the law. While the efficacy of the office on life beyond the office, or more generally the effects of bureaucratization, may be substantial, *in* the office, office work is consumed in repeating its own assumptions, verifying its own

---

[13] For the most influential reflection on the walls in Bartleby, see Leo Marx, "Melville's Parable of the Walls," *The Sewanee Review* 61.4 (1953): 602–627, although the "Marxist" aspect of his reading – that the walls appear as impenetrable, objective, even metaphysical limits, while being in fact constructs of society and history – is often overlooked.

[14] See Branka Arsić, *Passive Constitutions, or, 7 1/2 Times Bartleby* (Stanford: Stanford University Press, 2007), 1–10.

premises, and altogether validating the world view – which as we have seen is really a walled view – of the office. For this reason, office work in the nineteenth century and thereafter could be the occasion for critical or philosophical reflection as well as literary production; for it can be more or less true to the law or parody it, can take it in earnest or play with it. In all of these instances, it can serve as much to defer the law's execution as to render it effective. For the most part and above all, however, office work is the dull task of *just* copying the law – which is enough, no doubt, as with the lawyer's other scriveners, Nippers and Turkey, to give one indigestion or drive one to drink. In any case, whether it is carried out attentively or in fits and rages, sober or drunk, mechanically or maniacally, earnestly or ironically, office work does not immediately appear to produce anything more substantial than paperwork. Its efficacy, seen from the office, remains assumed or hypothetical, this being the contributing factor to its distinctive characteristics – the detachedness but also the intrusiveness, the easiness but also the frustration, the mildness but also the despair of the work.

## Insecurities

In *Bartleby* the office interior, as the lawyer presents it, provides a model also of the bureaucrat's interiority. Or rather his careful descriptions of the office space are implicated in, and therefore constantly betraying, his own preoccupations with boundaries and barriers that seem intended to further isolate him from the outside, without impinging on the possibility of efficient commerce. Notable in this respect is the way in which the office hierarchy is reflected in the ability to manipulate further enclosures or openings that in turn facilitate certain evasions in, as well as invasions into, the otherwise undifferentiated office space. The lawyer explains that semi-transparent, "ground glass folding-doors" divide his premises in two, one for the scriveners the other for himself, which he reports opening and closing according to his "humor" (19).

When Bartleby arrives, the lawyer decides to take "the quiet man" onto his side of the folding doors but brings another screen in to provide a further articulation of this more intimate space: "a high green folding screen, which might entirely isolate Bartleby from my sight, though not remove him from my voice. And thus, in a manner, privacy and society were conjoined" (19). A broader reflection on the workings of bureaucracy might be inferred from this dramaturgy of the office. For it is out of such official conjunctions and articulations, the folding screens that are introduced by, and subject to, bureaucratic discretion, that distinctions in "real life," like those between the public and the private, are defined. Inner and outer, private and public, here give way to the dialectical mechanisms of withdrawal and penetration of office government. Bureaucrats are taken up with putting up barriers that they can at any time, according to their humor, open or traverse but behind which they cannot be seen, nor called to account. The articulation of official authority depends on the private protection afforded by opacity – whether the opacity of walls or of recondite documents – and the force of penetration that, to use the lawyer's words, turns every seemingly private space into a "social" one, which is to say, a space subject to bureaucratic intervention. These are the conditions for the production of a bureaucratic attitude, at once punctilious and presumptuous, that in English is felicitously captured in the word, "officious." The structure of the office – also of the officiousness of the bureaucrat – shields itself according to the saying, "out of sight, out of mind," while expanding indefinitely (beyond any field of vision) the reach of official oversight.

The reach and ramifications of the lawyer's offices beyond the walls of Wall Street is only hinted at with the references to John Jacob Astor and, more generally, his "snug business among rich men's bonds and mortgages and title deeds" (14). And this oversight is structural to the extent that the story is constrained to the perspective of the narrator, which is to say, to the official view, even when, encountering Bartleby, this begins to break down. Not to suggest that his law office is employed in anything untoward, quite the contrary, he may just be doing his office. But precisely in attending to the law, the office contributes to the hegemony of a system, specifically a system of property relations, with which the narrator would, to judge from his difficulties dealing with Bartleby, be uncomfortable were he brought face to face with its realities.

One can thus speculate about the injustices, and perhaps even the inhumanity, with which the lawyer's legal avocations are complicit by paying attention to the particular context of historical New York,[15] or, more broadly, of mid-nineteenth-century political agitation in Europe and America and the accompanying critique of positive law.[16] *Bartleby*, however, is more illuminating in exhibiting

---

**15** See Barbara Foley, "From Wall Street to Astor Place: Historicizing Melville's 'Bartleby,'" *American Literature* 72.1 (2000): 87–116.
**16** In the American context, starting with Egbert Oliver's essay, "A Second Look at Bartleby," *College English* 6 (1945): 431–439, *Bartleby* has often been read as a response to Thoreau's "Resistance to Civil Government" (later "Civil Disobedience") which Melville might well have read in the *Aesthetic Papers*, ed. Elizabeth P. Peabody (Boston: The Editor, 1849). Michael Rogin is exemplary among those who see Bartleby as positively embodying Thoreau's "passive resistance," *Subversive Genealogy: The Politics and Art of Herman Melville* (New York: Knopf, 1983) 195; for a survey and criticism of such positions, see Dan McCall, *The Silence of Bartleby* (Ithaca: Cornell University Press, 1989), chap. 3. Marx's "On the Jewish Question" (1843) is probably the most famous and controversial expression of the critiques of law circulating in Europe; its emphasis on

the means and measures by which the office is – officially – insulated from the world of politics. The lawyer's contentment in the office is a function of its walls and other screening devices. He constantly betrays his preoccupations with ensuring his insulation from the effectivity of his office, his protection from the troubling consequences of action and the unsettling realities of the implementation of the law. It is in this respect above all that he is considered "an eminently *safe* man" (14).

Once again, the bust of Cicero comes to mind to the extent that "safe" in the lawyer's curious use of the term could be a translation, albeit a plaster-of-paris one, of the Latin word *salus*, which referred to physical safety (health and protection from harm) and played a crucial role in Roman political thought as the imperative of public safety which Cicero famously formulated, *salus populi suprema lex esto* (*De legibus*). Alternatively, the lawyer's term also suggests the Latin, *securitas*. And it was again Cicero in *De officiis*, although Melville may not have known this given the translation of the text he owned, who introduced the term "security" – which for him meant peace of mind – into the history of political thought.[17] In the first book, adapting the Stoic posture for the purposes of the office, he writes (in Cockman's translation): "We should free ourselves, in short, from all vehement passions and disorders of mind, not only those of desire and fear, but also of sorrow, of joy, and anger; that so the state of the mind may be calm and undisturbed [*securitas*]; which will make the whole life become graceful and uniform."[18] The condition of office-existence is to be free from cares (*curas*), for only in such a state of mind can one attend with prudence and honesty to determining and fulfilling one's duties.

In the nineteenth-century city office, in an almost parodic inversion of Cicero's text, security is the condition of office life insofar as it (ideally) enables the bureaucrat to attend to the duties determined by his office without any cares. In Cicero, security provides the "constancy and dignity" [*constantiam, dignitatem*] for the deliberation and execution of actions that one would stand by as one's own. In the translation Melville read, in contrast, security makes life "graceful and uniform," or simply, to use the lawyer's term, "snug," by retreating from the uncomfortable and unpredictable realm in which action takes effect. Thus the security of the office, as it is presented in *Bartleby*, means precisely

---

political/legal as opposed to human emancipation certainly resonates with *Bartleby*, see Karl Marx, *Early Political Writings*, ed. and trans. Joseph O'Malley with Richard A. Davis (Cambridge: Cambridge University Press, 1994).
**17** See John Hamilton, *Security: Politics, Humanity, and the Philology of Care* (Princeton: Princeton University Press, 2013), chap. 4.
**18** Cicero, *The Orations*, 3:43.

the opposite of Cicero's notion of dignity and constancy: it is the assurance that one will not have to stand by one's official actions – neither in the sense of assuming responsibility, nor in the sense of being on site to witness the effects.

On the basis of the narrator's presentation, a particular aspect of the typology of the modern office, which informs the type of the modern bureaucrat, begins to emerge: the office is the place where official persons are able to efficiently carry out their duties because they are shielded from, and so neither have to see nor face, the consequences of their avocations, as one says, "in person." The term "avocations" to describe the work particular to the office can thus be further specified: like the voice that penetrates the enclosure that keeps Bartleby out of sight, it is an action that is screened from its effect and the region of its efficacy. Or, to draw on another ambiguous term from the bureaucratic register, "oversight," the dialectic of seeing and screening in the office operates such that the more limited the perspective the greater the oversight. The greater the security, the easier it is for the office to be effective.

## Easiness

Bartleby never poses questions. He poses *as* a question. His very appearance, along with the iterations of his utterance, "I would prefer not to," present a perplexity that might be formulated in a question as to the grounds of the de-ontology of the office: "Why do something rather than nothing?" To this question, the lawyer, who introduces himself as "a man who from his youth upwards, has been filled with the conviction that the easiest way of life is the best" (14), would likely answer: because something is easier than nothing. It is, indeed, the notion of easiness – rather than the classic notions of safety and security – that articulates the snugness of the bureaucratic narrator. The modern office is the space that ensures that it is easy to do one's duty and that the execution of one's duty can be carried out with ease. In particular it means that one need not trouble one's conscience to carry out one's office duties and, furthermore, need not be troubled by the consequences of one's actions.

Not that the office is then a site of calm and tranquility. On the contrary, precisely on account of the easiness of the work, the most trivial of inconveniences – the height of a table, say – can drive the likes of Nippers to distraction. And to be sure, Turkey's alcoholism is a function of office work, enabling him to make of it something more challenging and exciting than it is. In drunken enthusiasm, he likens himself to an officer of a different sort: "'In the morning I but marshal and deploy my columns; but in the afternoon I put myself at their head, and gallantly charge the foe, thus!'—and he made a violent thrust with the ruler" (16). The office is the site in which, because of the easiness of the work, petty conflicts and trivial confrontations – with people and with things – seem to take on epic proportions. In the early twentieth century, this state of affairs will give rise to the phrase, "office politics." The exclusion of the political world from the office gives rise to its trivial recapitulation within.

Bartleby seems to be wonderfully unaffected by office politics. If he nonetheless wreaks havoc in the office, it is because his manner undermines the very principle of the office. Indeed, the calm figure proves to be nothing less than an anarchist. At first, Bartleby's "preferring not to," both his expression and his behavior, appear simply to be a failure to fulfill his office duties, prompting the lawyer on occasion to a rare imperative: "come forth and do your duty" (22). More perplexing to the lawyer is the fact that Bartleby appears to flout the principle of easiness on which the duties of the office are based. When Bartleby first says, "I would prefer not to," the task at hand is a "small affair," of the sort, indeed, for which the lawyer had expressly placed Bartleby "handy" to him on the other side of the screen, "to avail myself of his services on such trivial occasions" (20). The lawyer responds by presenting ever more trivial tasks to try and coax Bartleby into action according to the logic that the easier they are the less reason there is not to do them. According to the logic of the office: the more trivial, the more doable – and *therefore* the more it ought to be done.

It does not seem to occur to the lawyer that it may be the very triviality of the tasks that makes them not worth doing. Such a possibility is, however, at the center of Ralph Waldo Emerson's "The Transcendentalist" (1843). Part of a series of published lectures entitled "The Times," it may have influenced the composition of *Bartleby* and in any case diagnosed a historical condition with which Melville himself was all-too-familiar. Having listed the historical formations that gave rise to certain transcendentalist-type figures – from Stoics, to prophets, to protestants, to idealists – Emerson observes that in contemporary America such figures respond to the conditions of their epoch by preferring to do nothing:

> They hold themselves aloof: they feel the disproportion between their faculties and the work offered them, and they prefer to ramble in the country and perish of ennui, to the degradation of such charities and such ambitions as the city can propose to them. They are striking work, and crying out for something worthy to do! What they do is done only because they are overpowered by the humanities that speak on all sides; and they consent to such labor as is open to them, though to their lofty dream the writing of Iliads or Hamlets, or the building of cities or empires seems drudgery.[19]

---

**19** Ralph Waldo Emerson, "The Transcendentalist" (1841–1842/1843), in *The Essential Writings of Ralph Waldo Emerson*, ed. Brooks Atkinson (New York: Modern Library, 2000), 182. See C. W.

If Bartleby belongs, as he certainly seems to, among these nineteenth-century ranks of disaffected would-be "transcendentalists," it is because nothing appears "transcendent" any longer, the world is walled-in. The secret of their extraordinary inaction then is nothing mystical or profound; rather these are figures who have simply never found anything to do, in much the same manner that Kafka's "hunger artist," never found anything to eat that was "to his taste."[20] Thus where once one might have found an Achilles or a Hamlet, a Homer or Shakespeare, or simply the pathos of epic or tragedy, now there is just the more or less pathetic (in the modern sense) figure of Bartleby, the scrivener. Bartleby is the name of a figure who, in another life and times, might have been a hero or a genius or a prophet. He might, in short, have had a calling. Or it may be more precise to say that Bartleby is "called"; however the call – and this characterizes his "modernity" – is silent. He is called to nothing or to he knows not what, so that all he can do is respond to an advertisement for office help as a law-copyist.[21]

Of course the office is not a calling, rather it serves specifically to annul the need for a calling.[22] Nonetheless, Bartleby seems stubbornly to embark on his

---

Sten, "Bartleby the Transcendentalist: Melville's Dead Letter to Emerson," *Modern Language Quarterly* 35.1 (1974): 30–44.

**20** Franz Kafka, *Ein Hungerkünstler*, in *Ein Hungerkünstler. Vier Geschichten* (Berlin: Die Schmiede, 1924), 50.

**21** In "Up Wall Street towards Broadway: The Narrator's Pilgrimage in Melville's 'Bartleby, the Scrivener,'" *Studies in Short Fiction* 24 (Summer 1987): 263–70, Graham Forst suggests that Bartleby be understood as one of those "heroes" described in the work of Joseph Campbell who "refuses the call," and of whom Campbell writes: "Refusal of the summons converts the adventure into its negative. Walled in boredom, hard work, or 'culture,' the subject loses the power of significant affirmative action and becomes a victim to be saved," *The Hero with a Thousand Faces* (Princeton: Princeton University Press, 2004), 54. Here, around 1850 in New York, it does not, however, appear to be a matter of refusal, so much as a question as to what, under conditions of rationalization, the call, if there is something like a call at all, might be a call to do. This is the question of peculiar fanaticism.

**22** In "Politics as a Vocation" (1919), Max Weber reflects on the fate of "calling" as betrayed in the sense of the German word "Beruf." Of its "highest form," Weber writes with reference to the actual, or in any case effective, calling that defines charismatic authority: "Submission to the charisma of the prophet or warlord or of the great demagogues of the assemblies, the *ekklesia*, of ancient Greece or of Parliament means that such men are held to be the inwardly 'chosen' [called, *berufen*] leaders of humankind. People do not submit to them because of any customs or statutes, but because they believe in them," Max Weber, *The Vocation Lectures*, ed. David Owen and Tracy B. Strong, trans. Rodney Livingstone (Indianapolis: Hackett, 2004), 35. Although the call is still heard in the term, *Berufspolitiker*, it had come to mean simply "career," designating either a bureaucrat or a professional politician. The sociological but also existential

copying as if it were a calling, as if it might offer some sustenance for his transcendental hunger: "At first Bartleby did an extraordinary quantity of writing. As if long famishing for something to copy, he seemed to gorge himself on my documents" (19). He becomes, for a time, a monomaniacal copyist, preferring not to do anything else. But copying the law in the offices of the New York Master in Chancery in the 1840s is not going to inspire a religious transfiguration or a work of art. It opens no new vistas beyond, nor insight into, the bleak office in which it is carried out. Soon, therefore, Bartleby will give up copying altogether, the would-be transcendentalist falling instead into what the lawyer refers to as "dead-wall reveries" (29). There is, however, no writing on the wall, no revelation in the office, no transcendence whatsoever except perhaps that elusive line of flight indicated by the famous perlocution, "I prefer not to."

"To prefer not to" is to perform inaction. The utterance turns what would otherwise appear as mere inaction into the accomplishment of a preferred non-action. The phrase articulates a difference, albeit a barely manifest one, between sheer passivity (which could attest to indecision or inanimateness), and a sort of second-order passivity – a passivity that is performed in ineffectiveness, that executes an inactivity. This most difficult passivity – that acts by withdrawing from efficacy – is inexplicable from the point of view of the office, where it therefore generates considerable uneasiness.

Agamben, in his influential philosophical reading of *Bartleby*, suggests that one may understand this passivity as the being of potentiality. Bartleby is, accordingly, the exponent of potentiality in which its self-obfuscating character – that it can enter or not enter into actuality – is exhibited and, paradoxically, "actualized."[23] There is another more down-to-earth way, however, in which preference may be related to the possible. It relates to a distinction between preferring and willing, a distinction explicitly made in the text, that brings to mind discussions of the will in British empiricism and specifically a curious distinction made by Locke in his *An Essay Concerning Human Understanding* (1689), that is cited also in the opening pages of a book consulted by the lawyer in *Bartleby* (37): Jonathan Edwards' *Freedom of the Will* (1754).

Having defined willing in terms of preferring, "The word preferring seems best to express the act of volition," Locke suddenly adds, "it does it not precisely; for, though a man would prefer flying to walking, yet who can say he ever

---

problem, as Weber sees it, is therefore the question as to what keeps one, so to speak, careering along.
**23** Giorgio Agamben, "Bartleby, or on Contingency," in *Potentialities: Collected Essays in Philosophy*, trans. Daniel Heller-Roazen (Stanford: Stanford University Press, 1999).

wills it?"²⁴ Edwards takes exception to this distinction, it being as he sees it simply a matter of distance: willing is the more proximate and so immediate form of preferring. But Locke seems to suggest a qualitative distinction. Willing is what is preferred among courses of action or omission within the domain of what counts as possible: "Volition, it is plain, is an act of the mind knowingly exerting that dominion it takes itself to have over any part of the man, by employing it in, or with-holding it from, any particular action."²⁵ One can prefer what one knows one cannot do, but can only will what one *thinks* (with, presumably, a certain degree of accuracy or delusion) one can do. In other words, the will is defined and circumscribed by the capacities and capabilities of the human agent, while preference can apply beyond the limits of the human. And it is in this realm, insofar as he explicitly distinguishes it from willing, that Bartleby's negative preference situates itself – "'I would prefer not to.' 'You will not?' 'I prefer not'" (25).

Bartleby prefers not to do anything that falls under the dominion of what is considered possible. Around 1850, what counts as possible is no longer defined by the human agent but according to the institutional parameters of the office, or "agency" (a usage that, not coincidentally, was just becoming current). Bartleby thus rejects the easiness of the office, opting instead for the impossibly difficult. In preferring not to, Bartleby, whether in protest or pathology, prefers to prefer rather than to will. He thus presents an unexpected response to Locke's curious question, "for, though a man would prefer flying to walking, yet who can say he ever wills it?" In the manner of the *Schwärmer*, whose fiery elation he however lacks, Bartleby would prefer to fly and will not settle for walking – and so, in the meantime, although his attitude is patently futile and all for nothing (*vergeblich*), he prefers not to.

## What is to be done?

At one point, Bartleby mildly states about himself that he is "not particular" (41). He is, however, certainly peculiar, peculiar insofar as he fails to measure up to the norms of office life and ultimately even to what is taken to be human. The peculiarity of his preference expresses itself in a difficult, un-easy – indeed, fanatical – sort of passivity, or rather *impassivity*. This impassivity regularly "dis-

---

**24** John Locke, *The Works of John Locke* (London: Thomas Tegg (et al.), 1823), 1:244; Jonathan Edwards, *Freedom of the Will* (London: Thomas Nelson, 1845), 2.
**25** Locke, *Works*, 244.

arms," "unmans," and otherwise paralyzes or humiliates the lawyer, who finds in Bartleby's debilitating "mildness" that there is not "any thing ordinarily human about him" (21). Because Bartleby cares nothing for the office and appears wholly unaffected by its particular demands, the lawyer is at a loss as to what measure to take. "Poor fellow, poor fellow!" the lawyer repeats to himself at one point, "he don't mean any thing" (23). This is the very cause of his agitation: the insignificant figure, impoverished in his person and social standing, impassive in action, and inconsequential in his speech, means more and more to the narrator – and, apparently, without meaning to. Like Michael Kohlhaas before him, Bartleby – at least in the eyes of the law/lawyer – stands in vain (*vergeblich*) for nothing or next to nothing (*eine nichtige Sache*). The fanatical impassivity with which he insists on his apparently futile preference makes Bartleby a peculiar fanatic.

For although Bartleby seems to do nothing he has an extraordinary effect – or rather, generates extraordinary affect with disturbing consequences for the structure of the office and the person of the bureaucrat. It is particularly upsetting for the lawyer that he finds it difficult, indeed impossible, to assume Bartleby's air of indifference. That Bartleby's peculiar "passivity" does not appear to be of the order of passion at all, at least not of the sort the sentimental lawyer is in a position to recognize, seems to be all the more painful for him, as he finds himself, helplessly, passionate about Bartleby. Bureaucrat that he is, the lawyer seems obligated to interpret his feelings in terms of the logic of the office. He persistently and exclusively responds to the feeling Bartleby arouses with an office question, indeed, with the office question par excellence: *What is to be done?*[26] There is no unequivocal response: "What was to be done? He would do nothing in the office: why should he stay there? In plain fact, he had now become a millstone to me, not only useless as a necklace, but afflictive to bear. Yet I was sorry for him. I speak less than truth when I say that, on his own account, he occasioned me uneasiness" (32). His efforts to translate his conflicted and ambivalent feeling into a duty only generate an increasing tension and oscillation between attraction and repulsion. Having discovered that Bartleby lives in the office, a situation that attests not just to his poverty but to his "horrible" solitude, the narrator observes: "My first emotions had been those of pure melancholy and sincerest pity; but just in proportion as the forlornness of Bartleby grew and grew to my imagination, did that same melancholy merge into fear, that pity into re-

---

**26** This is a quintessential office-question but also, and not incidentally perhaps, presents *the* question of the nineteenth century extending from the canon of reason in Kant's *First Critique* (1781), via Marx's "Theses on Feuerbach" (1845), to Lenin's citation of Chernyshevsky's 1863 novel in his 1902 pamphlet, "What is to be done?"

pulsion" (29). Pity before misery that cannot be helped, about which nothing can be done, turns into repulsion. When the intensity of his feeling seems to run up against the limit of what can be done, against the walls of the office, as it were, it turns into its opposite. The ambivalent uneasiness that Bartleby occasions expresses itself in the lawyer's increasingly erratic behavior, going from one extreme to the other, as he attempts to arrive at a univocal determination of his duty.

The most prudent and easiest route would certainly have been to have recourse to the law, for, legally speaking, there is no question who is in the right. But it is immediately clear that the law cannot do justice in this case. For equity's sake, the Master in Chancery feels obliged therefore to look elsewhere, and in a certain sense beyond his office, when considering the right thing to do. One moment, he intends to "befriend" Bartleby (23), the next, burning "to be rebelled against," he seeks out hostility: "I felt strangely goaded on to encounter him in new opposition, to elicit some angry spark from him answerable to my own" (24). The impassive Bartleby, however, can be treated as neither a friend nor an enemy. Nonetheless, the escalating dialectic of attraction and repulsion that he occasions will very nearly make the office into the scene of a struggle-unto-death – and shortly thereafter, into something of a love affair.

After a particularly ineffectual exchange with Bartleby, the lawyer describes being in "such a state of nervous resentment" that he had to "check himself," being reminded of "the unfortunate Adams and the still more unfortunate Colt in the solitary office of the latter" (36). The reference is to the historical 1842 case in which John Colt (the brother of Samuel Colt, the firearms manufacturer) killed with a hatchet the printer, Adams, at his offices in New York (the confrontation was over costs relating to the printing of a textbook on bookkeeping!).[27] The lawyer tellingly presents the killing as one profoundly implicated in the office space, specifically suggesting that it was the fact that they were alone in the office that drove Colt into "wild" excitement in which the killing was committed "unawares" (so the lawyer). As the "safe man" explains in complete sympathy with Colt, who he sees as a victim of excessive office-affect: "It was the circumstance of being alone in a solitary office, up stairs, of a building entirely unhallowed by humanizing domestic associations—an uncarpeted office, doubtless, of a dusty, haggard sort of appearance;—this it must have been, which greatly helped to enhance the irritable desperation of the hapless Colt" (36). The office,

---

[27] On the extraordinary circumstances surrounding the historical case that also ended with a death in The Tombs, see William Dillingham, "Unconscious Duplicity: *Bartleby, the Scrivener*," in *Melville's Short Fiction, 1853–1856* (Athens: University of Georgia Press, 2008).

which is supposed to provide a context in which bureaucrats are insulated or alienated from ordinarily human concerns, becomes, by virtue of this inhuman desolation or desertedness, a place in which all irritation turns to wild excitement and the "fatal act" (36) is executed all-too-easily, without restraint or reflection.

The lawyer claims to have been "saved" from this particular office-destiny by recalling a higher duty, "the divine injunction: 'A new commandment give I unto you, that ye love one another,'" one that he, however, interprets in a typically office-like manner, "Aside from higher considerations, charity often operates as a vastly wise and prudent principle—a great safeguard to its possessor" (36). On the basis that "no man that ever I heard of, ever committed a diabolical murder for sweet charity's sake," he concludes that "mere self-interest," should lead "the high-tempered" to considerations of "charity and philanthropy" (36). This officious perspective on love gives way over the next few days to a more "salutary feeling," when in the light of a consultation of "Edwards on the Will," and "Priestley on Necessity," the lawyer "slips into the persuasion" that "Bartleby was billeted upon me for some mysterious purpose of an all-wise Providence, which it was not for a mere mortal like me to fathom" (37). In other words, the lawyer ventures – crucially, not without a certain irony that subverts the office logic he recites – to make sense of his conflicted feelings by interpreting Bartleby *as* his office by providence or predestination, "my mission in this world, Bartleby, is to furnish you with office-room for such period as you may see fit to remain" (37).

Assuming Bartleby as an unfathomable office "predestined from eternity" (37), the lawyer takes the logic of the office to a paradoxical extreme, where his duty is to do nothing and to leave Bartleby alone – and so is, in effect, no duty at all. His particular phrasing is more intriguing still, since to "furnish" Bartleby with "office-room," without asking questions or making demands, with room, in other words, in which Bartleby is left to prefer not to, is positively to engage in a kind of de-officing activity, the furnishing of a space of passivity and ineffectivity *in* the office in which no offices are in force. By taking on this unaccountable and possibly incoherent office, the lawyer enjoys, for a time, a happy relation with Bartleby that consists in embracing the obscure intimacy of his presence: "Yes, Bartleby, stay there behind your screen [...] I never feel so private as when I know you are here. At last I see it, I feel it; I penetrate to the predestinated purpose of my life. I am content" (37).

Earlier, when Bartleby stopped copying, "he did nothing but stand at his window in his dead-wall revery." The exasperated lawyer demands to know why: "'And what is the reason?' 'Do you not see the reason for yourself,' he indifferently replied" (32). Bartleby's obscure response leads the lawyer to reasonably conclude that, from all his copying, he might have "temporarily impaired his vision" (32). But if it is indeed impaired vision from which Bartleby suffers, it is the structural impairment of the office, concretized in the dead wall right outside the office window. The lawyer, however, tends to be too caught up in his office, or attending to his offices towards Bartleby, to look and see. If Bartleby's presence does indeed prompt him to reflect on and look beyond his offices, he neither looks beyond nor reflects on the limits or the limitedness of the office as such. He does not see, as a result, all that Bartleby makes him feel.

If the wall, as I have suggested, separates the action of office work from the sight, as well as the site, of its effects (that realm the landscape painters call "life"), then the "office-room" with which the lawyer takes it to be his duty to furnish Bartleby is, paradoxically, a space least penetrated by the rising rule of bureaucracy. Thus, it is in his office cubicle that the forlorn Bartleby is able to indulge his peculiarity, "loosed" from the escalating officiousness of everyday life. Bartleby's does indeed appear to be what Anglo-American philosophers and theologians called a "loose existence" that defies the nexus of causality,[28] or, in the contemporary terms of Continental thought, a figure of "potentiality" that "frees itself of the principle of reason."[29] But if he is an unaccountable apparition without the slightest cause or ground, who as he himself states, "would prefer not to be a little reasonable" (30), such a peculiar and *schwärmerisch* being can emerge in New York around 1850 only in an obscure retreat *within* the office, which, tellingly, he prefers not to leave. In short: the only space that eludes the proliferating oversight of the office is the "office-room" unofficially furnished within it.

By accepting, albeit as his preordained duty, Bartleby's occupation of an unofficial or non-official space inside his office walls, the lawyer comes to a happy though short-lived accommodation. The arrangement produces an intimate relation with only the mildest, most impassive mode of obligation or acknowledgment: "Yes, Bartleby, stay there behind your screen..." (37). And here the lawyer exclaims, as if in recognition of Bartleby's enigmatic question about seeing the reason for himself, "at last, I see it, I feel it" (37). If what he sees and feels is

---

**28** The phrase, which she draws from Edwards, is central to Branka Arsić's reading of *Bartleby*. "For if the event be not connected with the cause, it is not dependent on the cause; its existence is as it were loose from its influence, and may attend it, or may not...And to say that an event is not dependent on a cause is absurd," Edwards, *The Freedom of the Will*, 83, cited in Branka Arsić, *Passive Constitutions*, 23.

**29** Agamben, "Bartleby, or on Contingency," 254. For a summary and synthesis of many of the influential "Continental" readings of *Bartleby*, see Birgit Mara Kaiser, *Figures of Simplicity: Sensation and Thinking in Kleist and Melville* (Albany: SUNY, 2012), 88–96.

indeed love, and *Bartleby* then a curious kind of "office romance," it is not the sort of love that follows an injunction but, on the contrary, the attainment of a relation freed from every obligation: "I never feel so private as when I know you are here" (37). It would seem, however, that the lawyer never really sees it that way. Insofar as he continues to interpret the arrangement, however obtusely or even ironically, in terms of office-duty, his sight remains delimited by the perspective of the office, which is why the happy relationship cannot last.

## "Ah Bartleby! Ah humanity!"

Anne Smock writes that the encounter with Bartleby is "an unbearable burden – a duty that cannot be assumed nor is it possible to quit this debt."[30] This is indeed a precise characterization of the experience of the lawyer. It is the lawyer, after all, who ends up, for a time at least, assuming Bartleby as a predestined duty to whom he owes an infinite debt. There is something – or rather, as Smock accurately emphasizes, *nothing* – about Bartleby that leads to such a conclusion, as if he represented, as a number of commentators in different ways maintain, an inarticulable imperative or an impossible demand.[31] So interpreted, strange to say, Bartleby presents ultimately the very form of the office, the figure of duty.

It would seem therefore that there is an inclination, indeed, an *obligation,* to read *Bartleby* as an office encounter, one that poses but ultimately cannot resolve the question: *What is to be done?* which thus turns into an infinite task.[32] But this seems at once too office-like and too easy. Bartleby presents an "unheard-of perplexity" (35) in the office precisely because he neither responds to nor makes any demands. His impassive presence and impossible preference throw the fundamental assumptions of the office into confusion. And if the peculiar relationship the lawyer for a moment enjoys with him can be called an "ethical" relation, it is not one based on debt and demand but, on the con-

---

[30] Ann Smock, "Quiet," *Qui Parle* 2.2 (1988): 75.
[31] See J. Hillis Miller, "Who Is He? Melville's 'Bartleby the Scrivener,'" in *Versions of Pygmalion* (Cambridge, MA: Harvard University Press, 1990); Jeffrey Weinstock, "Doing Justice to Bartleby," *ATQ* 17.1 (March 2003): 23–42; Francesco Bigagli, "'And Who art Thou, Boy?': Face-to-Face with Bartleby; Or Levinas and the Other," *Leviathan* 12.3 (2010): 37–53.
[32] Hillis Miller, for one, insists that the "impossible task" presented by Bartleby extends to the text *Bartleby*: "Imperiously, imperatively, it says, 'Read me!'" to which, he claims, the reader can only respond in petrification, "Who Is He?" 175.

trary, on the paradoxical duty – which is in fact no duty at all – of leaving office-room.

It is in this light that the lawyer's final word or exclamation on his interaction with Bartleby is altogether illuminating: "Ah Bartleby! Ah humanity!" (45). The words follow his recounting of a "vague report" (13) that Bartleby had previously been employed as a subordinate clerk in the Dead Letter Office in Washington. For reasons that have to do with what he imagines to be the nature of that particular office, the thought makes the lawyer quite distraught, "When I think over this rumor, hardly can I express the emotions which seize me" (45). The Dead Letter Office is an office that documents and otherwise attempts to account for failed offices – those offices, whether public, private or commercial, borne by the post that missed their mark, arrived too late, or in any case could not take effect: "On errands of life, these letters speed to death" (45).

"Ah Bartleby! Ah humanity!" expresses this biopolitical conundrum: the official project of humanity addresses itself to death. It is fated to fail. All offices are ultimately Dead Letter Offices. There are always those elusive figures who cannot be helped nor cared for nor saved, whom offices, even "good offices" (but for the lawyer all offices are good offices), cannot reach, or reach only too late. Thus for the lawyer, "humanity" names an impossible demand – an impossible duty – it is the very motive and justification for office existence and the propagation of bureaucracy but, at the same time, the task that offices can never ultimately fulfill. Bartleby is the ghost-like apparition inside the office in which the limits of the office are presented, or felt, as an intractable obligation or infinite demand before which the lawyer's increasingly desperate avocations are hopeless – and his evasions equally so. "Ah Bartleby! Ah humanity!" is the despairing cry of the distraught bureaucrat torn between the unlimited imperative to officiate and the limited scope of specifically official efficacy.

So much for the obligatory, official reading. An unofficial reading might venture the following: Bartleby is the peculiarly fanatical figure in which humanity enters the office. Or rather: Bartleby is what humanity would look like in the office, were it ever to make such an appearance. It appears forlornly, with the calm, pale, composed impassivity of a statue. In short, humanity appears in the office as a figure about which there is nothing "ordinarily human" (21). And it announces itself as such by preferring not to, a gesture that elides and unsettles the biopolitical pretensions of the office to care for, and take care of, human life. Humanity, disconcertingly, asserts itself in the office not by making a demand, much less an absolute one, but by preferring not to acknowledge any offices at all. In short: the peculiar existence of humanity, if one can speak of such a thing, does not correspond to the ontology of the office and cannot be assimilated to the bureaucratic-humanitarian project.

The presence of Bartleby, the presence of humanity, is felt by the lawyer as a confusing combination of attraction and repulsion, of pain and pleasure, of awe and despair (29). This office-affect is very like that ambivalent feeling, at once painful and pleasurable, at once humiliating and uplifting, that Kant called *Achtung*. *Achtung*, on the one hand, discloses the immediate sense of duty before the law: "What I recognize immediately as a law for myself I recognize with respect [*Achtung*]...," and on the other – though for Kant these came to the same – the sense of humanity in our person: "Humanity in his person is the object of the respect which he can demand from every other human being ..."[33] In Kant, who Agamben takes to mark the culmination of the ontology of the office, *Achtung* discloses at once the office *and* humanity, the office *as* humanity. The *Achtung*-like feeling Bartleby elicits under the bureaucratic conditions of New York circa 1850 seems to betray an incommensurability between humanity and the office. The conflicting feeling no longer composes itself into the immediate conviction of a categorical duty but leads the lawyer again and again to pose the bemused, and perhaps misguided, question, *What is to be done?* In this light, the despairing exclamation, "Ah Bartleby! Ah humanity!" does not express an infinite demand on the office by humanity, but articulates an incongruence between the office and humanity, one that indicates a space (even if merely some provisional "office-room") for peculiar relationships and interactions – one could certainly also say "queer" ones – that would not be bound by duties.[34] Such relations would be human rather than official, instead of being human on account of being official.

Since humanity does no work in the office and its troubling effect does not seem to be of the order of a conflict of duties but the confounding of the assumptions of the office as such, its presence cannot be tolerated. The fact is: there is

---

**33** Immanuel Kant, *Groundwork of the Metaphysics of Morals* (1785), ed. and trans. Mary Gregor and Jens Timmermann (Cambridge: Cambridge University Press, 2012), 17, AA 4:401; Kant, *The Metaphysics of Morals* (1797), ed. and trans. Mary Gregor and Jens Timmermann (Cambridge: Cambridge University Press, 1996), 186, AA 6:435. For Agamben's discussion of Kant's notion of *Achtung*, which he sees as the irreducible theological moment in Kantian ethics, which in turn marks the culmination of the historical conversion of ethics into the office, see *Opus Dei*, 219–228. See also the discussion of *Achtung* in Chapter Two.

**34** For a related reading of Bartleby's "queerness" against the normative suppositions of the "humanities," see Lee Edelman, "Occupy Wall Street: 'Bartleby' Against the Humanities" *History of the Present* 3.1 (Spring 2013): 99–118. To Edelman's negative definition of queerness: "the queerness that every regime of 'what is' must construe as what is *not*, as the nothing, the negativity, or the preference for negation that threatens the normative order, whose name is always human community" (112), I would add: there is also a queerness that refuses or is refused by the official dispensation on account of preferring something other than "what ought to be."

no place for humanity in the architecture of the office, nor in the person of the bureaucrat. It is as if the concrete office space emerges to separate offices from humanity – and if humanity is still the end of the office in the sense that bureaucracy is an essentially humanitarian project, it has no place *in* the office but abides, as it were, in the officiated realm beyond its walls. Although, in the figure of Bartleby, humanity demands so very little, indeed nothing more than to be left some "office-room," it is ultimately too hard, if the office is to remain an office, to just leave humanity be. In the final instance, therefore, Bartleby will be cast out and treated as a kind of vagrant – which is to say, in nineteenth-century New York (and not only then), as the figure in which humanity appears when it fails to respond to the offices that structure ordinary life and which therefore has simply to be securely put away, taken "care" of.

## Vagrancy

The lawyer is soon embarrassed by his peculiar relationship with Bartleby. The rumors it generated in the circle of his "professional acquaintance […] worried him very much" (38). "Something severe, something unusual must be done" (38). Already, early on, the lawyer had posed the legal question to Bartleby who no longer copied and preferred not to do anything else: "What earthly right have you to stay here? Do you pay any rent? Do you pay my taxes? Or is this property yours?" (35). Once he is no longer employed and prefers, furthermore, not to accept any employment while also preferring not to leave the premises which he occupies night and day, Bartleby becomes increasingly vulnerable to the charge of vagrancy. In so doing, he comes to appear in the office as a figure of, and for, the urban underclass that was at the time in New York a significant political preoccupation and would likely, as the reference to John Jacob Astor emphasizes, have been subject to certain of the operations of the lawyer's private business as well as his office as the Master in Chancery.[35]

With regard to Bartleby, however, the lawyer goes to great lengths to evade the sort of conclusion that he had presumably reached often enough in cases that touched him less personally:

> And upon what ground could you procure such a thing to be done?—a vagrant, is he? What! he a vagrant, a wanderer, who refuses to budge? It is because he will *not* be a vagrant, then, that you seek to count him *as* a vagrant. That is too absurd. No visible means of support: there I have him. Wrong again: for indubitably he *does* support himself,

---

**35** See Robin Miskolcze, "The Lawyer's Trouble with Cicero," *Leviathan* 15.2 (2013): 43–53.

and that is the only unanswerable proof that any man can show of his possessing the means so to do. No more then. (38)

Itinerancy, a disregard (whether by necessity or preference) for borders and barriers, both public and private, and destitution, the lack of visible means of support, which principally defined vagrancy in nineteenth-century America (along with more specific definitions relating essentially to alcoholism and prostitution) gave rise to a figure that would, if under different names and criteria, be increasingly familiar to the rationalizing societies of the nineteenth century and beyond: a figure that had no other rights than those on account of being human. Or, put another way, vagrancy concerns people who are only bearers (as opposed to exercisers) of rights and are therefore in fact entirely subject to the "offices" of individuals, societies, or states.

Bartleby, as the lawyer's rather disconnected arguments show, does not fall easily into the category of vagrant. Although Bartleby has no rights to be where he is, he does not demand any rights nor seek to justify his position and although he has no visible means of support, he asks for no help and demands no care. What is difficult, and ultimately unbearable for the lawyer, is that Bartleby does not – prefers not to – submit himself to the de-humanizing indignity of soliciting good offices that would make officially treating him like a vagrant so much easier.

Rather than calling in the police, the lawyer moves out of the office himself and painfully leaves Bartleby behind, "and then, – strange to say – I tore myself from him whom I had so longed to be rid of" (39). Nevertheless, Bartleby's continued occupation of the office block (after the new tenant removes him to the stairwell) continues to cause ever greater consternation and even political fermentation: "Every body is concerned; clients are leaving the offices; some fears are entertained of a mob..." (40), and he is eventually removed to prison, namely, "the Tombs, or to speak more properly, the Halls of Justice" (42).[36]

Having occupied an office as an unemployable office worker, he now appears as a free prisoner, a non-criminal criminal, a passive outlaw:

> Being under no disgraceful charge, and quite serene and harmless in all his ways, they had permitted him freely to wander about the prison, and especially in the inclosed grass-platted

---

[36] For a reading of *Bartleby* and the concept of "civil death," in the context of the history of incarceration in the United States as well as of what he calls the *"poetics of the penitentiary,"* the "images and tropes that give meaning to the violence of detention," see Caleb Smith, "Detention without Subjects: Prisons and the Poetics of Living Death," *Texas Studies in Literature and Language* 50.3 (2008): 243–67.

> yard thereof. And so I found him there, standing all alone in the quietest of the yards, his face towards a high wall, while all around, from the narrow slits of the jail windows, I thought I saw peering out upon him the eyes of murderers and thieves. (43)

Here, Bartleby finds himself inside the walls he had looked out on from the office, that is to say, in a thoroughly administered space. This brings about a subtle and suicidal change in his aspect. Bartleby is no longer an office worker who prefers not to do his duty, he is an inmate who prefers not to be ministered to. While in the office he has lived on a diet of ginger-nuts, in prison, he "lives without dining" (45) – declining even the informal attentions of the grub-man, whom the lawyer calls an "unofficially speaking person" (43).

The change is also evidenced in his comportment towards the lawyer. When he first calls on Bartleby in the prison, no explicit reference to preference is made: "'I know you,' he said, without looking round, – 'and I want nothing to say to you'" (43). The lawyer, pained by this dismissive response and sensing an "implied suspicion," immediately feels obliged to offer an excuse: "It was not I that brought you here, Bartleby" (43). Bartleby's utterance, however, may not have been an accusation so much as a statement expressive of the indifference of incarcerated life. While it would be possible to paraphrase the utterance in more familiar terms, "I know you and I would prefer not to speak to you," for example, there is something decidedly different about his prison-language. In contrast with "to prefer not," which negatively implies a soaring sphere of preference beyond the immediate constraints of the context, "to want nothing" suggests a contracted field of immediate necessity (to be in want) or the urgent exigency of eminently executable activity (to want to). If Bartleby, in the office, could, in the manner of a peculiar fanatic, entertain a transcendent sphere of preference beyond the walls of what one could possibly will by preferring not to, in the prison he is reduced to the barren domain of want that is conventionally understood to be involuntary – and he responds by wanting nothing.

When the lawyer attempts to cheer him up, Bartleby gives him to understand that he is under no illusions as to where he is:

> "And to you, this should not be so vile a place. Nothing reproachful attaches to you by being here. And see, it is not so sad a place as one might think. Look, there is the sky, and here is the grass."
> "I know where I am," he replied, but would say nothing more, and so I left him. (43)

Prison life is no existence. It lacks the open and uncertain relation to the outside that characterizes even the trivial ecstasies of ordinary life. It is instead a place made up of bare statements of fact of the sort the lawyer unwittingly produces: "Look, there is the sky, and here is the grass," and a mode of being exhausted in the statement: "I know where I am." The reality of the prison is delimited to knowledge and want. Bartleby knows this and wants nothing.

The one time, therefore, that Bartleby does "prefer not to" in the prison – speaking to the grub-man: "I prefer not to dine to-day" (44) – the phrase is no longer of the same order as his peculiar office preferences. It no longer expresses the aspiration to a transcendent space of indeterminate or "impossible" activity (absolved of the assumptions of the office) but, on the contrary, the intention to reduce his condition of immanence to the zero degree. Bartleby wastes away in the prison. On a subsequent visit, the lawyer is directed by another prisoner to the "silent man," who he finds curled up in a corner, "But nothing stirred" (44). "Something," the lawyer notes, "prompted me to touch him. I felt his hand, when a tingling shiver ran up my arm and down my spine to my feet" (45). The involuntary and inofficious gesture of touching the lifeless body, that produces such a sensation through his whole body, is perhaps the lawyer's first human action. It is in any case the first time that he does not self-consciously accord with the office or react to its unfulfilled assumptions. Already recovering himself, he next performs that most seemingly humane of offices, "closing the eyes," before, to the grub-man's question whether he is asleep, he murmurs a formulaic but nonetheless resonant response, "With kings and counsellors" (45).

If Bartleby is, as the citation suggests, a nineteenth-century Job, in a similar way that the lawyer may be a nineteenth-century Cicero, he suffers not from the excessive afflictions that testify to an apparent abandonment by his god but rather from the milder, but all the more extensive, ministrations of a bureaucracy that will not let him be. And, if he is relieved of his earthly torment by withdrawing from the world into death, like those kings and counsellors who, in the King James Bible cited by the lawyer, "built desolate places for themselves,"[37] it is because there is no longer any "desolate place" left on earth that he could retreat to, except, for a brief time, behind the green screen in the lawyer's office at No.— Wall Street.

---

**37** "With kings and counsellors of the earth, which built desolate places for themselves," Job 3:14, *King James Version.*

# 4 Abstraction of the Earth: Conrad's *Heart of Darkness*

The New Imperialism at the end of the nineteenth century is the signature of the epoch of bureaucratic fanaticism. In Conrad's *Heart of Darkness* (1899/1902), this moment, the culmination of European domination of the globe, is presented as one of geopolitical displacement and disorientation of the sort Carl Schmitt, in his writings on the *nomos* of the earth, would call *Entortung*. This sense of displacement is articulated in *Heart of Darkness* in the dissonance of the "voice," which betrays the incommensurability between the bureaucratic-imperial attitude and the "earth" it presumes, abstractly, to organize. As the voice of the New Imperialism, Kurtz is the fanatical bureaucrat par excellence.

## Disfigurements

If the "great land appropriation," to use Schmitt's phrase,[1] commenced at the close of the fifteenth century with the colony Columbus planted on the part of an island that would become Haiti, then the end game of this appropriation played itself out in the "scramble for Africa"[2] at the end of the nineteenth century. A Papal Bull, Pope Alexander VI's *Inter caetera*, had first divided the globe between the earthly powers of Spain and Portugal in 1493 with a single line.[3] The terms that would complete the political mapping of the earth with the carving-up of the African continent were negotiated at a conference of European states (representatives from the Russian and Ottoman Empires as well as the United States government were also present) known as the Berlin Conference, or the Congo Conference, hosted from November 1884 to February 1885 by Chancellor Bismarck.

---

[1] Carl Schmitt, *The Nomos of the Earth: In the International Law of the Jus Publicum Europaeum* (1950), trans. G. L. Ulmen (New York: Telos Press, 2006), 101.
[2] The phrase was apparently coined in 1884. See Thomas Pakenham, *The Scramble for Africa, 1876–1912* (London: Weidenfeld & Nicolson, 1991), xix.
[3] Pope Alexander VI, Bull *Inter Caetera* (1493), in *International and United States Documents on Oceans Law and Policy*, ed. J. N. Moore (Buffalo: William S. Hein, 1986), 1:75–78. For a recent reflection on the significance of such lines with particular attention to the Bull and the Treaty of Tordesillas that followed it, see Philip Steinberg, "Lines of Division, Lines of Connection: Stewardship in the World Ocean," *Geographical Review* 89.2 (1999): 254–264.

A month earlier, in October 1884, on the other side of the Atlantic in Washington D.C., an International Meridian Conference had been convened by President Chester Arthur. Prompted by the increasing integration of the global economy, as well as by political and military considerations, the aim was to agree upon "a meridian to be employed as a common zero of longitude and standard of time reckoning throughout the world."[4] For reasons that were technical and historical, but also reflected the political situation of the day, Greenwich near London was recognized as the Prime Meridian.[5] The technical-scientific standardization of time and space around the globe thus coincided with the diplomatic concord of European imperialism, which concluded the political partition of the world according to the principle of sovereignty – dividing it into sovereign states and non-sovereign colonies or protectorates.

"Congo" and "Greenwich" thus stand as watchwords at the end of the nineteenth century to mark the culmination of the European, or in any case Eurocentric, epistemo-political project to dominate the earth. Conrad would call this new epoch, marked by the total mapping of the world, "geography triumphant."[6] Yet, as Conrad's literature attests, the victory was a hollow one. The moment was in fact characterized by a sense of groundlessness and disorientation, deception and demoralization. If the nineteenth century had seen, in European thought as in practice, the intensification and acceleration of its occupation of, and preoccupations with, the earth, then the pervasive mood as the century ended seems to have been a sense of alienation and despair, a sense captured by Marlow in *Heart of Darkness:* "the earth seemed unearthly."[7]

Emergent insecurities concerning space are evinced in the development of modern geopolitical thought. In one of its founding texts, the British geographer Halford Mackinder, writing in 1904, described a "geographical pivot in history," in which the imagery of an infinite Cartesian space that, he argued, had defined

---

4 *International Conference Held at Washington for the Purpose of Fixing a Prime Meridian and a Universal Day: October, 1884. Protocols of the Proceedings* (Washington: Gibson Bros, 1884), 1.
5 See Stephen Kern, *The Culture of Time and Space 1880–1918* (Cambridge, MA: Harvard University Press, 1983), esp. 10–14.
6 In "Geography and some Explorers," Conrad compares the adventurous open horizons and "blank spaces" of the exploration of the globe that characterized the epoch he calls, "geography militant," starting with Columbus, reaching its peak with Cook and ending with Livingstone, with the new epoch, "geography triumphant," which, however, is somehow disappointing, Joseph Conrad "Geography and some Explorers," *National Geographic*, March 1924. See Con Coroneos, *Space, Conrad, and Modernity* (Oxford: Oxford University Press, 2002).
7 Joseph Conrad, *Heart of Darkness*, in *Youth, Heart of Darkness, The End of the Tether*, ed. Owen Knowles, *The Cambridge Edition of the Works of Joseph Conrad* (Cambridge: Cambridge University Press, 2010), 79. Subsequent citations in-text.

the political consciousness of the preceding era, had given way to a new epoch characterized by what he called "closed space" and a concomitant "closed political system." Henceforth, he observed, political life would have to be understood in terms of the finite dimensions of the earth in which there are no strictly isolated events: "Every explosion of social forces, instead of being dissipated in a surrounding circuit of unknown space and barbaric chaos, will be sharply re-echoed from the far side of the globe, and weak elements in the political and economic organism of the world will be shattered in consequence."[8]

The emergence of a realist political theory, premised on the inherent limitations of the earth and its resources, coincided with the sudden and radical escalation of European colonial politics known as the "New Imperialism," which exhibited an altogether unrealistic attitude to space. What was "new" about the New Imperialism was captured in a fanatical declaration made by Cecil Rhodes, which Arendt would use as the epigraph to her study of imperialism in the second part of *The Origins of Totalitarianism* (1951): "I would annex the planets if I could."[9] The phrase, Arendt argues, at once expresses the doctrine of the New Imperialism, that "expansion is everything," as Rhodes put it, and betrays "its inherent insanity and its contradiction to the human condition."[10] An imperialism premised on expansion for expansion's sake is insane and inhuman – it is, indeed, fanatically bureaucratic – insofar as it aims for space in the abstract without reference to any geographical measure of the earth nor to the realities of human finitude. The application of what Arendt took to be the principle of economic speculation – "expansion" – to the inherently finite sphere of geopolitics would prove catastrophic for the earth and its peoples: "the human condition and the limitations of the globe were a serious obstacle to a process that was unable to stop and to stabilize, and could therefore only begin a series of destructive catastrophes once it had reached these limits."[11] What was most devastating about this historical moment – from which it is far from clear that we have emerged – was a paradoxical failure to acknowledge the finitude of the earth, combined with an ungrounded but altogether destructive determination to wield dominance over it.

In *The Nomos of the Earth* (1950), Schmitt would also relate the imperialism of the period 1890 to 1914 to an *Entortung* that heralded the collapse of the Euro-

---

[8] H. J. Mackinder, "The Geographical Pivot of History," *The Geographical Journal* 23.4 (April 1904): 422. On Conrad and Mackinder, see Coroneos, *Space, Conrad, and Modernity*, esp. 21–38.
[9] Cited in Hannah Arendt, "Imperialism," *The Origins of Totalitarianism* (New York: Harcourt, 1966), 124.
[10] Arendt, *Origins*, 124.
[11] Arendt, *Origins*, 144.

centric spatial order of the earth he considered to have been in place since the appropriation of the New World. In this regard, the Congo Conference marked the culmination, but also the fateful tipping point, of the Eurocentric *nomos*, premised on the appropriation and domination of non-European territories and their exclusion from the order of sovereignty: "The triumph symbolized by the word 'Congo' was short-lived."[12] In Schmitt's theory, the *nomos* of the earth in any given epoch is defined by a particular concrete relation between order (*Ordnung*) and place (*Ortung*). *Entortung* indicates not simply a change in the relation between a given order and place, but a practical and epistemological perplexity regarding the very understanding of order and place. It is, in other words, no longer clear how to "take" space, in the sense of appropriating it (*nehmen*), but also in the sense of perceiving it, as in the German term, *wahrnehmen*. For the norms supposed to constitute the concrete unity of order and place no longer hold. The consequence is a geo-political crisis, beyond Schmitt's resolutely Eurocentric assessment, for which the word "Congo" may be taken to be paradigmatic.

Although the Congo basin was a significant source of slaves for the Atlantic slave trade,[13] it had not become the object of European colonial interests in part because the Congo River, with its strong currents, had not been navigable by European craft and so was not viable for commercial exploitation. Nineteenth-century developments in steam-power, with the railway and more significantly the steamboat, changed that.[14] When Bismarck called the Berlin Conference, it was in order to secure "free trade" in central Africa and to prevent competition among European powers on the continent from descending into conflict. Although the imperial map of Africa was not explicitly drawn up at the conference, the negotiations over the "conventional basin of the Congo" defined the terms and set the precedent.[15] The convention room itself would be depicted in the

---

12 Schmitt, *Nomos*, 226.
13 On the penetrating effects of the Atlantic slave trade (and of the slave trade more generally) on societies within Africa, see Patrick Manning, *Slavery and African Life: Occidental, Oriental and African Slave Trades* (Cambridge: Cambridge University Press, 1990): "with the allure of imported goods and the brutality of capture, slave traders broke down barriers isolating Africans in their communities. Merchants and warlords spread the tentacles of their influence into almost every corner of the continent. By the nineteenth century, much of the continent was militarized; great kingdoms and powerful warlords rose and fell, their fate linked to fluctuations in the slave trade," 147.
14 Henry Morton Stanley brought the first river steamers in sections to the navigable part of the river in 1879.
15 William Roger Louis, in *Ends of British Imperialism: The Scramble for Empire, Suez, and Decolonization* (New York: I.B. Tauris, 2006), describes the "wild geographical interpretation of the

media with a large map of Africa hanging on the wall,[16] and the conference became the symbol for the imperial mapping of the continent with borders that bore little relation to the social, political, and demographic contours of Africa.[17] The map defines the territory.

The implementation of the abstract projection of order onto place, with little regard for local particularities, was facilitated by the notion, central to the conference, of "effective occupation," which was supposed to distinguish a legitimate colonial possession from mere flag-raising. The premise, drawing on a tradition of natural law that links property to use, was the capacity to demonstrate the effective exploitation of the land. Since the continent of Africa was patently occupied and put to use, the possibility of African sovereignty had to be acknowledged – whence the emergence of the term "protectorates." But by and large, the tacit understanding operative at the Berlin Conference was that, although African territories were owned, in the sense that they were used and governed by their present occupants, they were nonetheless not *sovereign* in the significant sense of the term.[18] Legal scholars, sympathetic to the imperial project, sought to distinguish sovereignty (*imperium*) from simple ownership of the land or its use (*dominium*). Sovereignty was defined instead as having an administra-

---

Congo basin," which one British Foreign Office official observed would be comparable to drawing the Rhine in the basin of the Rhône. It turns out, Louis adds, that it was in fact the British ambassador who had secretly proposed the line in order to maximize the eventual zone open for "free trade," 102–103. His chapter, "The Berlin Conference and the (Non-) Partition of Africa," presents a detailed account of the diplomatic negotiations.

16 The large map, drawn by Heinrich Kiepert, features prominently in many of the representations of the conference as it did in the conference room itself. See, for example, Adalbert von Rößler's illustration in *Über Land und Meer. Allgemeine illustrierte Zeitung* 7 (1884): 1233.

17 Writing in the context of the independence movements of the 1960s, Kwame Nkrumah would state, "the original carve-up of Africa [was] arranged at the Berlin Conference of 1884," *Challenge of the Congo* (New York: International Publishers, 1967), x. While not strictly speaking true, the claim is testament to the historical significance of the conference that is indeed symbolic of the "arrangement" of the African continent. For recent historical accounts of the conference and its relation to the "scramble," see Pakenham, *The Scramble for Africa*, esp. 239–255; H. L. Wesseling, *Divide and Rule: The Partition of Africa, 1880–1914* (Westport: Praeger Publishers, 1996), esp. 71–130; A. I. Asiwaju, "The Conceptual Framework," as well as J. D. Hargreaves, "The Making of the Boundaries: Focus on West Africa," in *Partitioned Africans*, ed. A. I. Asiwaju (New York: St. Martin Press, 1985); for a brief survey of the history of "African boundaries" with particular attention to early efforts to address the problem of colonial borders in post-independence Africa, see Adekunle Ajala, "The Nature of African Boundaries," *Africa Spectrum* 18.2 (1983): 177–189.

18 On the history of the re-definition of (European) sovereignty on the basis of colonial encounters, see Antony Anghie, *Imperialism, Sovereignty, and the Making of International Law* (Cambridge: Cambridge University Press, 2005).

tive structure requisite for entry into the community of nations and so of international law.[19] Although it would also, and in every case, completely disregard the rights of indigenous people to their property, the New Imperialism understood itself as essentially an administrative enterprise.

In the very midst of the formal proceedings of the conference, which prepared for the carve-up of ostensibly non-sovereign African territories among sovereign European states, a series of backroom bilateral arrangements were reached that facilitated the emergence of an extraordinary "sovereign" state. Representatives of King Leopold II of Belgian managed to convince the nations present that the aims of neutrality and free trade in the Congo basin could best be guaranteed by a private enterprise, Leopold's "philanthropic" *Association Internationale Africaine*, which should furthermore be recognized as a "neutral" sovereign power. By the end of the conference, every country in the community of states present (except Turkey) had recognized what would become Leopold's own *État indépendant du Congo* or, as it equally ironically became known in English, the Congo Free State.[20] In absolute contrast to the words in its name and the philanthropic discourse that surrounded its establishment, the Congo Free State would soon refer to a regime of forced labor and "primitive accumulation" that committed some of the most egregious humanitarian and ecological excesses and atrocities of the New Imperialism.[21] "Congo" thus names an overdetermined but for that reason exemplary site of the *Entortung* that defined the epoch.

Certainly for Conrad this was the case. In "Geography and some Explorers" (1924), Conrad presents the "Congo" as the very threshold in the history of geography that he felt he had lived through. Echoing almost verbatim certain of Marlow's lines in *Heart of Darkness*, Conrad remarks on how the mysterious "blank spaces" in central Africa on the map that he had found so alluring as a child had become instead the source of "…the distasteful knowledge of the vilest scramble

---

**19** On the role of the *Institut de droit international* where the legal grounds for the new imperialism were discussed, see Andrew Fitzmaurice, "The Genealogy of Terra Nullius," *Australian Historical Studies* 38.129 (2007): 1–15.
**20** See Fitzmaurice, "The justification of King Leopold II's Congo enterprise by Sir Travers Twiss," in Ian Hunter and Shaunnagh Dorsett, eds., *Law and Politics in British Colonial Thought* (New York: Palgrave, 2010). Merwin Crawford Young sees the Congo Free State as exemplary of the deleterious consequences of the contemporary doctrine of sovereignty, *The African Colonial State in Comparative Perspective* (New Haven: Yale University Press, 1994), 78.
**21** See Adam Hochschild, *King Leopold's Ghost* (London: Pan Macmillan, 1998). Frances Singh notes that the phrase, "state of terror," in the sense of state or state-sponsored terrorism, was coined with regard to the Congo Free State by A. E. Scrivener writing critically of Leopold's regime in 1903, "Terror, Terrorism, and Horror in Conrad's *Heart of Darkness*," *Partial Answers: Journal of Literature and the History of Ideas* 5.2 (2007): 200.

for loot that ever disfigured the history of human conscience and geographical exploration."²² If the word "Congo" does not appear in *Heart of Darkness*, this is because the novella is not a story about, much less a record of, events of a concrete historical place, but explores rather the contours of the disfiguration Conrad alludes to. If Conrad's "disfigurement" can be read as a possible translation of *Entortung*, referring to the spatial perplexities that emerged at the turn of the century, it complicates Schmitt's strictly geopolitical category by bringing it into relation with questions of conscience. Indeed, in Conrad's usage, the word "conscience" carries the association of the French and so also means "consciousness"; the disfigurement relates geo-political concerns to phenomenological ones. This complex of space and conscience is subject to literary treatment in *Heart of Darkness* and the remarkable figure at the heart of the historical disfiguration – of the earth and of the conscience – is the fanatical bureaucrat, Kurtz.

## Ivory, Ideas, Rivets

Around the time that Conrad published *Heart of Darkness* (1899/1902), Husserl published his *Logical Investigations* (1900/1901). This would just be a coincidence were it not for the fact that Conrad's novella revolves around the uncertain status of "the Intended," just as Husserl's investigations seek to clarify the obscurities of "intentionality" (*Intentionalität*). To be sure, the respective use of these "intentional" terms is not the same: Conrad's novel is as little a "logical investigation" as Husserl's study is a work of modern literature. But in their invocation of intention, each writer elaborates, I would argue, a shared source of anxiety that relates to the geographical and political *Entortung* they were living through.²³

If Husserl's work is an investigation of intentionality understood to be the "correlation between being and consciousness,"²⁴ *Heart of Darkness* describes a journey into a space of tensions, incongruities and conflicts that seem to undermine such correlation and to lie at the heart of the imperial enterprise as a

---

22 Conrad, "Geography and Some Explorers," 272.
23 Looking back in 1919, Husserl would refer to the sense of "distress" (*Not*) out of which the *Investigations* emerged. See Edmund Husserl, Letter to Arnold Metzger, September 4, 1919, *Philosophisches Jahrbuch der Görres-Gesellschaft* 62 (1953): 195.
24 Edmund Husserl, *Introduction to Logical Investigations: A Draft of a Preface to the Logical Investigations*, written in preparation for its republication in 1913, ed. E Fink, trans. P. J. Bossert and C. H. Peters (The Hague: Martinus Nijhoff, 1975), 29.

bureaucratic undertaking. Referring to the novella as "a drama of officialdom,"[25] Levenson reads *Heart of Darkness* as a critique of imperial bureaucracy insofar as it presents "the empty assertion of an institutional formalism in the face of violently anti-institutional facts." It thus situates itself in a broader Conradian problematic: "the incongruity between value and fact, between the system of meanings that we devise and the world reluctant to accept them."[26] Kurtz, whose name means "short," would be the *reductio* of the imperial attitude. I would argue, however, that the reduction presented in the figure of Kurtz is not adequately captured in the fact-value distinction drawn from British empiricism. It resonates rather with the contemporary concerns of phenomenology. Indeed, the phenomenological analysis of intentionality was supposed to overcome the spurious distinction between values understood as ideal, psychical attributes projected onto material facts. For intentionality denoted the way in which consciousness is always consciousness *of* something, such that it made no sense to separate an ostensibly neutral given fact from a supposedly projected value; fact and value were rather always already integral to the intention of a given object.[27]

Husserl's phenomenological "breakthrough" in the *Investigations* drew a radical consequence for epistemology: truth was no longer the question of *adaequatio rei et intellectus*, as intentionality was nothing other than the consciousness-object correlation; rather, truth became a matter of "fulfillment."[28] Truth relates to the degree to which an intention found fulfillment in intuition (*Anschauung*): "*the full agreement* of what is meant [*Gemeinten*] with what is

---

**25** Michael Levenson, "The Value of Facts in the Heart of Darkness," *Nineteenth-Century Fiction* 40.3 (1985): 266. Setting his discussion in the context of Max Weber, Levenson is one of the few commentators who emphasizes the centrality of bureaucracy in *Heart of Darkness*: "after Kafka, Conrad is our most searching critic of bureaucracy," 269.
**26** Levenson, "Value of Facts," 267.
**27** Bruce Johnson addresses the relation between Husserl's project and *Heart of Darkness*, although the question of intentionality with regard to the Intended is not expressly discussed, "Conrad's Impressionism and Watt's 'Delayed Decoding,'" in *Conrad Revisited: Essays for the Eighties*, ed. Ross Murfin (University: University of Alabama Press, 1985).
**28** See Ernst Tugendhat, *Der Wahrheitsbegriff bei Husserl und Heidegger* (Berlin: de Gruyter, 1967). In Tugendhat's account this sea change in the conception of truth led ultimately to the trivialization of truth in analytic philosophy and the indiscriminate expansion of truth in Heidegger's notion of *aletheia*. He proposes a return to Husserl's phenomenological approach to truth as a process of adequation towards fulfillment. Tugendhat does not address the emergence of "existential" epistemology that may also be traced to this moment and with which the Heidegger of *Being and Time* is certainly associated.

*given as such.*"²⁹ It is, ideally, given "in the flesh" (*leibhaftig*) as full living presence.³⁰ The *Investigations* thus register the beginning of a shift: out of the nineteenth-century tension between scientific truth and humanistic value-thinking emerges an existential epistemology in which authenticity and inauthenticity would assume priority over truth and falsehood. Although it is not yet a discourse of authenticity, Husserl's theory of truth as fulfillment speaks to a perceived need that is betrayed in the contemporary "nihilistic" idiom of hollowness and vacuity paradigmatically recited in *Heart of Darkness* – along with the appetite for something more substantial. One recalls Marlow's description of his first sighting of Kurtz: "I saw him open his mouth wide – it gave him a weirdly voracious aspect, as though he had wanted to swallow all the air, all the earth, all the men before him" (106). Told belatedly, *Heart of Darkness* is the account of a longing for fullness in which the longing itself proves to be suspect – empty and inauthentic.

In *Heart of Darkness* the search for fullness or fulfillment is oriented around two "values" that govern the imperial attitude: ivory and ideas. In a much-cited reflection before he starts his tale, Marlow observes, not without irony, that the conquest of the earth is redeemed by "...the idea only. An idea at the back of it, not a sentimental pretense but an idea; and an unselfish belief in the idea – something you can set up, and bow down before, and offer sacrifice to..." (47). While at the Inner Station and elsewhere, he remarks on the ivory-worship of the "faithless pilgrims" in the employ of the Company: "The word 'ivory' rang in the air, was whispered, was sighed. You would think they were praying to it" (65). If ivory takes on a fetish-character among the "pilgrims," at once present and absent, it is significant that it is always also a token of death or decay.³¹ Even when it is actually present in a veritable heap in front of you, it rings hollow: "I don't deny," explains the manager after they have found Kurtz, "there is a remarkable quantity of ivory—mostly fossil" (108). The truth of ivory is that there is never enough.

---

**29** Edmund Husserl, *Logical Investigations, Volume 2*, trans. J. N. Findlay from the second German edition of *Logische Untersuchungen*, ed. Dermot Moran (New York: Routledge, 2001), Investigation VI, 263.
**30** Husserl, *Logical Investigations, Volume 2*, Investigation V, 608–609.
**31** For a range of valences of ivory in *Heart of Darkness*, see Darrel Mansell, "Trying to Bring Literature Back Alive: The Ivory in Joseph Conrad's 'Heart of Darkness,'" *Criticism* 33.2 (1991): 205–215; Jeffrey Myers, "The Anxiety of Confluence. Evolution, Ecology, and Imperialism in Conrad's *Heart of Darkness*," *Interdisciplinary Studies in Literature and Environment* 8.2 (2001): 97–108; Stephen Ross, *Conrad and Empire* (Columbia: University of Missouri Press, 2004).

Set on "ivory," one extreme of the imperial attitude is exemplified in the Eldorado Exploring Expedition, which exhibits a buccaneering attitude of unadulterated expropriation: "To tear treasure out of the bowels of the land was their desire, with no more moral purpose at the back of it than there is in burglars breaking into a safe" (73). In her study of imperialism in *Origins of Totalitarianism*, which treats Conrad's novella as a sourcebook on the imperial character, Arendt cites Marlow's description of the Eldorado Expedition – "reckless without hardihood, greedy without audacity and cruel without courage" (73) – to identify what she calls the "superfluous type."[32] In her account, those who engaged in the imperial enterprise at the end of the nineteenth century differed from the individuals and explorers who had characterized the earlier period of exploration by the fact that they had been rendered superfluous in their own societies, "the refuse of all classes," and so were driven to such extremes: "they had not stepped out of society but had been spat out by it; [...] they were not enterprising beyond the permitted limits of civilization but simply victims without use or function."[33] The superfluity of the "mere wealth" they craved corresponded to their own superfluity and the ruthless, risky, and thoughtless manner in which they sought to attain it.[34] It would be characteristic of the New Imperialism – and nowhere more so than in the Congo Free State – that the economy of pure extraction of the value of the earth (of ivory and subsequently, wild rubber) would be pursued with abandon, without concern for more lasting institutions of government and the appropriation, partitioning, and investment characteristic of more sustainable land-use.

Arendt takes the figures in *Heart of Darkness*, including Kurtz, as exemplary of the "worst elements in Western civilization," particularly disposed to racism, which she identifies as one of two "devices" that characterized and facilitated the propagation of the New Imperialism.[35] She overlooks in her reading of Conrad the fact that the novella is equally about that other device of imperialism which, she argues, "attracted the best, and sometimes even the most clear-sighted, strata of the European intelligentsia," namely, bureaucracy.[36] Arendt reads *Heart of Darkness* as a racism novella, but it is also a bureaucracy novella and Kurtz is not, or not only, a "superfluous" type but also the imperial bureaucrat

---

[32] Arendt, Origins, 189. On Arendt's reading, see Christopher GoGwilt, "Joseph Conrad as Guide to Colonial History," in *The Oxford Historical Guide to Joseph Conrad*, ed. John Peters (Oxford: Oxford University Press, 2010).
[33] Arendt, *Origins*, 155, 189.
[34] Arendt, *Origins*, 188.
[35] Arendt, *Origins*, 186.
[36] Arendt, *Origins*, 186.

par excellence. His clear-sighted bureaucratic attitude propels him to fanaticism in a manner that is altogether in keeping with Arendt's assessment of the "insanity" of the New Imperialism.

In contrast to the superfluous greed exhibited by the Eldorado Expedition, the so-called "new gang—the gang of virtue" (68), among whom Kurtz and Marlow are numbered, are motivated not only by ivory but also by ideas. While it is tempting, and certainly consoling, to dismiss the idealism of imperialism as mere cynicism, it is more challenging to treat it as authentic. In *Heart of Darkness* an abstract or abstracted idealism reveals itself to be the motor, and not just the justification, of certain horrors of the New Imperialism. In contrast to ivory, which proves always to be somehow hollow and fossil-like, ideas *as* ideas can be both living and full. That ideality could be intuited in its own right and was therefore not simply psychological nor "abstract" in the derivative sense proposed by British empiricism, was Husserl's discovery in the *Logical Investigations*. In the Sixth Investigation Husserl showed that what is given in intuition are not only sensible objects (such as ivory) but also ideal objects (ideas). Although this was not his express intention, Husserl's expansion of intuition to include "categorial intuition" showed why ideas could present themselves as living and full – often even more so than the sensible reality they are supposed to articulate. Measured by intuition, the abstraction of ideality could be more fulfilling than the concreteness of reality. This phenomenon serves in part to explain the fanaticism of Kurtz – and the swarming he inspires. In *Heart of Darkness* the allure of ideality is presented in women back in Europe, most notably the Intended, who is the very impersonation of the "idea" and, as such, at the very "heart" of the novella, even if she appears in person only, and anonymously, at the very end. Her attitude is prefigured in the enthusiasm of Marlow's aunt, who, brushing aside the fact that the Company is run for profit, talks about her nephew as if he were "something like an emissary of light, something like a lower sort of apostle," setting out to wean "'those ignorant millions from their horrid ways'" (53).

Although he is not himself an idealist – his account is rather one of disillusionment – Marlow's manifest intention, by no means successful, is to maintain a distinction between ivory and idea, to prevent, on the face of it, the thoroughgoing contamination of idealism by avarice. At a more fundamental level, however, his intention can be understood as securing the survival – if not the full living presence – of the ideal intention, namely, the Intended. This will be the significance of his final gesture, the lie, in which he despairingly attempts to preserve the Intended from the ivory with which she, her forehead "smooth and white" (123) in the evening light of her home in a European capital, threatens to become confounded.

As the closing scene makes explicit, the Intended lives for, lives on account of, Kurtz's "voice." The fascination with Kurtz, is the fascination with his voice: "The man presented himself as a voice" (92). Throughout *Heart of Darkness* the fullness of ideality is expressed, or rather it is promised and betrayed, in his voice. For this "gift," Marlow observed "carried with it a sense of real presence" (92). It is remarkable that, although it is not a central term in the *Investigations*, Derrida would use the word "voice" to characterize the "metaphysics of presence" that, based on his reading of the First Investigation, informs Husserl's project.[37] If *Heart of Darkness* presents a critical treatment of the phenomenon of the "voice" that resonates with the concerns Derrida identifies in the contemporary work of Husserl, it does so by situating the logical and epistemological preoccupations of the *Investigations* in the expressly geopolitical context that is captured in the word "Congo." In the reading I propose, *Heart of Darkness* presents a reflection on the colonizing violence attendant to the metaphysics of the "voice." To judge from Marlow's account of Kurtz, the ideal of the voice, the ideality expressed in the voice, becomes all the more alluring and compelling under historical conditions of *Entortung*.

Schematically: in *Heart of Darkness* the emptiness of the idle talk of "ivory" in the Inner Station is opposed to the fullness of Kurtz's voice. Or to concentrate the reduction further, the tension can be described between Kurtz's voice and his "invoice" – the invoice that, much to the manager's irritation, he sends down the river with the ivory he has acquired (74). The colonial world is constituted out of this tension of imperialist intentionality between the unfulfilling extraction of ivory and the fullness promised in the abstractions of the voice. This world, which is really only a series of "stations," is throughout the novella set in relief against a third entity that betrays the phantom-like unreality of all imperialist activity. Of the Inner Station, Marlow observes: "By Jove! I've never seen anything so unreal in my life. And outside, the silent wilderness surrounding this cleared speck on the earth struck me as something great and invincible, like evil or truth, waiting patiently for the passing away of this fantastic invasion" (65). "Outside" the discourse of imperialism, the frenetic discoursing of imperialism, lies a silence that it does not address and indeed ignores: "It was the still-

---

[37] Jacques Derrida, *Voice and Phenomena: Introduction to the Problem of the Sign in Husserl's Phenomenology*, trans. Leonard Lawlor (Evanston: Northwestern University Press, 2011). For a Derridean reading of *Heart of Darkness* that, however, does not consider the contemporaneity of *Heart of Darkness* and the First Investigation from which Derrida's deconstruction of phonocentrism departs, see Vincent Pecora, "Heart of Darkness and the Phenomenology of Voice," *ELH* 52.4 (1985): 993–1015.

ness of an implacable force brooding over an inscrutable intention" (77).[38] This silence is not simply nothing; on the contrary, it is fully given. And if its meaning cannot be discerned, it nonetheless presents itself as meaningful as if it were for its part "intentional." What is exposed here is an incongruity more fundamental than that between fact and value. On the one hand, there is the imperialist agency. It either seeks the superfluous fulfillment afforded by the unreserved extraction of ivory, or, altogether convinced of its ideals, "lives" like Kurtz in a state of abstraction. On the other, there is a given reality, "the wilderness" or simply "the earth," that disappoints and conflicts with imperialist intentions. Furthermore the earth exceeds imperialist intentions in fullness as if withholding an inscrutable meaning. Out of this incongruity between the intentionality of imperialist agency and the givenness of the earth follows the disfigurement of the earth and of the conscience.

Staged in *Heart of Darkness* is a problem that is indicated, but unresolved, in Husserl's *Logical Investigations* and that led to a break among early phenomenologists between realist and transcendental phenomenology: is "the given" a determination of the intuiting subjectivity by means of its sheer transcendental intentional activity (epitomized in the "voice") or is there a givenness that transcends and escapes the objectifying intentions of the subject ("the earth" or "the wilderness")?[39] Insofar as this question retrieves and reframes aspects of the polemic between Fichte's subject-centered philosophy and Schelling's *Naturphilosophie*, *Heart of Darkness* articulates the contours of a latent *Schwärmer* problematic around 1900. For what here emerges as the question of the "earth" relates to the problem of "nature." The question becomes: is the articulation of the earth a human imperative, as Fichte would have it, the imposition of order

---

**38** On the problem of this kind of presentation insofar as it is taken to be an "image of Africa" and contributes therefore to a tradition of what Rob Nixon has called the "figurative arrest" of the African continent, see Chinua Achebe's influential polemic, "An Image of Africa," *Research in African Literatures* 9.1 (Spring 1978), 1–15; Rob Nixon, *London Calling: V.S. Naipaul, Postcolonial Mandarin* (Cambridge, MA: Harvard University Press, 1992), 91; also Patrick Brantlinger, *Rule of Darkness: British Literature and Imperialism, 1830–1914* (Ithaca: Cornell University Press, 1990); Peter Firchow, *Envisioning Africa: Racism and Imperialism in Conrad's Heart of Darkness* (Lexington: University Press of Kentucky, 2000).

**39** For phenomenological realism inspired by the *Investigations* that resisted Husserl's transcendental turn, see Adolf Reinach, "Concerning Phenomenology" (1914), trans. Dallas Willard, *The Personalist* 50 (1969): 194–221; Roman Ingarden, *On the Motives which led Husserl to Transcendental Idealism* (Dordrecht: Kluwer, 1975). For a recent assessment of the question of realism and Husserl's phenomenology in the light of the emergence of "speculative realism," see Dan Zahavi, *Husserl's Legacy: Phenomenology, Metaphysics, and Transcendental Philosophy* (Oxford: Oxford University Press, 2017).

onto place, or does the primacy belong rather to the self-ordering of, or an order implicit in, the earth itself? If Fichte's "voice of conscience" finds its articulation a century later in Kurtz's "gift of expression," then Schelling's "nature" is obscurely expressed in the silent intentionality surrounding the imperial stations. To take up terms developed in my reading of *Michael Kohlhaas*, what is called "the wilderness" is *given* in the mode of *das Vergebliche* – that is to say, in a manner that is intuited but cannot, for all its fullness and presence, be intended, at least by the agents of imperialism, as anything other than negligible or *nichtig*.

For his part, Marlow attempts to inoculate himself from the *Schwärmerei* of the Inner Station by maintaining a "hold on the redeeming facts of life" (65) and concentrating on the work of repairing the steamer. But precisely those "facts" are missing – the rivets – that would make possible the fastening of the boat and of reality. Instead, the caravans from the coast bring only superfluous goods of sham value to be traded for ivory: "And several times a week a coast caravan came in with trade goods – ghastly glazed calico that made you shudder only to look at it, glass beads value about a penny a quart, confounded spotted cotton handkerchiefs. And no rivets" (71). Imperialist commerce, parodied here, is not just in bad faith, it exhibits a structural aversion to communicating with "the wilderness."

## The Outraged Law

Marlow's account of his journey from Europe is presented as a successive disintegration of reality, described in relation to the phenomenon of the voice. As his ship engages in the process of depositing customs-house clerks and soldiers at tin shed stations along an unchanging African coast, Marlow remarks: "The voice of the surf heard now and then was a positive pleasure, like the speech of a brother. It was something natural, that had its reason, that had a meaning" (54). The voice of the surf provides the last coherent sense of an otherwise disintegrating reality: "For a time I would feel I belonged still to a world of straightforward facts" (55).

This sense would disintegrate at the sight of a French man-of-war firing into the wilderness, a scene memorably adapted and updated with napalm in the opening of *Apocalypse Now* (1979): "In the empty immensity of earth, sky, and water, there she was, incomprehensible, firing into a continent" (55). Upon reaching the outer station, Marlow is reminded of the French ship and the "enemies" it was supposed to be targeting when he encounters a chain gang of forced-laborers working on the railway at the coastal station, a project that appears to be consumed in little else than an "objectless blasting" (56) of the landscape:

> Another report from the cliff made me think suddenly of that ship of war I had seen firing into a continent. It was the same kind of ominous voice; but these men could by no stretch of imagination be called enemies. They were called criminals, and the outraged law, like the bursting shells, had come to them, an insoluble mystery from the sea. (57)

In contrast to the pleasingly matter-of-fact voice of the surf, the "ominous voice" of imperialism is strikingly duplicitous: it arrives as a senseless blasting of the recalcitrant earth, and it articulates a legal-bureaucratic discourse which imposes itself, incongruously and therefore all the more ruthlessly, upon all it encounters. Indeed, the outrage of the "outraged law" is a function of the lack of correlation between the law and the territory, between order and place, which it only exacerbates. The more the imperialist voice insists on its geopolitical intentions, the more the "earth seemed unearthly."

In Marlow's account of his journey, the problematic of the voice is brought into relation with that of space. The journey, in which the voice is progressively dislocated from the earth, becomes an exploration of the "heart" of the historical and political *Entortung* of the epoch. And this finds its most radical expression in the topology constituted by Marlow's narrative itself – the topology of *Heart of Darkness* – which, drawing attention to its own uncertain status and incommensurability, envelopes his listeners, generating a sense of "faint uneasiness" that cannot be disentangled from the moral and political uneasiness of the story. "For a long time already [Marlow] had been no more to us than a voice," the narrator observes, "I listened, I listened on the watch for the sentence, for the word, that would give me the clue to the faint uneasiness inspired by this narrative that seemed to shape itself without human lips in the heavy night-air of the river" (70). The listener awaits the word that would provide the metric to translate Marlow's story into established conventions, to localize and legalize it according to the habitual norms of genre, as well as geography, so that the tale might be enjoyed by its listeners on the pleasure boat on the Thames, in splendid isolation – aesthetic and political – from the feverish space into which they are unwillingly drawn by Marlow's account. It is precisely such distance and security that Marlow's tale, which begins with a reflection indicating the Thames, Greenwich, London – "And this also [...] has been one of the dark places of the earth" (45) – fails to provide. It ends with the telling of a "lie" that, in mapping Kurtz's expiring cry, "The horror! The horror!" onto the name of the Intended, repeats the very disfiguration to which the story gives voice.

If Marlow's account in *Heart of Darkness* performs the *Entortung* that he seeks to describe, it thereby discloses the implication of the geopolitical with the phenomenological – of geography with conscience or consciousness. While the "ominous voice" of the "outraged law" articulates an imperialist topology that is increasingly abstracted from the "earth" it is supposed to map, the earth for its part appears monstrous: "We are accustomed to look upon the shackled form of a conquered monster, but there – there you could look at a thing monstrous and free" (79). Indicated here is the most elemental geo-political – and also ecopolitical – concern about the relation of human activity to the earth. If, in Schmitt's account, the word "Congo" marked the culmination and decline of the *nomos* of the earth premised on a European appropriation (*Landnahme*) of the non-European world, *Heart of Darkness* presents a more elemental question that corresponds to the tension between realist and transcendental phenomenology and again recalls the *Schwärmer*-debates between Fichte and Schelling. If there is a *nomos* of the earth, understood as the reciprocal mapping of *Ordnung* and *Ortung*, does it emerge from the imposition of order onto place or the other way round? Even if, as Schmitt does, one premises the *nomos* on an originary taking (*nehmen*) of the land, the phenomenological question imposes itself: what are the concrete human and geographic conditions that would determine the measure of this geopolitical taking?

The question of measure is central to *Heart of Darkness* to the extent that any metric that would congruously or coherently map order onto place is missing. Far from establishing a concrete, much less a legitimate, order, the appropriation of the land takes the form of measureless violence that is, to be sure, accompanied by the rampant exploitation of human life. These geopolitical and "necropolitical" dimensions of the imperialist enterprise are concentrated in Marlow's description of the coastal station.[40] Against the backdrop of "objectless blasting," he picks his way through a scarred landscape – "I avoided a vast artificial hole somebody had been digging on the slope, the purpose of which I found it impossible to divine. It wasn't a quarry or a sandpit, anyhow. It was just a hole" (57) – and comes upon a "grove of death" (61). Here, displaced and exhausted African laborers, "brought from all the recesses of the coast in all the legality of time contracts, lost in uncongenial surroundings, fed on unfamiliar food," withdraw to die (58). To his horror, he encounters: "Black shapes crouched, lay, sat between the trees leaning against the trunks, clinging to the earth, half coming out, half effaced within the dim light, in all the attitudes of pain, abandonment, and despair" (58). Marlow describes, "more bundles of acute an-

---

[40] See Achille Mbembe, "Necropolitics," *Public Culture* 15.1 (2003): 11–40.

gles [...] all about others were scattered in every pose of contorted collapse, as in some picture of a massacre or a pestilence" (59). Yet, there has been neither a plague nor a mass killing. This scene is instead the outcome of an abstract and incongruent, and so in fact fanatical, implementation of bureaucratization. It could be called an "administrative massacre," if the phrase was not to be reserved to refer to intentional imperialist policy.[41]

Subsequently Marlow will look out from the accountant's office and survey the curious mapping of death at the imperial outpost: "In the steady buzz of flies the homeward-bound agent was lying finished and insensible; the other, bent over his books, was making correct entries of perfectly correct transactions; and fifty feet below the doorstep I could see the still tree-tops of the grove of death" (61). It is the accountant's dubious achievement to make "perfectly correct" entries while scrupulously discounting the death and dying all about him. This state of official abstraction he seeks to cultivate is exhibited by his pristine clothing that makes no concession to the environment. The realities of colonial life cannot, however, be altogether shut out of the colonial office, a situation that provokes an unaccountable outburst. When the tumult that accompanies the arrival of a caravan from the interior disturbs his work, the accountant declares, in words that anticipate those of Kurtz: "When one has got to make correct entries, one comes to hate those savages – hate them to the death!" (60–61). When the calculus based on the exploitation of black life and the discounting of black death can no longer ignore the horrific realities of this operation, the despairing representatives of the law respond with a veritably genocidal outrage. Here is laid bare the implication of race and bureaucracy, the two imperialist "devices" Arendt identifies.[42]

Death is commonplace – also among the "whites." Already on the French man-of-war that was shelling the coast, Marlow reports, "men in that lonely ship were dying of fever at the rate of three a day" (55). Less than half of the agents, Marlow suggests, sent out to "the wilderness" by the Company return to Europe.[43] The otherwise completely unexceptional General Manager of the

---

[41] Arendt sees the emergence of the notion (coined by Carthill) of "administrative massacre" proposed by bureaucrats in India as the culmination of the logic of imperial bureaucracy, *Origins*, 186. Massacre-like outcomes had, however, long been "collateral" to the imperial project.
[42] See Arendt, *Origins*, chap. 7.
[43] On Conrad's own journey to the Congo, a fellow passenger Prosper Harou informs him that sixty percent of the company's employees resigned their posts before six months were up; others who could not stand the tropical climate were quickly sent back to Europe; still others died from various tropical diseases and exhaustion; and only seven percent completed their three-year contracts, such as that signed by Conrad/Korzeniowski. From Zdzisław Najder, *Joseph Conrad:*

Inner Station owes his position of power simply to the fact that he is able to survive the climate: "Once when various tropical diseases had laid low almost every 'agent' in the station, he was heard to say, 'Men who come out here should have no entrails'" (64). As Marlow approaches the coastal station, the Swedish ferry captain mentions he had transported a man recently who had hanged himself on the road. When Marlow incredulously asks why, he replies, "Who knows? The sun too much for him, or the country perhaps" (56).

But before the travails of the heat and the environment, before the communication of any specific tropical disease – malaria, typhoid, dysentery – the very entry into the "heart of darkness" is already a sign or a symptom that they have abandoned themselves to the "sickness unto death." If Kierkegaard's pseudonymous book on despair throws light on the new imperialist project a half century later, it is because the psychology it describes is disposed to being oblivious to, or in denial about, the finitude of the human condition. What is always potentially *schwärmerisch* about despair – and no less so about its antidote, "faith" – is that death loses all significance. Where death no longer means anything, nothing is worth dying for – and one is ready therefore to die for next to nothing. Whence the reckless risk-taking, the ruthlessness, the racism, and finally the horrific indifference to the deaths of others exhibited by the representatives of the imperialist undertaking.

It is this particular type of "madness" that the doctor is interested in studying when he asks to measure Marlow's head, "I always ask leave, in the interests of science, to measure the crania of those going out there," adding, however, that "the changes take place inside, you know" (52). The "fever" that all the agents of imperialism exhibit is first of all the one they bring with them and that drives them on: despair. And although it certainly cannot be physically measured – a lack of measure is precisely what despair attests to – it is not clear it takes place "inside" as a strictly psychological phenomenon. For far from being an interior state, despair is the fundamental mood that "corresponds" to the historical and geopolitical *Entortung*, disclosing an incongruous space of anomie that is

---

*A Life* (Rochester: Camden House, 2007), 147. Conrad himself would fall ill in the Congo and return to Europe after only a few months; the recurrent bouts of illness that he suffered for the rest of his life have been attributed to this episode in the Congo. Marianna Torgovnick writes, "The African *is* death in the novella," arguing that along with "the feminine" and "the primitive," these figures indicate or concentrate the sort of acephalous white-male fantasies of extraordinary and irresponsible license of which *Heart of Darkness* is a not particularly critical catalogue, *Gone Primitive: Savage Intellects, Modern Lives* (Chicago: University of Chicago Press, 1990), 155. Death in *Heart of Darkness* is for the most part, however, banal and commonplace and so devoid of such allure.

particularly disposed to the exigencies of imperialism. The "superfluous men," as Arendt referred to them, made redundant by the European economic system, are compelled to become its piratical proponents abroad, in the sense that they are handed over to risk: willing to risk their lives and squander those of others (they value neither), in the hope of disproportionate rewards in the form of "percentages." When Marlow asks his sick and overweight traveling companion, who can barely make the journey to the Inner Station, why he had come, he replies scornfully: "'To make money, of course. What do you think?'" (62). Shortly thereafter the heavy man is dropped by his black bearers. Marlow finds him, "very anxious for me to kill somebody...," at which point, Marlow, recalling the doctor, observes: "I felt I was becoming scientifically interesting" (62).

Thus even that existential and biopolitical metric – "death" – is missing in *Heart of Darkness*. The absence of this limit, along with established epistemological and political norms, would not however be sufficient to explain the positive lack of "restraint" that Marlow describes as the unsettling signature of the imperialist enterprise. On the Company steamer this imperialist attitude stands in stark contrast with the restraint of the supposed "cannibals" working on board. Although they by definition have no cultural and furthermore "no earthly reason for any kind of scruple" (86), and despite their gnawing hunger, they do not turn on the Europeans whom they significantly outnumber. Marlow observes: "And I saw that something restraining, one of those human secrets that baffle probability, had come into play there" (85). The common name for this human secret of restraint is "conscience."[44] Nietzsche, some years before *Heart of Darkness*, and Freud, some years later, would argue that the mysterious restraint of "conscience" was a function of the internalization of the external constraints – spatial and affective – necessary for life in settled political communities.[45] These theories were arguably themselves occasioned by the dramatic

---

**44** In a letter to Roger Casement, written in 1903, Conrad wrote: "It is an extraordinary thing that the conscience of Europe which seventy years ago has put down the slave trade on humanitarian grounds tolerates the Congo State to day," Joseph Conrad, *The Collected Letters of Joseph Conrad*, ed. Frederick Karl and Laurence Davies (Cambridge: Cambridge University Press, 1983–2008), 3:96. The letter was written with a propaganda purpose in mind to contribute to the growing English and international movement opposing the abuses in Leopold's Congo. Indeed, the passage was taken up by Edmund Morel, who cited it in parliament in July the following year and quoted from it (as well as referring to *Heart of Darkness*) in his book, *King Leopold's Rule in Africa* (London: William Heinemann, 1904), 174. The letter would become one of the most memorable and effective indictments of the Congo Free State.

**45** See Friedrich Nietzsche, "Second essay: 'Guilt,' 'Bad Conscience' and Related Matters," in *On the Genealogy of Morality*, ed. Keith Ansell-Pearson, trans. Carol Diethe (Cambridge: Cambridge University Press, 1994); Sigmund Freud, "The Ego and the Id" (1923), in *The Standard Edition of*

transformations and upheavals in communal life and the accompanying dislocation of the "conscience" throughout the nineteenth century, conditions that were all the more starkly pronounced in the colonial context.

The lack of restraint of the agents of imperialism is not due to a loss of conscience under historical and political conditions of displacement, but rather to a distorted or, to use Conrad's term, disfigured conscientiousness. It is the outcome of an institutional culture that disrupts and dislocates communities, operating according to abstract laws that, indifferent to death, recognize no earthly or human measure. It is not enough then that "out there there were no external checks" (64),[46] the fanatical lack of restraint is a positive product and incentive of the imperialist undertaking. The "voice of conscience" that determines the agent of imperialism, that lends him sense and orientation, is the abstracted voice of the "outraged law." If one poses here the hackneyed question as to how supposedly "good Europeans," or in any case ordinary ones, could do such horrific things in the colonies, it would not – not always – be because they found themselves freed from conventional social constraints as well as those of the conscience. Rather, under the dislocated institutional conditions of imperialism, they were all too conscientious.

This disfigurement is exemplarily captured in the story of Marlow's predecessor, a Danish captain named Fresleven. In what sounds like a parody of *Michael Kohlhaas*, he had died in a "scuffle with the natives" that arose over a misunderstanding concerning "two black hens" (49). Although Fresleven was allegedly "the gentlest, quietest creature that ever walked on two legs," Marlow observes dryly:

> he had been a couple of years already out there engaged in the noble cause, you know, and he probably felt the need at last of asserting his self-respect in some way. Therefore he whacked the old nigger mercilessly, while a big crowd of his people watched him, thunderstruck, till some man—I was told the chief's son—in desperation at hearing the old chap yell, made a tentative jab with a spear at the white man—and of course it went quite easy between the shoulder-blades. (49)

---

*the Complete Psychological Works of Sigmund Freud*, vol. 19, trans. James Strachey and Anna Freud (New York: Vintage, 2001).

**46** It is often assumed that *Heart of Darkness* presents the "anarchic" condition that follows from the removal of external constraints giving rise to licentious and indeed terroristic behavior. See, for example, Ian Watt, *Conrad in the Nineteenth Century* (London: Chatto & Windus, 1980), 92.

To paraphrase Anti-Climacus: in despair, the Dane wills to be a self.[47] The attitude brought to the continent under the title "noble cause" presupposes an "indigenous" respect for sovereign European selfhood that, on the ground, turns out not to measure up to such expectations, either in terms of selfhood or self-respect. But, of course, the superfluous figures attracted to imperialism found no more self-respect in Africa than they had in Europe. Fresleven's attempt to insist on the self-respect that he lacks and to shore up the self that he feels to be in dissolution, culminates in his excessive display of violence – and the "desperate" response it solicits. It is not the anarchic or atavistic absence of institutional markers, as if in a state of nature, but rather the violent imposition of an abstract institutional apparatus that transforms the "captain," the word suggesting a model for sovereign individuality, into a deranged and eventually outraged "agent" of an impersonal imperialist enterprise. The paradigm of such "agency" is provided with a fitting image when Marlow finally arrives at Kurtz's station, which turns out to be surrounded by a fence of decapitated heads on posts, their gaze turned inwards. The heads do not stake out a limit, certainly not the limit of "death," but rather, as Marlow concludes: "They only showed that Mr. Kurtz lacked restraint" (104).

In her discussion of bureaucracy, Arendt relates the unrestrained character of the imperialist bureaucrat to the unrestrained process of imperial expansion. The exigency to efface the self for the sake of the institutional process culminates, paradoxically, in an inflated sense of self-righteousness and the pretensions of a kind of sovereignty by proxy. At his most fanatical, the bureaucrat, in the name of the system he serves, takes himself for a god:

> No matter what individual qualities or defects a man may have, once he has entered the maelstrom of an unending process of expansion, he will, as it were, cease to be what he was and obey the laws of the process, identify himself with anonymous forces that he is supposed to serve in order to keep the whole process in motion; he will think of himself as mere function, and eventually consider such functionality, such an incarnation of the dynamic trend, his highest possible achievement. Then, as Rhodes was insane enough to say, he could indeed "do nothing wrong, what he did became right. It was his duty to do what he wanted. He felt himself a god – nothing less."[48]

Kurtz expresses similar sentiments in his pamphlet for the International Society for the Suppression of Savage Customs:

---

[47] "In despair to will to be oneself," is the third kind of despair in Kierkegaard's typology, *Sickness unto Death*, 13. I discuss Kierkegaard's book in more detail in Chapter 1.
[48] Arendt, *Origins*, 215.

> He began with the argument that we whites, from the point of development we had arrived at, 'must necessarily appear to them [savages] in the nature of supernatural beings – we approach them with the might of a deity,' and so on, and so on. 'By the simple exercise of our will we can exert a power for good practically unbounded,' etc., etc. (95)

As anticipated by the words of the accountant, in Kurtz's text Arendt's two devices – race and bureaucracy – coincide. Both devices, in her account, propagate a kind of extremism that corresponds to the aporias of conscience brought about by imperialism: "Race, in other words, was an escape into an irresponsibility where nothing human could any longer exist, and bureaucracy was the result of a responsibility that no man can bear for his fellow-man and no people for another people."[49] Where this leads is captured in the "postscript" to Kurtz's report, scrawled in pencil on the bottom of the last page: "Exterminate all the brutes!" (95).[50]

At its extreme in the colonies, office-duty turns into sheer capriciousness. To recite Rhodes: "It was his duty to do what he wanted."[51] And what the bureaucrat does becomes right. This is not simply a reiteration of the old adage, *might makes right* – although it may indeed amount to the same. Rather, the "ground" of duty, in a manner that recalls, but also disfigures, Fichte's *The Determination of the Human*, is taken to be sheer subjective activity in its willfulness and wantonness – without regard for what is usually considered to be reality. Out of the conviction that the world is exhaustively constituted by bureaucratic acts, the bureaucratic fanatic assumes that, in principle and by right, "Anything – anything can be done in this country" (75). This is the articulation of a completely abstract order without regard for place, of an order in the midst of *Entortung*.

Arendt's argument that the apotheosis of the bureaucrat owes to the power of abstract processes is only persuasive if one can also account for the power of abstraction. If the imperialist attitude operates, as she argues, on an abstract concept of infinite expansion detached from human and geographic realities, under what conditions can such an inhuman and unearthly ideal become convincing? This brings us back to the perplexity of Kurtz's voice. For indeed, if Kurtz feels like a god, it is on account of his voice.

---

49 Arendt, *Origins*, 207.
50 For a reading of *Heart of Darkness* as a "subtle" re-writing of R. B. Cunninghame Graham's "Bloody Niggers" (1897), a critique of the political theology of imperialism, which he showed was capable of sanctioning genocide in the name of a "divine" mission, see Michael Lackey, "The Moral Conditions for Genocide in Joseph Conrad's *Heart of Darkness*," *College Literature* 32.1 (2005): 20–41.
51 Rhodes cited in Arendt, *Origins*, 207.

## Voice, Voices

Marlow's tale, which recounts a journey to "the farthest point of navigation and the culminating point of my experience" (47–48), is also a work of mourning. He mourns the loss of what he sought to find – Kurtz, or rather the "gift" of Kurtz, for Kurtz had always been "little more than a voice" (93). And since it turns out that this voice never existed as such – "It echoed loudly in him because he was hollow to the core..." (104) – Marlow mourns the idea of Kurtz's voice. Throughout, Marlow's own voice betrays the loss that he recounts. At one point, Marlow interrupts himself to ask his listeners on the Thames: "Do you see him? Do you see the story? Do you see anything? It seems to me I am trying to tell you a dream – making a vain attempt – [...]" (70).[52] What has been lost, what turns out to have been always already lost, is the ability to convey full living presence, such that one may see what is meant, intuit what is intended: "... No, it is impossible; it is impossible to convey the life-sensation of any given epoch of one's existence—that which makes its truth, its meaning—its subtle and penetrating essence" (70).

Kurtz's voice is the life of the Intended. Marlow's failure to mourn Kurtz is betrayed in his futile attempts to save the ideality of the Intended from the "darkness," not to mention the "horror," with which she becomes confounded. Of Kurtz's lost voice, of the way that Kurtz's voice now gets lost among other voices, Marlow reflects:

> And I was right, too. A voice. He was very little more than a voice. And I heard – him – it – this voice – other voices – all of them were so little more than voices – and the memory of that time itself lingers around me, impalpable, like a dying vibration of one immense jabber, silly, atrocious, sordid, savage, or simply mean, without any kind of sense. Voices, voices – even the girl herself – now – –  [53]

---

[52] A certain dislocation (*Entstellung*) of order and place, as well as of sense and signification, is of course characteristic of the experience of a dream, as Freud suggested at the same historical moment, Sigmund Freud, *Die Traumdeutung* (Leipzig: F. Deuticke, 1900 [1899]). For the classic psychoanalytic reading of *Heart of Darkness* departing from the notion of Kurtz as Marlow's "secret sharer," see Albert Guérard, *Conrad the Novelist* (Cambridge, MA: Harvard University Press, 1958), chap. 1. Conrad had apparently not read any Freud and refused to do so when lent copies of his work by H. R. Lenormand on a visit to Corsica in 1921. See Najder, *Joseph Conrad*, 535.
[53] The Cambridge Edition reads " – even the girl herself – now ..." (93). Here I follow the 1921 Heinemann edition of Conrad's *Collected Works*, the last version of the text that Conrad approved although the status of certain emendations is debated, as reprinted in Joseph Conrad, *Heart of Darkness: Complete, Authoritative Text with Biographical and Historical Contexts, Critical History, and Essays from Five Contemporary Critical Perspectives*, ed. Ross C. Murfin (Boston:

Marlow's own voice descends in this passage into a kind of "jabber." Lamenting his failure to convey the phenomenon of the voice in a meaningful expression, the chain of adjectives to which Marlow resorts performs the very truncating of meaning to become something "simply mean" that he stutteringly describes.[54] But just because it is "simply mean," that is to say, lacking the ideal of meaning to which Marlow still aspires, it is not "without any kind of sense" – as the text of *Heart of Darkness* demonstrates. On the contrary, the simple meanness of sense betrays the way that significance transcends, or subtends, any particular conscientious intention and can for that reason never be exhaustively fulfilled "in person." It is not a coincidence therefore that with the deictic eruption of the " – now – –" into his increasingly inarticulate recollections, "the girl herself" is unintentionally announced or announces herself as the Intended. Marlow insists – despite her being there in the here and now of his narration – that she is or has to be "out of it" (93), as if attempting to preserve a realm of ideal meaning beyond the simple meanness of the voices, first of all his own, that constitute his account. At the end of his tale, however, Marlow cannot overhear the thoroughgoing contamination, "the sound of her low voice seemed to have the accompaniment of all the other sounds, full of mystery, desolation, and sorrow, I had ever heard –" (124).

Kurtz's voice is the life of the Intended because his "gift of expression" consists in the ability to lend abstract ideals life and fullness. The ideas themselves are of little consequence, it is ideality that he is able to authentically convey. Marlow, who can remember few of the details of Kurtz's pamphlet, can nonetheless recall its effect: "From that point he soared and took me with him. The peroration was magnificent, though difficult to remember, you know. [...] It made me tingle with enthusiasm. This was the unbounded power of eloquence—of words – of burning noble words" (95). This is not rhetoric. Kurtz himself is convinced by his own voice – and ultimately only by his voice. A journalist colleague would later say to Marlow, Kurtz "'could get himself to believe anything – anything.'" The conversation proceeds: "'He would have been a splendid leader of an extreme party.' 'What party?' I asked. 'Any party,' answered the other. 'He was an – an – extremist'" (120). Kurtz's fanaticism thus discloses a dialectic of despair – a need for belief along with the acknowledgement that there is nothing in particular to believe in. Kurtz's voice achieves just this: conviction in the abstract. As such, it is the *schwärmerisch* expression of a complete *Entortung*:

---

Bedford Books of St. Martin's Press, 1996), 64. See the editor, Owen Knowles' "The Texts: An Essay," in the *Cambridge Edition*, esp. 300–302.

**54** For an influential criticism of Conrad's "adjectival insistence," see F. R. Leavis, *The Great Tradition* (New York: New York University Press, 1969), 180.

"There was nothing either above or below him [...]. He had kicked himself loose of the earth. Confound the man! he had kicked the very earth to pieces" (113).

The gift of Kurtz's voice is the conviction that he owns what he means, a duplicity captured in the German terms for first-person possession, *mein* (mine), and for intentional meaning, *meinen* (to mean). As Marlow observes when he finally meets or rather hears Kurtz in person: "'My ivory.' Oh, yes, I heard him. 'My Intended, my ivory, my station, my river, my – ' everything belonged to him" (94). This is the colonizing disposition of the voice. Of course, these claims are true, which is to say, full only insofar as they are purely ideal expressions of the voice and only as long, therefore, as they remain "little more than a voice." As soon as more mundane fulfillment is sought, the incongruities become apparent and the problem of the ideality of meaning (*meinen*) betrays itself to be "simply mean" (*gemein*) and solicits the concrete problem of taking (*nehmen*), with all the far from ideal violence and real meanness that entails. That one understands oneself to be taking what one already owns, that one takes what one intends, that ultimately one operates under the assumption that the order of the world corresponds to the order of one's intentions: this all serves to explain the horrifying combination of idealism and terrorism in which conscientious agents of imperialism seem capable of engaging.

Kurtz is so patently convinced by his own voice, so possessed by what he claims to be in possession of, that Marlow remarks as they head back down river: "It made me hold my breath in expectation of hearing the wilderness burst into a prodigious peal of laughter that would shake the fixed stars in their places" (94). The question of the incongruities between meaning and taking, or owning and appropriating, is brought into focus in the scene of departure when a black woman appears on the bank of the river. The "savage and superb" woman, adorned in "the value of several elephant tusks," prompts a highly exoticized description by Marlow of "fecund and mysterious life," which he holds as counterpoint to the morbidity and fossilization he, in disillusionment, associates with Kurtz and the rest of the imperialist enterprise (107). Regarding her performance on the water's edge, Marlow asks: "'Do you understand this?' Kurtz made no answer, but I saw a smile, a smile of indefinable meaning, appear on his colourless lips that a moment after twitched convulsively. 'Do I not?' he said slowly, gasping, as if the words had been torn out of him by a supernatural power" (114–115).

It is not by chance that the ambivalence of Kurtz's relation to the woman on the riverbank revolves around a form of the most paradigmatic of those speech acts in which intention – the intention indeed to "take" a person – is supposed to be fulfilled in the expression: *I do.*⁵⁵ If the question formulated in the negative – *Do I not?* – seems defiantly to attempt to lay claim to what it cannot understand by placing the onus on the interlocutor, it nonetheless concedes that there is a sphere of significance that exceeds his intentions and does not correspond to the order of his possession. Thus, the *Do I not?*, which seems to be "torn out of him" rather than expressed, is not in his "own" voice. It betrays rather the limits of his intentional expression in the face of a living presence that does not correspond to his objectifying possessive intentions. However much he may want her, may intend to have her – the woman is not his. He cannot "take" her, he cannot even "name" her.⁵⁶ The failure of imperialist intentionality and the impotence of imperialist agency, when brought face to face with an inscrutable intentionality that it cannot make its own, prompts senseless violence: the performance ends with her figure disappearing in a hail of gunfire.

Marlow will see the shadow of this woman again in the gestures of the Intended whom he sees as a "tragic and familiar Shade" (125), when he visits her on his return to Europe. The Intended had, in her own words, "survived" (122), but declaring the need for something "to live with," she demands to know Kurtz's last words drawing the lie out of Marlow: "The last word he pronounced was – your name" (125). If Marlow sought with the lie to keep the Intended "out of it" – "Oh, she had to be out of it. You should have heard the disinterred body of Mr. Kurtz saying, 'My Intended.' You would have perceived directly then how completely she was out of it" (93) – the gesture maps Kurtz's expiring cry, "The horror! The horror!" onto the name of the Intended and so into the very "heart of darkness."

The lie furthermore unfolds Marlow's tortured relation to Kurtz. Earlier he stated: "There is a taint of death, a flavour of mortality in lies—which is exactly what I hate and detest in the world—what I want to forget" (69). The flavor of mortality that Marlow wants to forget is related to the loss of truth in the sense of full living presence. With the "lie" to the Intended, he paradoxically attempts to keep the Intended as intended alive (that is, in her full ideal presence) but can only do so on the condition of acknowledging the "flavour or mortality" that is already in any case betrayed by her spectral aspect.

---

**55** See J. L. Austin, *How to Do Things with Words: Second Edition*, ed. J. O. Urmson and Marina Sbisà (Cambridge, MA: Harvard University Press, 1975).
**56** On the relation between naming, taking and the *nomos*, see Carl Schmitt, "Nomos-Nahme-Name," in *Staat, Großraum, Nomos: Arbeiten aus den Jahren 1916 bis 1969* (Berlin: Duncker & Humblot, 1995). In a curious attempt to restitute certain forms of constitutive naming and taking, which he acknowledges have fallen into disrepute in the contemporary contexts of decolonization, Schmitt proposes marriage as a model of non-violent "taking."

It would be a lie, however, to conclude that there is a fundamental truth in Marlow's lie, as if "The horror! The horror!" were the moral of the story. That would be far too fulfilling an ending. There is a way, however, in which Marlow seems to believe his own lie: ultimately he appears convinced that Kurtz's final intention is expressed in his dying expiration and that therefore the "The horror! The horror!" is indeed, although he cannot bring himself to say so outright, the "name of the Intended." Of Kurtz's dying utterance Marlow declares: "He had summed up—he had judged. 'The horror!' He was a remarkable man" (118). Whether these are indeed to be treated as Kurtz's "last words," and whether they do indeed present a conscious let alone a judicious conclusion, is, however, less than clear. Even Marlow's pronounced insistence on the fullness of their meaning seems, given his own account, to ring hollow: "this was the expression of some sort of belief; it had candour, it had conviction, it had a vibrating note of revolt in its whisper, it had the appalling face of a glimpsed truth – " (118). He has been once again seduced by the voice, even as Kurtz's fabled voice fails him – "a cry that was no more than a breath" (117). For "The horror! The horror!" need not be, and in all likelihood is not, an intentional expression in the meaningful, possessive sense beloved of Kurtz. It is not, in other words, an expression of the sort: *My Intended*. It is in all likelihood rather an expression of Marlow's more or less conscious intention that Kurtz should meaningfully conclude. In other words, Marlow took the significance of Kurtz's expiration in a determinate way by naming it and fulfilling his own ideal intention: *the horror!* What he cannot accept, but what his lie betrays, is that Kurtz's dying expression was "simply mean."

In a recent attempt to "name contemporary violence," Adriana Cavarero has proposed the term "horrorism" to describe a certain experience of violence understood from the perspective of vulnerability rather than perpetration. Cavarero's neologism does not appear to have been in the first instance inspired by *Heart of Darkness*, although she takes up the book along with *The Secret Agent* and *Under Western Eyes* in her epilogue observing that "it is above all Joseph Conrad who supplies interesting material on the topic of horrorism."[57] That Cavarero finds in Marlow's closing line a sheer vacuity – "*Heart of Darkness* opens onto a void or, as the (telling) expression goes in Italian, onto *un orrido* (a deep and gloomy gorge)" [58] – indicates the very different intention that informs her gesture of naming, which is by no means supposed to provide closure

---

[57] Adriana Cavarero, *Horrorism: Naming Contemporary Violence* (2007), trans. William McCuaig (New York: Columbia University Press, 2008), 116.
[58] Cavarero, *Horrorism*, 119.

but on the contrary to look into the abyss. The reading of *Heart of Darkness* I have proposed seeks to contribute to the understanding of the kind of violence Cavarero examines by relating horrorism to bureaucratic fanaticism. Horrorism would be the experience, the subjection to a bureaucratic fanaticism that proceeds, on the basis of intentional activity, to impose an ideal significance without regard for people or parameters that may not measure up to such abstract intentions. And bureaucratic fanaticism would be the horrifying way in which, under certain historical and political conditions of disfigurement, the attraction of a full life of abstraction can appear more intuitive and compelling than the simple meanness of reality.

# 5 Poverty of Agency: Conrad's *The Secret Agent*

The "secret" in *The Secret Agent* (1907) lies in the ambiguity of the word "agent." The story reflects on its historical moment by playing on the slippage from "agency," expressive of a capacity for individual action, to "agency," in the functional or institutional sense.[1] London, characterized by a proliferation of competing and overlapping agencies, agents and agendas, emerges as the site of this slippage. At a time when the imperial metropolis was the hub of diverse political, economic, and techno-scientific developments and notably a significant "liberalization" of political life, *The Secret Agent* conveys an impression of the city as one of poverty and incapacity, of destitution and dependence, devoid of the sort of concerted, concentrated, and committed action associated then as now with social engagement and political life. In thus presenting the contours of the crisis of "agency" in the city of London at the turn of the twentieth century, Conrad's novel sets in relief the apparently paradoxical implication of liberalism and bureaucracy in modern mass societies, in a manner that, furthermore, illuminates the disaffection, not to mention the outrage, to which liberal governmentality can lead.

## Temperature

If heat and fever form part of the atmosphere, but also of the telling, of *Heart of Darkness* and contribute to the spectral and dream-like character of the novella, an experience not unlike the delirium of moving in and out of consciousness, then *The Secret Agent* discloses a considerably colder context conveyed by a quite different narrative approach. Indeed, Albert Guérard remarks that the book "may well be the coldest" of Conrad's novels.[2] This change in temperature, or in any case in temperament, was acknowledged by Conrad himself, who wrote in the "Author's Note" to the 1920 reprinting, "I rather think that a change in the fundamental mood had already stolen over me unawares" (4). Characteristic of this change seems to have been a sense of inauthenticity and futility, and above all a disappointment and deception, indeed a despair, over the possibility of gen-

---

[1] Joseph Conrad, *The Secret Agent: A Simple Tale* (1907), ed. Bruce Harkness and S. W. Reid, *The Cambridge Edition of the Works of Joseph Conrad* (Cambridge: Cambridge University Press, 1990). Citations in-text.
[2] Albert Guérard, *Conrad the Novelist* (Cambridge, MA: Harvard University Press, 1958), 231.

uine expression, as Conrad recollects: "It was a period, too, in which my sense of the truth of things was attended by a very intense imaginative and emotional readiness which, all genuine and faithful to facts as it was, yet made me feel (the task once done) as if I were left behind, aimless amongst mere husks of sensations and lost in a world of other, of inferior, values" (4).

This mood Conrad describes, as well as the tone of *The Secret Agent* to which it gave rise, recalls a famous passage in a letter to R. B. Cunninghame Graham (with whom he carried out, over many years, the most political of his correspondences), written almost a year before the composition of *Heart of Darkness* and after the torturous completion of his first London-based short story, "The Return" (1897/1898). In the letter, Conrad gives vent to a particularly cold mood in the form of a manifesto of despair:

> The attitude of cold unconcern is the only reasonable one. Of course reason is hateful – but why? Because it demonstrates (to those who have the courage) that we, living, are out of life – utterly out of it. The mysteries of a universe made of drops of fire and clods of mud do not concern us in the least. The fate of a humanity condemned ultimately to perish from cold is not worth troubling about. If you take it to heart it becomes an unendurable tragedy. If you believe in improvement you must weep, for the attained perfection must end in cold, darkness and silence. In a dispassionate view the ardour for reform, improvement for virtue, and knowledge, and even for beauty is only a vain sticking up for appearances as though one were anxious about the cut of one's clothes in a community of blind men. Life knows us not and we do not know life – we don't know even our own thoughts...[3]

This bitter passage on the reasonableness of cold unconcern in life as in politics is over-determined to the point of parody by the chilling rhetoric of late nineteenth-century science and in particular by Kelvin's notion, based on the second law of thermodynamics, of the heat death of the universe. The passage thus expresses despair by betraying a lack of faith, if not a disdain, for the scientific grounds for despair it expounds. Despair, after all, is the chronic experience of a fundamental lack of grounds. And it is on this note that Conrad breaks off in his letter to Cunninghame Graham, despairing over the possibility of authentically expressing despair, for any attempt to do so, however deeply felt, seems condemned to descend into platitude, as Conrad jokingly concedes: "– only the string of my platitudes seems to have no end."[4] The irreducibility here between the expression of despair and despairing over expression that pro-

---

[3] Letter to Cunninghame Graham, January 14, 1898, Joseph Conrad, *The Collected Letters of Joseph Conrad*, ed. Frederick Karl and Laurence Davies (Cambridge: Cambridge University Press, 1983–2008), 2:16–17.
[4] Conrad, *Collected Letters* 2:17.

duces, perhaps despite itself, a dark humor, anticipates the "fundamental mood" as well as the "ironic treatment" that Conrad referred to in describing the origin of *The Secret Agent* in the "Author's Note" (7). Returning from, and deceived by, the desperate tropics of *Heart of Darkness* and *Nostromo* (1904), as well as from the reflections on his youth in the *Mirror of the Sea* (1906), Conrad finds himself exposed once again to the cold.

In *Politics and the Novel*, Irving Howe expresses exasperation at the tone of the tale: "Conrad's ironic tone suffuses every sentence, nagging at our attention to the point where one yearns for the relief of a direct statement almost as if it were an ethical good... So peevish an irony must have its source less in zeal or anger than in some deep distemper."[5] Conrad himself acknowledged, *The Secret Agent* did indeed emerge out of a "deep distemper," one that, however, was not so much the expression of a personal indisposition as a fundamental mood – the cold condition of despair. And out of this bleak mood emerges the city of London. Conrad records how, as opposed to the "crude sunshine and brutal revolutions" of the Latin American climes he had been exploring in *Nostromo*, "the vision of an enormous town presented itself, of a monstrous town more populous than some continents and in its man-made might as if indifferent to heaven's frowns and smiles... Irresistibly the town became the background for the ensuing period of deep and tentative meditations" (6). If Kelvin's speculation involved the eventual dying out of the sun, then Conrad's London is "a cruel devourer of the world's light" (6), as if the heat death of the universe were not a cosmic but a metropolitan occurrence. London is not simply the setting for "a desperate tale"[6] but the impoverished space of life and experience disclosed by such a despairing mood.

Howe's frustrated yearning for direct statement in his study on political novels fails to acknowledge the novel political environment that Conrad presents in *The Secret Agent*. London is not a place for clearheaded statements nor is it a feverish tropical landscape. It is instead presented as a smoggy entropic topography.[7] London consumes, it does not conserve, heat. But as in Conrad's letter, here thermodynamics concern temperament rather than temperature. In *The Se-*

---

5 Irving Howe, *Politics and the Novel* (New York: Horizon Press, 1957), 95–96.
6 The subtitle given in the book-edition to Conrad's earlier short story, "An Anarchist" (1906/1908), republished in *A Set of Six*, Dent Collected Edition (London: Dent, 1954).
7 For readings that focus on the allusions to entropy, see Jill Clark, "A Tale Told by Stevie: From Thermodynamic to Informational Entropy in *The Secret Agent*," *Conradiana* 36.1–2 (2004): 1–31; Alex Houen, *Terrorism and Modern Literature from Joseph Conrad to Ciaran Carson* (Oxford: Oxford University Press, 2002), 53–4; also Bruce Clarke, "Allegories of Victorian Thermodynamics," *Configurations* 4.1 (1996): 67–90.

*cret Agent*, London presents precisely the sort of chill, alienating and de-sensitized space that can give rise to heated bursts of outrage – especially among those of a more political disposition.

## A Period of Reaction

The temper of the work can also be related to more explicitly political developments. *The Secret Agent* was written in 1906 on the eve of a series of social reforms by the newly elected Liberal government – most notably the introduction of national health insurance and social security, but also changes to the system of education, measures supplementary to the outdated Poor Laws and changes in the administration of mental deficiency – that in certain crucial respects anticipated, and in the long term facilitated, the establishment of the welfare state in Britain.[8] Although these policies were only the most dramatic in a long-running process of piecemeal reform and increasing government (as well as civic) intervention, they reflect a significant transformation in the self-understanding of British liberalism, which had, as a party as well as an ideology, suffered major setbacks since its loss of power in 1886. While classic liberalism had promoted a policy of minimal government, individualism, private property and a *laissez-faire* economy, the "new liberalism" was forced to come to terms with the emergence of a new and amorphous political entity, "the social," which could not be reduced to the sum of its parts and operated – this being the subject of the emergent field of sociology – according to a logic of its own. Specifically, liberalism was obliged to confront those aspects of the social – issues relating to poverty, public health and education, un- and underemployment, housing and sanitation, child care, old age, and disability – that together constituted the so-called "social problem" (also the title of J. A. Hobson's 1901 book).[9] In the light of accelerating urbanization, the economic depression of the 1880s, and the rising specter of socialism across Europe, these issues could no longer credibly be said to resolve themselves progressively according to classical liberal principles.

Most unsettling of all, perhaps, was the fact that the "period of reaction" (from 1886), as one of the more significant "new liberals," L. T. Hobhouse, would describe it, owed in some degree at least to the liberalization of the fran-

---

[8] See Michael Freeden, *The New Liberalism: An Ideology of Social Reform* (Oxford: Clarendon Press, 1978); and Maurice Bruce, *The Coming of the Welfare State. With a Comparative Essay on American and English Welfare Programs* (New York: Schocken Books, 1966).
[9] J. A. Hobson, *The Social Problem: Life and Work* (London: J. Nisbet & Co., 1901).

chise beyond determinations of property brought about by Gladstone's legislation, most notably the "Representation of the People Act" (1884).[10] Apart from demanding further social protections, the "people," it turned out, were susceptible, especially at the hands of the mass media, to the worst kinds of demagoguery and jingoism, supporting such illiberal policies as imperialism and the Boer War. Of course, such developments were by no means simply a liberal problem in the party-political sense but were rather indicative of broader tendencies and transformations in the nature of "the political" in the late nineteenth century. As the social emerged over the course of the century as a distinct, albeit difficult to define, agency that was widely perceived to harbor a potentially decisive political force, politics in general, in Europe at least, became a matter of addressing, conforming to, or representing it. As the liberal politician, William Harcourt, caricatured as the Home Secretary in *The Secret Agent*, famously declared in 1888, "we are all socialists now."[11] What this meant for the new liberalism, as espoused by the likes of Hobhouse and Hobson, was an acknowledgement that the individual subject presupposed by classic liberalism was in fact a social construct and that the social had accordingly to be intensively managed by state and private initiatives in order to facilitate the production of liberal subjects. The "social liberalist" reforms undertaken after 1906 can be understood as the crystallization of this realization.

Thus the liberalization of the field of eligible political participation was complemented by an institutional apparatus that, on the one hand, was concerned with the constitution of the sort of autonomous individuals inclined to adopt liberal conduct and, on the other, with delimiting participation according to new criteria of dependence and disability that would exclude those elements of the population taken to be disposed to illiberal and illegitimate behavior. As was being realized across Europe, a state built on the principle of individual liberal agency required, in fact, the intervention of countless bureaucratic agencies in order to mediate and mitigate the latent agency that seemed to invest the social and to be inextricably connected with the social question.

*The Secret Agent* is a product of this period of reaction. Conrad himself would write, again in the "Author's Note," that the tale originated in a "period of mental and emotional reaction" (3). To be sure, he intends something quite different from the reaction that was the occasion of Hobhouse's 1904 book, *Democracy and Reaction*. Nevertheless the two books share a common preoccupa-

---

[10] L. T. Hobhouse, *Democracy and Reaction* (1904), ed. P. F Clarke (Brighton: Harvester Press, 1972).
[11] See Avrom Fleishman, *Conrad's Politics: Community and Anarchy in the Fiction of Joseph Conrad* (Baltimore: Johns Hopkins Press, 1967), 213–214.

tion or "fundamental mood," namely, a profound anxiety about the locus and scope of agency. For Conrad, in particular, this mood betrays a disappointment with, perhaps even despair over, the possibility of authentic action.

Insofar as *The Secret Agent* presents what many, not without reason, consider Conrad's "reactionary" literary engagement with the social question, it is striking that the poverty in question appears first of all to be characterized by a destitution of individual agency.[12] In the story, the apparently autonomous agents – above all those who call themselves anarchists – appear thoroughly dependent, unable to act at all except under the influence of impersonal agencies. And the only action that might be said to take place in the story – Stevie's outrage, Verloc's killing, and Winnie's suicide – is of the pathetic sort that is, as it were, too poor, and too desperate, to count as such in the emergent liberal-democratic political space.

A period of reaction is one in which the character and possibility of action is in question. Writing in 1887, in *On the Genealogy of Morality*, Nietzsche provided a scathing account of the social origins of the liberal conception of agency. It was, he argued, the result of a long history of the institutionalization of reaction, or *ressentiment*, that sought to valorize an attitude of powerlessness and thereby to normalize human conduct in both the moral and the statistical sense. The "rationalization" of agency was in the first instance the result of a "seduction of language" that assigns a grammatical agent which is then mistaken for a substantive being: "But there is no such substratum; there is no 'being' behind the deed, its effect and what becomes of it; 'the doer' is invented as an afterthought, – the deed is everything [*das Tun ist Alles*]."[13] In Nietzsche's account, actions are simply the "innocent" expression or outburst of impersonal forces or what he calls affects (*Affekte*). And such action was felt to be more than ever suppressed and disparaged by the reactive homogenizing agencies – those of modern science as well as liberal politics – that governed modern life. In short, Nietzsche responded to the contemporary crisis of action with a polemical attempt to liberate action from the responsible liberal subject, affect from rational agency.

---

12 For varying assessments of the reactionary conservatism of *The Secret Agent*, see, amongst others, Howe, *Politics and the Novel*; Fleishman, *Conrad's Politics*; Terry Eagleton, "Form, Ideology and *The Secret Agent*," in *Against the Grain: Essays 1975–1985* (London: Verso, 1986); and Daphna Erdinast-Vulcan, "'Sudden Holes in Space and Time': Conrad's Anarchist Aesthetics in *The Secret Agent*," in *Conrad's Cities: Essays for Hans van Marle*, ed. Gene Moore (Amsterdam: Rodopi, 1992).

13 Friedrich Nietzsche, *On the Genealogy of Morality*, ed. Keith Ansell-Pearson, trans. Carol Diethe (Cambridge: Cambridge University Press, 1994), 26. Translation modified.

Some twenty years later, in what amounts to a commentary on the sociopolitical changes underway in Europe over the period, *The Secret Agent* presents a reflection on the state of agency in early twentieth-century London where action seems to have lost both its power and its potential. The pervading sense is one of the hopelessness of all "direct action." The reaction of Conrad's text is thus characterized not first of all by the demonstration of the "criminal futility" (5) of ostensibly anarchist activities but, more pressingly, by the way that the field for action of the spontaneous and passionate sort Nietzsche describes has been reduced to futile and criminal agitations on the edge of society. *The Secret Agent* presents a moment in which Nietzsche's declaration, "the deed is everything," has lost whatever force it may have had. Instead, considering such "acts" as those in which Stevie and Winnie are engaged, it would seem that the deed has in fact become nothing. For there is no socially significant act without the acknowledgement or mediation of an institutional agency, except insofar as it may be written about and at the same time written off with such "ready-made" (30) journalistic phrases as "*An impenetrable mystery seems destined to hang for ever over this act of madness or despair*" (228).

Nonetheless, even if *The Secret Agent* is forever, coolly, "explaining away" (30) the mystery that hangs over it, namely, the condition and the very possibility of anarchist action, the desperate violence exhibited by Stevie and Winnie presents ciphers for an unaccountable politics – a non-codifiable and imprescriptible, and therefore apparently futile, politics of and for ineligibles. It indicates a political praxis that neither follows an abstract program nor orients itself towards an idea, thus a praxis that does not aspire to, and is not even capable of, the abstract intentionality required to consent to, or contract with, the prevailing dispensation as a responsible agent. In short, such destitute action, hopelessly rejects the impoverished terms in turn-of-the-century London of what counts as "the political" – of what prescribes the conditions for rightful political participation, on the one hand, and of what defines the conditions of state-administered dependency prohibiting participation, on the other.

## Dulling Agents

It is not incidental that London would impose itself as the site for Conrad's literary reflections on the shifting sense of agency. For it was in the big cities of the early twentieth century that the atrophy of the individual, over and against a hypertrophy of objective socio-historical and institutional processes, found its paradigmatic instantiation. Such, in any case, was the argument presented by Georg Simmel in his 1903 essay, "The Metropolis and Mental Life." According to Simmel, the *"intensification of nervous life"* in the modern metropolis requires a corresponding dulling on the side of the subject if the individual is not to be "leveled and consumed in the social-technical mechanism."[14] Amidst and against the impersonal objective agencies of city life, a new kind of individual emerged: reserved, intellectual, indifferent to the intense and diverse sensory stimulation of the city and habituated to treating all inter-actions as abstract transactions. This individual was "blasé."[15]

The city, for Simmel, is thus the site of the subordination of affect. There is a structural trade-off between blasé autonomy and substantial affective attachment, whence the characteristic urban sense of loneliness and alienation. The blasé is the consciousness that corresponds to the modern city insofar as it cultivates an intellectual distance, a state of abstraction, that filters the manifold impressions of the city as well as the emotion such impressions may otherwise excite. Such an attitude fittingly describes the "ironic" narration of *The Secret Agent*, which at the same time betrays a keen awareness of the affect it lacks, whence the fascination – which never, however, achieves a sense of genuine empathy – with the likes of such passionate figures as Winnie and Stevie.[16]

Whereas Simmel is concerned with the ideal-type of the urban individual, the figures in *The Secret Agent* are those that emerge – if they emerge as individuals at all – along the edges of mainstream city-life and in its backwaters. While Simmel addresses the mode of conscious life proper to the city, Conrad's figures present its sub-conscious, semi-conscious and in some cases simply non-conscious elements. Here, the dulling that Simmel identifies as necessary for city life is characterized not by an intellectual filter of the sensory, as in the case of the blasé, but by recourse to anaesthetizing or depressive substances (or agents) that reduce sensitivity to the violence of the everyday and disclose a dreary, apathetic urban environment.

One substance or agent that dulls the senses is alcohol. This is the case with such figures as the maimed night-cabby who heads for the pub after complaining to Stevie of the burden of his wife and children; or the charwoman, Mrs Neale, who, as the narrator blithely describes, "exhaled the anguish of the

---

**14** Georg Simmel, "Die Großstädte und das Geistesleben," in *Die Großstadt. Vorträge und Aufsätze zur Städteausstellung* (Dresden: v. Zahn & Jaensch, 1903), 188, 187. Translations mine.
**15** Simmel, "Die Großstädte und das Geistesleben," 193.
**16** On Simmel's (and Walter Benjamin's) analysis of modern city life in relation to *The Secret Agent*, see Jonathan Arac, "Romanticism, the Self, and the City: *The Secret Agent* in Literary History," *boundary 2* 9.1 (1980): 75–90; see also Christina Britzolakis, "Pathologies of the Imperial Metropolis: Impressionism as Traumatic Afterimage in Conrad and Ford," *Journal of Modern Literature* 29.1 (2005): 1–20.

poor in a breath of soap-suds and rum, in the uproar of scrubbing, in the clatter of tin pails" (138). As opposed to the exaltation of a certain kind of drunkenness, they are driven to the depressive effects of alcohol and the dulling of alcoholism. Although both cabby and charwoman are ostensibly gainfully employed, the extreme of destitution and poverty they exhibit corresponds to an existence that is not altogether human. The narrator describes Mrs Neale "on all fours amongst the puddles, wet and begrimed, like a sort of amphibious and domestic animal living in ash-bins and dirty water" (140). Drowned in drink and just keeping their heads above water in the "slimy aquarium" of the city (114), such amphibious figures present one cold extreme of London's life.[17]

At another extreme, a very different agent, equally chilling and intoxicating in its way, is dynamite (patented in 1867). Revolutionary in conventional warfare and of course in mining and construction, dynamite seemed also to present a potentially revolutionary transformation in the politics of the nation-state – lending minorities and oppressed nationalities (notably the Irish), as well as political radicals of every persuasion, the potential to turn the tables on established, conventional powers.[18] In 1884, *The Anarchist*, exemplary among anarchist publications for its praise of the possibilities of the new weapon, would publish an article entitled "Dynamite: The Modern Agent of Revolution."[19] Indeed, it is as if dynamite emerged in reaction to the consolidation of the state monopoly on violence and the expansion at home and abroad of the bureaucratic and biopolitical techniques of governmentality. Such, in any case, was the argument of the libertarian, Auberon Herbert, in his 1894 polemic, "The Ethics of Dynamite." Suddenly, an individual armed with "the power that can be carried in the pocket of any ragged coat" (213) could present a substantial threat to the state. Herbert went on to argue that such fanatical minorities, and the violence to which they had recourse, were in fact the product of the increasingly invasive and impersonal character of the state.[20] Dynamite is the condition but also the expression of

---

**17** In the late nineteenth century, the spread of alcoholism was the subject of intermittent public concern and consternation. Under the title, "The Slavery of Drink," the *Daily Telegraph*, for example, was forum to a discussion on alcohol, August-September 1891. See Susan Marjorie Zieger, *Inventing the Addict: Drugs, Race, and Sexuality in Nineteenth-Century British and American Literature* (Amherst: University of Massachusetts Press, 2008).
**18** On dynamite in the public imagination, see Deaglán Ó Donghaile, *Blasted Literature: Victorian Political Fiction and the Shock of Modernism* (Edinburgh: Edinburgh University Press, 2011).
**19** Unsigned article, "Dynamite: The Modern Agent of Revolution," *The Anarchist*, March, 1885; discussed by Ó Donghaile, *Blasted Literature*, 143.
**20** Auberon Herbert, "The Ethics of Dynamite," *Contemporary Review* 65 (1894): 667–687, reprinted in *The Right and Wrong of Compulsion by the State and Other Essays* (Indianapolis: Liberty Classics, 1978), 213.

"the force of the minority" in a world of increasing homogenization that turns people, in Herbert's words, into "administration material."[21]

In *The Secret Agent*, the Professor, who sees himself as a "moral agent" (66), impersonates an "ethics of dynamite" not unlike that sketched by Herbert. In response to the oppressive multitude of oblivious people and the impersonal social structures of the city, his extreme *ressentiment* finds its perfect formulation in a chemical reaction. Indeed his whole "philosophy" is expressive of a sort of dynamite-dependence; he carries a bomb on him at all times, stroking the rubber detonator in his pocket. Reciting Blanqui, the "perfect anarchist" as the narrator refers to the Professor, declares, "My device is: No God! No Master" (227); this device *is* dynamite. Destruction is "the agent of his ambition" (66). Indeed, his very sense of agency and autonomy, the exalted distinction from the swarm-like city crowd in which he fears to be consumed, is paradoxically dependent on the possibility of the incommensurate destruction that would accompany his suicide bombing: "Lost in the crowd, miserable and undersized, he meditated confidently on his power, keeping his hand in the left pocket of his trousers, grasping lightly the india-rubber ball, the supreme guarantee of his sinister freedom..." (67). His desperate exaltation is entirely a function of his dynamite fixation: "By exercising his agency with ruthless defiance he procured for himself the appearances of power and personal prestige" (67). And his political vision is likewise constrained to that of mere explosion; he hopes by his "home industry" (159), carried on in the lonely precincts of his room, to deliver "a telling stroke" – "a blow fit to open the first crack in the imposing front of the great edifice of legal conceptions sheltering the atrocious injustice of society" (66). In truth, the mode of his existence is ultimately little different from that of the substance in his pocket – latency without futurity. As the blasé narrator observes: "He had no future. He disdained it. He was a force. His thoughts caressed the images of ruin and destruction" (231).

---

**21** Herbert, "The Ethics of Dynamite," 213, 192. On the other hand the dread of dynamite in the public sphere was in large part attributed, at the time at least, to its terrifying impersonality. In "The Fear of Dynamite," for example, an article in *The Spectator*, April 14, 1883, 9–10, the commentator observes, "We expect to find a man behind the rifle or the revolver, and have a courage for the combat with man, however armed, which does not exist for a combat with a blind, speechless, and, so to speak, natural force. To face dynamite is to the imagination to face the lightning, or an earthquake, or a lava-stream, or any other death-giver, before which fortitude is useless, and retreat not dishonourable." Referring to this article among others from the period, Alex Houen argues dynamite revealed to the public that they were "*already living as potential statistics,*" *Terrorism and Modern Literature, from Joseph Conrad to Ciaran Carson* (Oxford: Oxford University Press, 2002), 25. On the contemporary literary engagement with terrorism, see also Barbara Arnett Melchiori, *Terrorism in the Late Victorian Novel* (London: Croom Helm, 1985).

## Outcast London

The attitude of barely concealed disdain for the likes of Mrs Neale and the Professor betrays a more general anxiety on the part of the narrator about the particular social stratum of the London population with which the story is engaged and that is epitomized in the description of the novel's central figure, Mr Verloc:

> There was also about him an indescribable air which no mechanic could have acquired in the practice of his handicraft however dishonestly exercised: the air common to men who live on the vices, the follies, or the baser fears of mankind; the air of moral nihilism common to keepers of gambling halls and disorderly houses; to private detectives and inquiry agents; to drink sellers and, I should say, to the sellers of invigorating electric belts and to the inventors of patent medicines. (16)

Especially during the depression of the 1880s, also the time of the most significant wave of dynamite "outrages" in London, but equally in the early years of the twentieth century in the wake of the Boer War, the notion of an "outcast London" – and the attendant vision of an "urban residuum" composed of hapless, physically, socially, morally and even biologically unredeemable types variously labeled "unemployables," the "undeserving poor," and even the "degenerate" – was increasingly visible as a phenomenon and prevalent in public discourse as well as public policy.[22] Unrespectable without being infamous, illicit without necessarily being criminal, such figures, where they were not entirely dependent on charitable or state institutions, subsisted for the most part in the shadow-economy. Altogether they present that "superfluous" category of the late nineteenth-century European population, identified by Arendt, that proves to be as characteristic of metropolitan as of colonial life.[23]

The disparate figures in *The Secret Agent* emerge from – or attend to – this underclass or, rather, since it does not compose itself into a class, this outcast residuum. It was such *déclassé* elements – and the volatile, reckless, and ultimately impoverished agency with which they, individually and en masse, were

---

[22] See Gareth Stedman Jones, *Outcast London: A Study in the Relationship between Classes in Victorian Society* (Oxford: Clarendon Press, 1971); and Daniel Pick, *Faces of Degeneration: A European Disorder, c.1848–c.1918* (Cambridge: Cambridge University Press, 1989).

[23] The parallels between outcast London and the colonies was very much a part of public discourse from the period, emblematically presented by William Booth's *In Darkest England and the Way Out* (1890), which cited Stanley's best-seller, *In Darkest Africa* (1890). On the colonial resonances of Conrad's presentation of London, see Rebecca Stott, "The Woman in Black: Race and Gender in *The Secret Agent*," *The Conradian* 17.2 (Spring 1993): 46–50; also Robert Hampson, "Conrad and the Idea of Empire," *L'Epoque Conradienne* (1989): 9–22.

seen to be invested – that generated the urgent impetus across the formal political spectrum for addressing "the social question." Perceived as a danger to society, such types were considered ineligible to engage as responsible participants in the social contract and therefore presented an unsettling complication to the liberalizing society which was anxious to make sure that they remained outside the expanding political franchise and passive, if not altogether dependent, with regard to the policies designed for them. Thus, addressing the social question in the context of efforts to produce a more inclusive, stable and above all productive political dispensation was a matter of making distinctions among the poor, enabling those considered amenable to socially responsible conduct (above all work) and disabling the disaffected elements who were considered dangerous or incorrigible.[24] These processes were intended to defuse recourse by the poor to more radical action.

While there was a consensus across the political spectrum that poverty generated, at least potentially, a radical kind of agency and fostered a disposition to extremism, the particular agency to be assigned to poverty, or rather the kind of poverty likely to generate revolutionary agency, was at the center of debates between socialists and anarchists in the late nineteenth century. Marx insisted on the immiseration thesis, namely, that the progressive economic impoverishment inherent to capitalism would be the motor of revolutionary action so long as it was supplemented by the organization of the working class. In contrast, Bakunin's concept of the power of impoverishment emphasized the affective basis of radical social action, which led Marx to write dismissively of the unscientific character of his theory: "*Willpower*, not *economic conditions*, is the basis of his social revolution."[25] Far from denying the significance of economic conditions, however, Bakunin was suspicious of the authoritarian bureaucratic tendencies that Marxists seemed to be uncritically appropriating from the bourgeois social order, for he considered organization itself – whether socialist or capitalist – to risk defusing the spirit of revolt.

In *Statism and Anarchy* (1873), Bakunin drew a distinction between the complacency of the employed working class and the militantism of the outcast residuum that Marx considered susceptible to reaction and dismissed as the *Lumpenproletariat*. Among the working class, Bakunin claimed, "even the impulse to give way to despair is smothered by a complete insensibility toward his own rights, and an imperturbable obedience [...] They would rather die than rebel."

---

**24** See Michel Foucault, "About the Concept of the 'Dangerous Individual' in 19th-Century Legal Psychiatry," *International Journal of Law and Psychiatry* 1.1 (1978): 1–18.
**25** Karl Marx, "Notes on Bakunin's Book: *Statehood and Anarchy*" (1874), *Marx & Engels Collected Works* (New York: International Publishers, 1975–2004), 24:518.

In contrast to the "hopeless" poverty to which workers as a class remained bound, the poverty of the *Lumpenproletariat* was fueled by "despair": "But when a man can be driven to desperation, he is more likely to rebel. Despair is a bitter, passionate feeling capable of rousing men from their semiconscious resignation if they already have an idea of a more desirable situation, even without much hope of reaching it."[26] It was not the working class therefore but the disordered and disorderly – and above all truly desperate – poor whom Bakunin considered capable agents of the revolutionary cause. Without their "negative passion," the destructive passion of desperation, "revolution would be impossible."[27] These sentiments, which in fact only reflect the anxieties of the establishment (even the Marxist establishment) about the potential agency of the *Lumpenproletariat*, are expressed by the Professor who declares at the end of *The Secret Agent:* "'Madness and despair! Give me that for a lever, and I'll move the world'" (230).

If the assumption that the urban poor were repositories of a formidable kind of agency undergirded both anarchist hopes and establishment anxieties, what is striking in *The Secret Agent* is the lethargy and lack of passion regarding all "activism." Not only does outcast London present a site devoid of the kind of individual agency that could pass for conscious self-possession or willed intentional action, but it is even more strikingly characterized by the absence of the agency associated with any sort of passionate engagement and even of the "negative passion" that is the minimal condition for the most negligible or nihilistic of actions. As agency in the big city has been dispersed and distributed by and among impersonal agencies, its inhabitants, even those most abused and disaffected, have been rendered apathetic. Poverty in the city is not just economic. London is poor in passion.

## Institutional Agencies and Dependencies

Mr Verloc, the secret agent of the tale, is thoroughly in his element in underground London – and is for that reason altogether apathetic. Obese, idle, unhygienic, his consciousness is imprinted not by the superfluous multiplicity of its sensations but by the inertia of its sheer materiality. Like "a soft kind of rock" (17), Verloc is corporally and psychologically unmoved by the bustle of the

---

[26] Mikhail Aleksandrovich Bakunin, *Statism and Anarchy* (1873), *Bakunin on Anarchy*, ed. and trans. Sam Dolgoff, preface by Paul Avrich (London: Allen & Unwin, 1973), 334–335.
[27] Bakunin, *Statism and Anarchy*, 334.

city and is correspondingly idle, idle "with a sort of inert fanaticism, or perhaps rather with a fanatical inertness" (16).

Verloc acts as mediator between the anarchists and the legal agents responsible for their oversight, since he is, in fact, a double agent – and more than that. He is not just a delegate to various anarchist societies, he is also a police informant, an *agent provocateur* and an agent of foreign governments, as well as a seller of "shady wares" (pornography and anarchist literature). He is thus, for all his inertness, or rather precisely because he does nothing, an exemplary agent at the nexus of London's confounding web of agencies and institutions of which he even sees himself to be the "protector" (11). Above all, however, he is representative of, which is also to say dependent on, the domestic institution of marriage: "He was thoroughly domesticated" (11).

At the embassy Vladimir observes that an anarchist ought to be neither fat nor married (20). Verloc is both – and the two are by no means unrelated. Indeed, Verloc's indolent obesity is not only a satirical commentary on his attenuated agency but also a function of his institutional attachments.[28] This relation is complemented by the history of the other obese anarchist, Michaelis, the "ticket-of-leave apostle" who lives writing his memoirs under the protection of a lady-patroness, having emerged from prison irredeemably overweight: "He had come out of a highly hygienic prison round like a tub [...] as though for fifteen years the servants of an outraged society had made a point of stuffing him with fattening foods in a damp and lightless cellar" (37). Feeding stands in the text for the institutionalization of dependence. The relation of the individual – whether as an "agent" or a "patient" – to the agencies that govern social life is figured as a kind of substance abuse. Even privileged agents – police commissioners, home secretaries, embassy first secretaries – seem to get their fix out of institutional affiliations. For the rest, the mediation of institutional agencies, much like that of certain chemical agents, facilitates their coping with the rapacious violence of everyday metropolitan life.

In an early meeting of the anarchists at Verloc's house, Karl Yundt, "the terrorist, as he called himself," splutters: "Do you know how I would call the nature of the present economic conditions? I would call it cannibalistic. That's what it is! They are nourishing their greed on the quivering flesh and the warm blood of the people—nothing else" (44). Stevie seems to take the venomous remark literally: "Stevie swallowed the terrifying statement with an audible gulp, and at

---

**28** On the political resonance of the comic figure of the fat anarchist, see James English, "Anarchy in the Flesh: Conrad's 'Counterrevolutionary' Modernism and the *Witz* of the Political Unconscious," *MFS* 38.3 (1992): 615–630.

once, as though it had been swift poison, sank limply in a sitting posture on the steps of the kitchen door" (44). Shortly thereafter, however, Stevie, all worked up, will start "prowling round the table like an excited animal in a cage" (47).

Coming from Yundt's toothless mouth, such talk is not meant as a statement of fact but as a "terrifying statement," characteristic of his brand of terrorism:

> He seemed to sniff the tainted air of social cruelty, to strain his ear for its atrocious sounds. [...] The all but moribund veteran of dynamite wars had been a great actor in his time—actor on platforms, in secret assemblies, in private interviews. The famous terrorist had never in his life raised personally so much as his little finger against the social edifice. He was no man of action [...]. (42)

Unlike the others who are in one way or another desensitized, Yundt has a nose for suffering, indeed he is in a certain sense dependent on it. He copes by directing such passion into an act, a "speech-act" of outrage, which, however, does not so much express as defuse the outrage – the affect as well as the political or terroristic action. Inflationary rhetoric proves to be just another depressive agent.

An actor, not a man of action – this could be said of just about all the figures in the book. London, for all its hustle and bustle, is the site of a crisis of action. And if *The Secret Agent* is about agency,[29] it is about the way that it is secreted in those obscure arrangements and institutions, the substances and abstractions, that excite, instigate, or provoke movement or movements in collective and individual bodies but also, more pressingly, dull and sedate such bodies. As Guérard observes, "The particular chill humor of *The Secret Agent* derives from such elevation of passion and suffering to abstraction and from such reduction of the human being to a function or formal status."[30] The apathetic types catalogued in the book are indeed either functionaries of certain agencies or functions of certain agents.

The perplexity of agency emerges – both for the protagonist and the reader – at moments at which individuals break down, when the dulling mechanisms that defend against the onslaught of the outside cease to function. Suddenly they find themselves brought into an emotional encounter with startling and unfiltered impressions. Here, the case of Verloc is once again exemplary for he loses his cool at the very mention of the word *activity*. When Vladimir states: "What we want

---

**29** For other considerations of "agency" in *The Secret Agent*, see Ruth Kolani, "Secret Agent, Absent Agent? Ethical-Stylistic Aspects of Anarchy in Conrad's *The Secret Agent*," in *The Ethics in Literature*, ed. Dominic Rainsford, Andrew Hadfield, and Tim Woods (Basingstoke: Macmillan, 1999); John Attridge, "Two Types of Secret Agency: Conrad, Causation, and Popular Spy Fiction," *Texas Studies in Literature and Language* 55.2 (2013): 125–158.
**30** Guérard, *Conrad the Novelist*, 227.

now is activity—activity," the narrator observes: "Every trace of huskiness disappeared from Verloc's voice. The nape of his gross neck became crimson above the velvet collar of his overcoat. His lips quivered before they came widely open" (23). If "activity," the demand for "facts – startling facts" (25), puts Verloc into a hot distemper, it is because neither he nor any other "anarchist" in the story, for all their talk of *what is to be done*, is in a position to do anything. Structurally, the demand for action – in the emphatic sense, an action "in fact, mad" (30) – made by Vladimir, who is himself acting in his bureaucratic capacity as embassy agent, is paradoxical because he cannot expect of his secret agent anything other than agency "work" – namely, functionalism. If Vladimir cynically demands propaganda by the deed, Verloc, for his part, responds in effect that he can only "do" propaganda, which is to say, like Yundt, can only "act." The unhappy situation – not unconnected with the propagation of newspaper discourse in *The Secret Agent*[31] – is such that there are no "felicitous" acts, not even speech-acts, for language has lost any pragmatic reference to reality.[32] "Propaganda par le fait," the famous call central to anarchist discourse from the 1870s onwards, would be symptomatic of this historical disjunction: an expression of the failure of its own speech-act – propaganda – that calls in fact for the action that it cannot in itself accomplish.[33] Action and speech-action alike have been relegated, or rather delegated, to the perfunctory. Language – as *The Secret Agent* itself demonstrates – has become the agency-language of bureaucracies and news agencies.

In this sense, Vladimir's ironic "philosophy of bomb-throwing" which targets the "sacro-sanct fetish" of science enshrined in the Greenwich Observatory, rightly posits anarchism against science (30). If anarchism sought to insist on an agency that was felt to be lost or threatened by modern metropolitan life, then the nascent social sciences over the same period seemed increasingly preoccupied with explaining away, if not effacing, such "anarchic" agency, both individual and collective. They did so with such mystifying terms as "degeneration" but equally, if from the opposite end of the spectrum, "progress." Thus

---

[31] See Patricia Pye, "Hearing the News in 'The Secret Agent.'" *The Conradian* 34.2 (2009): 51–63; Peter Nohrnberg, "'I Wish He'd Never Been to School': Stevie, Newspapers and the Reader in The Secret Agent," *Conradiana* 35.1–2 (2003): 49–62.

[32] On "infelicities" and other failed performances in speech-act theory, see J. L. Austin, *How to Do Things with Words*, ed. J. O. Urmson and Marina Sbisà, 2nd edition (Cambridge, MA: Harvard University Press, 1975).

[33] On the anarchist apology for terrorism as a compensation for its theoretical impotence, see Marie Fleming, "Propaganda by the Deed: Terrorism and Anarchist Theory in Late Nineteenth-Century Europe," *Terrorism* 4.1–4 (1980): 1–23.

the "anarchists" in *The Secret Agent* – the medical student, Ossipon, who cites Lombroso (41), and the scientific-materialist, Michaelis, who, on the basis of "cold reason," predicts the "logical" end of all private property (38–39) – distinguish themselves amidst the urban outcasts amongst whom they belong only insofar as they recite the "scientific" discourses that explain, but also confirm, their own indolent, indigent and politically impotent conditions.[34] The "anarchists," in short, are mouthpieces for the contemporary meta-discourses that expound their own crisis of agency.

When Vladimir proposes an outrage on Greenwich, "the blowing up of the first meridian" (32), he argues that it would be the only convincing demonstration of a truly anarchist conviction: "You anarchists should make it clear that you are perfectly determined to make a clean sweep of the whole social creation" (30). Greenwich is the *arche* of the "social creation" inasmuch as it presents an overarching agency assigned to project a global organization of space and synchronization of time. It is the representation of global order as such, and moreover and more menacingly, of an ostensibly scientific and so "apolitical" one that fundamentally occludes the scope for "anarchist" action. To attack Greenwich would, Vladimir argues, be truly anarchist in that it is, by any reasonable measure, completely senseless: "But what is one to say to an act of destructive ferocity so absurd as to be incomprehensible, inexplicable, almost unthinkable; in fact, mad?" (30).

Vladimir's cynical ratiocinations on "anarchism" do, however, amount to a compelling theory of terrorism: An "outrage" is all the more outrageous if it is a senseless and a futile one. The very poverty of the action is a measure of its fanatical intensity and of the terror it inspires. Vladimir's "philosophy of bomb-throwing" thus articulates one extreme response to the impoverishment of political agency under bureaucratization: fanaticism as sheer, empty activism. Need-

---

[34] On Lombroso in *The Secret Agent*, see Norman Sherry, *Conrad's Western World* (Cambridge: Cambridge University Press, 1980), 276; Allan Hunter, *Joseph Conrad and the Ethics of Darwinism: The Challenges of Science* (London: Croom Helm, 1983), chap. 5; Rebecca Stott, "The Woman in Black"; Robert Jacobs, "Comrade Ossipon's Favorite Saint: Lombroso and Conrad," *Nineteenth-Century Fiction* 23.1 (1968): 74–84. Ironically, Lombroso concluded that anarchists were not generally criminal-types and should be categorized under "crimes of passion." "Like religious fanatics, they like suffering for its own sake," he observes in "The Status of Anarchism To-Day in Europe and the United States," *Everybody's Magazine* 6 (1902), reprinted in David M. Horton and Katherine E. Rich, eds., *The Criminal Anthropological Articles of Cesare Lombroso published in the English Language Periodical Literature during the late $19^{th}$ and early $20^{th}$ Centuries*, (Lewiston, NY: Edwin Mellen Press, 2004), 165–168. On the critical presentation of the nineteenth-century scientific world in *The Secret Agent*, see Ludwig Schnauder, *Free Will and Determinism in Joseph Conrad's Major Novels* (Amsterdam: Rodopi, 2009), chap. 7.

less to say, this kind of fanatical action lends itself to cynical exploitation for political purposes. In *The Secret Agent*, in contrast, a peculiar kind of "outrage" occurs that fails to be registered as an action at all. Such destitute action, which cannot be distinguished from pure passion, exposes the limits of the institutional constitution of agency. This "futile" (*vergeblich*) agitation, which, however, is not for nothing, is characteristic of the passive activism of the peculiar fanatic.

## "Poor Stevie"

One can be an agent in London, that is to say, emerge as a more or less conscious individual, only insofar as one represents, or is under the influence of, one or more agencies. Approaching the extremes of poverty or destitution, one is increasingly relegated to a less-than-conscious state of near total dependence on institutional agencies or chemical agents. In *The Secret Agent*, Stevie seems to present the limit-figure of "complete dependence" (120) and therefore does not appear as an agent at all. Perhaps it is for precisely this reason that it is possible for him to become involved in an anarchist action.

Stevie, usually referred to as "poor Stevie," but also as "luckless Stevie," "innocent Stevie," as well as, by Ossipon, a "degenerate," is always characterized in terms of the deprivation of his situation: "Stevie was destitute – and a little peculiar" (120). His mother, herself physically incapacitated, worries that Stevie, who has "nothing in the world" and no "sufficient standing" in the world (125), would, without the precarious protection of his sister, Winnie, and her husband, his brother-in-law, Verloc, end up in the "workhouse infirmary" (127). Indeed, his destitute state is such that it would be difficult to speak of him as even having so much as a distinct individuality – if this means being in possession of oneself. Stevie's peculiarity is characterized by his inability to discern or function according to the habitual distinctions of city-life – he is forever *losing himself* in the city, a fact that costs him his first job:

> But as errand-boy he did not turn out a great success. He forgot his messages; he was easily diverted from the straight path of duty by the attractions of stray cats and dogs, which he followed down narrow alleys into unsavoury courts; by the comedies of the streets, which he contemplated open-mouthed, to the detriment of his employer's interests; or by the dramas of fallen horses, whose pathos and violence induced him sometimes to shriek piercingly in a crowd, which disliked to be disturbed by sounds of distress in its quiet enjoyment of the national spectacle. When led away by a grave and protecting policeman, it would often become apparent that poor Stevie had forgotten his address—at least for a time. A brusque question caused him to stutter to the point of suffocation. (13)

Distracted, without a fixed sense of time, and disoriented, without a determinate sense of place or direction, he is incapable of assuming a useful function in city life: "He is difficult to dispose of, that boy" (13).

Dull though he may be, he is too sensitive to the sensations of city-life to operate as an effective agent of any of its agencies. Indeed, his peculiarity is premised on a heightened memory of, or rather an inability to forget, sensations: "Stevie, though apt to forget mere facts, such as his name and address for instance, had a faithful memory of sensations" (129). It is his memory of sensations that prevents him from developing a faculty of abstraction and this in turn prevents him from distancing himself, as an individual, from the flood of sensations and inarticulate suffering of the city.

In the cabby-ride scene Stevie is moved by compassion at the sight of the cabby's maimed and abused horse:

> Stevie was staring at the horse, whose hind quarters appeared unduly elevated by the effect of emaciation. The little stiff tail seemed to have been fitted in for a heartless joke; and at the other end the thin, flat neck, like a plank covered with old horse-hide, drooped to the ground under the weight of an enormous bony head. The ears hung at different angles, negligently; and the macabre figure of that mute dweller on the earth steamed straight up from ribs and backbone in the muggy stillness of the air. (128)[35]

The sight of this silent suffering body again puts Stevie into a state of agitation, one heightened when the cabby reproaches him for worrying about the horse when he himself, as a poor night-cabby, has to survive and provide for his indigent children, a remark that solicits an image of London in which one kind of suffering literally feeds on another. Stevie is indignant at the prospect of "the poor cabman beating the poor horse in the name, as it were, of his poor kids at home" (132).

The relation between suffering and consumption finds its succinct expression in the skinny Stevie's open or vacant mouth. The sight or the image of

---

[35] Conrad seems to have been unsettled by the sensitivity – his own and others' – to the plight of London's horses. In both "Autocracy and War" (1905) and the Congo letter to Roger Casement (December 21, 1903), cited in the previous chapter, he brings up fallen horses as instances of exaggerated sentimentality at home, while the public remains desensitized to the human atrocities carried out abroad: "An over-worked horse falling in front of our windows, a man writhing under a cart-wheel in the streets awaken more genuine emotion, more horror, pity, and indignation than the stream of reports, appalling in their monotony, of tens of thousands of decaying bodies tainting the air of the Manchurian plains...," "Autocracy," 84. See Michael DiSanto, "'Dramas of Fallen Horses': Conrad, Dostoevsky, and Nietzsche," *Conradiana* 42.3 (Fall 2010): 45–68.

pain makes his mouth drop open. And in fact his mouth is always somewhat open. The only external sign of Stevie's peculiarity is the "vacant droop of his lower lip" (13). Averse to both the abuse and the substance-abuse that makes it possible, Stevie does not take part in the cruel consumption of the city – he is in this sense something of a hunger-striker, or perhaps rather a hunger-artist, to take up the ironic reference to Stevie as "the artist" devoting himself attentively to drawing circles at the kitchen table (40).

Although he struggles with words and thoughts, he is forever seeking adequate means of expression – of which his circles are perhaps the most memorable. Stevie's open mouth is not mute. He may be no good as a messenger and stutters when addressed, but he exhibits a particularly passionate relation to language: "Stevie was no master of phrases, and perhaps for that very reason his thoughts lacked clearness and precision. But he felt with greater completeness and some profundity" (131–132). The strength of his feeling comes at the expense of his articulacy. Faced with the case of "human and equine misery," he had initially stammered: "Poor! Poor!" (131) and subsequently: "Shame!" "Beastly!" (132) – exclamations communicating little beyond the explosive intensity of his compassion. When he does eventually formulate an articulate utterance, he arrives at the pathetic, all but tautological expression: "Bad world for poor people" (132). In *The Secret Agent*, relayed by the blasé narrative voice, this sentence can be no more than a sentimental phrase in the pejorative sense – a truism demeaning to the truth it expresses. Deprived not of force but of meaning, indeed, deprived of significance as a function of their affective intensity, Stevie's "poor utterances" convey the poverty of passionate expression.

Winnie's characteristic response to her brother's distress, "Nobody can help that" (132), expresses an apathy borne of powerlessness that infuriates the compassionate Stevie who recognizes himself in the negated "agent" of the phrase. His compassion gives way to righteous fury: "The anguish of immoderate compassion was succeeded by the pain of an innocent but pitiless rage. Those two states expressing themselves outwardly by the same signs of futile bodily agitation…" (130). Ultimately, deprived of sympathetic outlets for significant expression, Stevie will literally explode in outrage.

At the "mercy of his righteous passions" (132), Stevie brings to mind Michael Kohlhaas. But also, in a different tenor, he is reminiscent of the "poor fellow," Bartleby. While neither can, nor will, become implicated as blasé agents, officers or employees, messengers or scriveners in the various agencies and bureaucracies that envelop them, the impassive Bartleby and the impassioned Stevie present two "peculiar fanatical" responses to the institutional injustices of nineteenth-century life. The contrast is perhaps most pronounced in a reference to Stevie's earlier life, when as an office-boy he had been found "busy letting off fire-works in the staircase" causing panic in the building, "Wild-eyed, choking clerks stampeded through the passages full of smoke, silk hats and elderly business men could be seen rolling independently down the stairs" (13). The motives, the narrator drily observes, for "this stroke of originality" on Stevie's part were difficult to establish: "It seems that two other office-boys in the building had worked upon his feelings by tales of injustice and oppression till they had wrought his compassion to the pitch of that frenzy" (13). The image of Stevie in the stairwell evokes the stairwell to which Bartleby retreats, causing by his impassive presence such agitation that he is eventually forcibly removed by the police. While Bartleby "prefers not to" and impassively withdraws, Stevie in contrast is profoundly moved, he is indeed little more than emotion. And his final convulsive movement will be the desperate expression of his anguished compassion and pitiless rage (130). As such, Stevie becomes the peculiar fanatical exponent of otherwise neglected figures of suffering – such "mute dweller[s] on the earth" (128) as the maimed cabman and his emaciated horse.

## Actions and Accidents

The narrator as well as the characters in *The Secret Agent* seem persuaded that the "attempted bomb outrage in Greenwich Park" (144) as it is referred to, was an accident. The Professor speculates that the bomber set the detonator but "forgot the time," adding later, "He either ran the time too close, or simply let the thing fall" (63). The constable on the scene "announces positively" that he must have "stumbled," adding, "I stumbled once myself... Them roots do stick out all about the place. Stumbled against the root of a tree and fell, and that thing he was carrying must have gone off right under his chest, I expect" (72). Verloc, for his part, is quick to label the event "a pure accident" (193). Readers and critics, too, tend to toe the official line: there was no "outrage," just an accident occasioned by the hapless and irresponsible Stevie who was operating under the influence and instructions of his brother-in-law, the secret agent, Verloc, who was acting in turn on Vladimir's orders to carry out the attack on the Greenwich Observatory.[36] Stevie therefore "figures merely as a prepared victim" – albeit of society writ large – and his death can only be treated as some kind of mistake or misstep.[37]

---

[36] The Greenwich Outrage on which *The Secret Agent* is loosely based was "the sole [anarchist] outrage that occurred in London, a bomb explosion outside the Greenwich Observatory in February 1894, killed the man carrying the bomb," Hermia Oliver, *The International Anarchist Movement in Late Victorian London* (London: St. Martin's Press, 1983), 90.

[37] Howe, *Politics and the Novel*, 97.

It is reasonable to suspect that Stevie, who stammers when speaking, would stumble when acting. But the fact that Stevie stammers does not, as we have seen, mean that he has nothing to say. In fact, "poor Stevie," may have been making his own kind of peculiar statement. In 1901, as it happens, across Europe in Vienna, another cosmopolitan and imperial metropolis, such "action," which is to say, the curious convergence of failure and accomplishment, was given the name *Fehl-leistung* (literally: a failed-achievement), in what has become known in English as a "Freudian slip."[38] The explosion is perhaps a *Fehlleistung* but for that reason need neither have been a failure nor failed to be an action. Perhaps, on the contrary, the "futile bodily agitation" of the outrage in Greenwich Park should be considered a truly selfless action or "altruistic exploit," to use a phrase with which the narrator described one of Stevie's earlier inscrutable incendiary reactions (14).

Stevie may have stumbled over a root or he may be a radical activist – and these are by no means mutually exclusive. Rather than a failed attempt on the Greenwich Observatory, *there was* an anarchist outrage in Greenwich Park. Since, however, by the measure of any rational agency, it appears to be a senseless act and, serving no determinate political end, a futile one, it is written off as a mere accident or slip-up. As such, however, it evokes a different kind of intervention altogether – the artistic or aesthetic. Stevie's could then be seen as a very avant-garde performance engaging the tortured question of bodies and agencies in the modern metropolis. Indeed, the action would be political insofar as it is aesthetic, since it touches on the anaesthetized or unacknowledged "aesthetic" violence on which the metropolis relies.

Whether he is to be treated as an anarchist, artist,[39] cosmopolitical activist or simply a suicide[40] or simpleton,[41] Stevie's is an annihilating expression of

---

**38** Sigmund Freud, *Zur Psychopathologie des Alltagslebens* (Berlin: Karger, 1901).
**39** On the sympathy of avant-garde artists with anarchism, which was more often aesthetic than political for they were fascinated by the spectacle of the explosion, which, like their own work, tested the limits of representation, see Uri Eisenzweig, "Poetique de l'attentat: Anarchisme et littérature fin-de-siècle," in *Anarchisme et création littéraire*, ed. Société d'histoire littéraire de la France (Paris: Presses universitaires de France, 1999).
**40** Émile Durkheim's 1907 study, *Le Suicide* (Paris: Presses universitaires de France, 1993), opens with a discussion of the question of the "sphère d'action," which presents itself immediately when one seeks to arrive at a sociological definition of that category of deaths of which "le patient est l'auteur," 3.
**41** On the problem of "idiocy" and "mental deficiency," see Matthew Thomson, who argues that it became a particular preoccupation around 1900 because of anxieties about the changing terms of political citizenship with the end of a property-based franchise, *The Problem of Mental Deficiency: Eugenics, Democracy, and Social Policy in Britain c.1870–1959* (Oxford: Oxford Univer-

despair, that "negative passion" Bakunin associated with extreme poverty. As such Stevie's deed is the most impoverished expression, the zero degree of the agency understood as an outburst of affect, that Nietzsche had sought to validate with the line, "the deed is everything." In London, however, such agency has been consigned to the margins of the field of sanctioned social action by the overarching agencies of administration – to the extent that Stevie's destitute action is disavowed on all sides as a pure accident. The incendiary demonstration, in its very futility, negatively indicates a field of passionate political engagement that has been, in the interests of social utility, systematically closed off. In London, even Stevie's outrageous demonstration promptly becomes a matter for the apathetic consumption of the city, not least the media-consumption of the newspapers: "The trade in afternoon papers was brisk, yet, in comparison with the swift, constant march of foot traffic, the effect was of indifference, of a disregarded distribution" (65).

## Winnie, "the free woman"

If the questionable action with which Stevie is associated goes unacknowledged on account of his assumed lack of agency, his sister, Winnie, is, in contrast, attributed a certain agency. However, her "actions" equally exclude her from the field of recognized political engagement by being written off as *both* criminal and mad. Like her brother, the constraints of institutional life – epitomized in married life insofar as marriage presents *the* representative institution in the story – leaves only a destitute space for free action in the form of the desperate gestures, unhesitatingly identified by the authorities as murder and suicide, with which the story ends.

Like her brother, Winnie is a passionate figure, but her passion is exclusively focused on Stevie. Indeed, it is only for the sake of the otherwise destitute Stevie that she compromised her passion and entered into the cold world of social relations at all. Specifically, she enters the contract of marriage as if it were "a contract with existence" (189), confining herself to a domestic life providing "wifely attentions" to Mr Verloc (11). When, therefore, this contract that was meant to protect Stevie has, in fact, lead to his cruel obliteration, she suddenly finds herself a "free woman" (189, 191, 198). Verloc, however, only notices, "something peculiar in the blackness of his wife's eyes" (189).

---

sity Press, 1998), 297; Joseph Valente, "The Accidental Autist: Neurosensory Disorder in *The Secret Agent*," *Journal of Modern Literature* 38.1 (2014): 20 – 37.

Winnie, who has always been cleaning and dusting Mr Verloc's house, is suddenly in a position, in Vladimir's words defining anarchism, "to make a clean sweep of the whole social creation" (30). The killing of Verloc falls therefore not simply under that questionable category of irresponsible, unpremeditated, crimes, so-called *crimes passionnels,* but presents a second, clandestine, anarchist "outrage." Be it act or accident, "hazard has such accuracies,"[42] the narrator observes cryptically during the event (197), the killing is as much an attack on the institutionalism that Verloc as exemplary agent represents, as on his person, which is in any case so thoroughly defined by institutional affiliations, above all the domestic one.

The violence implicit in the domestic contract – "No system of conjugal relations is perfect," the narrator had dryly observed just previously (186) – is concentrated in the moments preceding the killing.[43] Winnie seizes the carving knife to kill Mr Verloc at the very moment that he, oblivious to her righteous fury, calls her over to the sofa – brutally – for conjugal relations: "'Come here,' he said in a peculiar tone, which might have been the tone of brutality, but, was intimately known to Mrs Verloc as the note of wooing" (196–197). Schematically, the scene traces the breakdown of Mrs Verloc, the reserved agent of the social contract, into the free woman, Winnie, whose liberation culminates in an anarchist gesture against patriarchy.

The complete dissolution of her contractually instituted personhood gives rise to a death-like freedom that is absolved, as the narrator with some discomfort observes, of all the trappings of agency: "She had become a free woman with a perfection of freedom which left her nothing to desire and absolutely nothing to do [...] She was a woman enjoying her complete irresponsibility and endless leisure, almost in the manner of a corpse. She did not move, she did not think" (198). If, following Stevie's death, Winnie becomes something of a feminist activist, she does not do so in the manner, to refer to the contemporary movement, of the suffragette. While the suffragettes asserted themselves as political actors intent on participating in the institutions of liberal democracy, Winnie's short-lived freedom is, in contrast, thoroughly irresponsible, anti-

---

[42] In the serial edition the phrase reads, "Accident has such accuracies," cited in Harold Davis, "Conrad's Revisions of *The Secret Agent:* A Study in Literary Impressionism," *Modern Language Quarterly* 19.3 (1958): 251. On the extensive re-writing of the killing scene in order to heighten the ambiguity of agency in question, see Kolani, "Absent Agent?," 94–96.
[43] On "violent domesticity" in *The Secret Agent*, see Rischona Zimring, "Conrad's Pornography Shop," *Modern Fiction Studies* 43.2 (1997): 321.

institutional and, for these reasons, emphatically anarchistic.[44] While the authorities would doubtless agree with Mr Verloc, "His wife had gone raving mad – murdering mad..." (197), the accidental, destitute act is the despairing assertion of a desolate freedom that abandons the agencies that structure and ratify social existence.

Such anarchist abandonment, however, leaves her without protection or solidarity, "a very friendless woman" (202) in the heart of London. She can think only of Mrs Neale, herself the "victim of her marriage with a debauched joiner" (138), and feels compelled to throw herself on the womanizer Ossipon in an attempt to re-establish an ad-hoc contract of protection: "'I won't ask you to marry me,' she breathed out in shame-faced accents" (216). Robbed and abandoned by Ossipon, her fate is reported in a newspaper article he finds the next day under the title, "Suicide of Lady Passenger from a cross-Channel Boat" (228). A wedding ring is found "left lying" (230) – dropped or discarded? – on a chair.

Ossipon, nicknamed "the Doctor," deduces from the article that "behind that white mask of despair [...] there was struggling against terror and despair a vigour of vitality, a love of life that could resist the furious anguish which drives to murder and the fear, the blind, mad fear of the gallows" (229). This expert diagnosis, which proceeds on the assumption of the agency of a certain biological "vitality," ignores the fact that Winnie "freely" dies rather than return to the precincts of the institutions to which she had been bound for so long. There is no place where she might be free. Committing herself anonymously to the anarchic waves, she elides all further regulation, as if asserting a "right" *not* to exist for the processing of socially contracted agencies – "She was nowhere" (230). The newspaper, whose reach she cannot altogether escape, has the last word: "*An impenetrable mystery seems destined to hang for ever over this act of madness or despair*" (228).

## Blank Wall

When she first learns of Stevie's death, she stares at the wall: "Mrs Verloc gazed at the whitewashed wall. A blank wall—perfectly blank. A blankness to run at and dash your head against" (184–185). The wall, devoid of its social inscrip-

---

[44] On *The Secret Agent* in the context of the "suffragette disturbances," at a time when "the feminists were the anarchists" (36), see Karen Piper, *Cartographic Fictions: Maps, Race, and Identity* (New Brunswick: Rutgers University Press, 2002), esp. 21–40. For a contemporary feminist-anarchist dismissal of "Woman Suffrage," see Emma Goldman, *Anarchism and Other Essays* (New York: Mother Earth Publishing, 1910).

tions and functions – the articulation of domestic and public space, of inside and outside, of private and public intercourse – appears as the sheer confinement of her "contract with existence" (189). But the sight of the wall as such, the "whitewashed wall with no writing on it" (182), is also the opening or invitation to leave a mark on the world. As he was groomed by his brother-in-law for the Greenwich attack, Stevie too turned from his circle-drawing to the wall: "At odd times he clenched his fists without apparent cause, and when discovered in solitude would be scowling at the wall, with the sheet of paper and the pencil given him for drawing circles lying blank and idle on the kitchen table" (142–143).

The blank wall thus presents the image of the anarchist problem of agency. It figures for the problem of unscripted action, the question of a non-functionalist agency outside the agencies of governmentality, and the scope for the peculiar sort of activity that declines the demands of intentional projects, programs, and ideals. Put blankly, at stake is the thoroughly concrete distinction between the inside and the outside of the political. And it further presents the question as to the place and character of destitute actions and agents, even if they are just "futile bodily agitation" (130) that do not so much as make an impression on the social edifice. "As to the outer wall of the Observatory," Conrad writes in the "Author's Note" of the historical Greenwich outrage, "it did not show so much as the faintest crack" (5).

Indeed, if, as has been suggested, the blank wall also presents a "primal scene" of writing,[45] Conrad's "Author's Note," written in 1920, acknowledges the implication of the activity of writing in the crisis of agency the story presents. For the "period of mental and emotional reaction" (3) was, Conrad recalls, characterized by a sense of the failure of his activities as an author and a despair over the possibility of authentic expression. Rather than attempting to recuperate the authenticity he felt his writing lacked, in *The Secret Agent* Conrad reflects instead on this "fundamental mood" (4). Thus a story that some take to be in bad faith is indeed, by the author's own admission, a reflection on his sense of bad faith as a writer. The inauthenticity of the subject matter – "anarchists or rather anarchist activities" (4) – is imbricated in the inauthenticity of the narration – of the writer or rather his literary activities. In *The Secret Agent* the field of literature, as much as the political field it represents, seems to be evacuated of the "sense of the truth of things," and rendered instead a site in which one is left "aimless

---

[45] Michael Fried, "'A Blankness To Run At and Dash Your Head Against': On Conrad's *The Secret Agent*," *ELH* 79.4 (2012): 1039–1071.

amongst mere husks of sensations and lost in a world of other, of inferior, values" (4).

The "Note" concludes therefore with an appropriately ironic re-citation of *The Secret Agent*'s repeated newspaper byline: "But still I will submit that telling Winnie Verloc's story to its anarchistic end of utter desolation, madness and despair, and telling it as I have told it here, I have not intended to commit a gratuitous outrage on the feelings of mankind" (8). For, in the dim light of *The Secret Agent*, nothing is less certain than talk of "intentions," of the gratuitousness of allegedly committed "outrages," and especially the authority of any "author's note."

It is only in such peculiar, unliterary, and apolitical figures as Stevie and Winnie that the "genuine and faithful" (4) expression, otherwise missing, seems to be achieved – if only in desperate, barely accountable, and annihilating form. Their destitute action presents a commentary on the poverty of the field of political as well as of literary representation. Thus, in the closing line of the "Author's Note," Conrad is neither denying committing a gratuitous outrage nor implying the expression of a necessary one. Rather the remark conveys the sense that a "gratuitous outrage on the feelings of mankind" is not, or is no longer, possible within the institution of literature – and may no longer be possible within the carefully regulated precincts of an increasingly disaffected mankind. If *The Secret Agent* is anything more than a conservative recitation of the "writing on the wall," it achieves this by staging, and perhaps also generating, futile agitation that attests, albeit negatively, to an attenuated field for the expression of passionate agency, those "feelings of mankind," for which – beyond the newspaper appeal of a melodramatic sentimentality – the walled-in agencies that govern social life make neither place nor time.

# 6 State of Embarrassment: Kafka's *In the Penal Colony*

If *Heart of Darkness* presents an exploration of what Conrad, with reference to events in the Congo at the turn of the twentieth century, called the disfigurement of the human conscience and of geography, Kafka's *In the Penal Colony*, drafted on the eve of, and published following, the First World War, articulates a related sense of displacement and disorientation. I call this "embarrassment," translating the German words *Peinlichkeit* and *Verlegenheit*. The state of embarrassment disclosed in Kafka's text corresponds to a political and epistemological crisis of authority that is in fact the crisis of so-called "bureaucratic authority." While bureaucratization owes its efficacy to "disenchanting" all modes of authority but its own, it understands and legitimates its own authority as delegated. Thus, bureaucracy runs into the paradox of exercising an authority that is authorized by an instance it is ultimately unable to credit, or positively discredits. This situation is cause for considerable embarrassment – and gives rise to despairing, but for that reason all the more excessive, gestures of self-assertion and assurance.

## The Postcolonial Moment

The question of Kafka and colonialism is inescapable in a reading of *In the Penal Colony*. For as postcolonial theory and history have shown, the "colony" in general can be defined as a site of punishment, where European fantasies of sovereignty and technologies of domination are explored and experimented with. The colony is the metropole's permanent state of exception in which its normative structures are rehearsed in excessive and grotesque, but altogether exemplary, theatrics. On account of the convergences of Kafka's text with contemporary and historical discourses of colonialism, as well as its resonance with anti-colonial and postcolonial writers in the late twentieth century, *In the Penal Colony* has been read, notably by Paul Peters, as a "master narrative of the 'primal scene' of colonialism itself."[1]

---

[1] Paul Peters, "Witness to the Execution: Kafka and Colonialism," *Monatshefte für deutschsprachige Literatur und Kultur* 93 (2001): 403. See also Margaret Kohn, "Kafka's Critique of Colonialism," *Theory and Event* 8.3 (2005); John Zilcosky, "Savage Travel: Sadism and Masochism in Kafka's Penal Colony," in *Kafka's Travels: Exoticism, Colonialism, and the Traffic of Writing* (New York: Palgrave Macmillan, 2004); Rolf Goebel, "Kafka and Postcolonial Critique: *Der Verschol-*

Kafka's Prague, of course, presented its own constellation of imperial, minoritarian, and subaltern problems that might have provided ample context for the postcolonial resonances that pervade his work.² An avid reader of popular exotic and colonial travel literature set in distant lands, Kafka was in fact personally not far removed from the frontlines of the New Imperialism. His uncle, Joseph Loewy, was involved in the construction of the Congo railway that Conrad witnessed and Marlow describes in *Heart of Darkness*.³ It is doubtful therefore that he would have been unaware of the humanitarian uproar in the European press around the Congo at the turn of the century. Nor are the controversies and atrocities that attended the nascent German colonial project likely to have escaped his attention.⁴

Nonetheless, *In the Penal Colony* is, I suggest, not in the first instance a thematic engagement with the *topos* of colonialism. It presents, rather, the articulation of what I would call the postcolonial moment in European history marked, at the height of the New Imperialism, by the outbreak of the First World War. Already in *Heart of Darkness*, the dislocation (*Entortung*) that contributed to what Schmitt refers to as the European order of states is indicated in the fanatical excesses of its imperial agents. In *In the Penal Colony*, Kafka draws the political and epistemological conclusion, so to speak, of this sea change, which solicited nothing less than the "provincializing of Europe" as a geopolitical configuration and as a discursive formation.⁵ The attempt to articulate this historical moment,

---

lene, 'In der Strafkolonie,' 'Beim Bau der chinesischen Mauer,'" in *A Companion to the Works of Franz Kafka*, ed. James Rollston (Rochester: Camden House, 2002); Karen Piper, "The Language of the Machine: A Postcolonial Reading of Kafka," *The Journal of the Kafka Society of America* 20 (1996): 42–54.

**2** See Scott Spector, *Prague Territories: National Conflict and Cultural Innovation in Franz Kafka's Fin de Siècle* (Berkeley: University of California Press, 2000); Gilles Deleuze and Félix Guattari, *Kafka: Toward a Minor Literature*, trans. Dana Polan (Minneapolis: University of Minnesota Press, 2012).

**3** Anthony Northey, *Kafkas Mischpoche* (Berlin: Wagenbach, 1988), 19–30. For a literary reflection on this coincidence, see W. G. Sebald, *The Rings of Saturn*, trans. Michael Hulse (London: Vintage, 2002), 121.

**4** In 1909, General Lothar von Trotha justified his refusal to negotiate with the Herero, whose uprising in German Southwest Africa he had been sent to put down, because to do so would have been to show "weakness and embarrassment" (*Ohnmacht und Verlegenheit*). It was so as not to be embarrassed that in 1904 the German army proceeded to engage in what was effectively the genocide of the Herero people. Lothar von Trotha, "Politik und Kriegführung," *Berliner Neueste Nachrichten*, February 3, 1909, 1. See Isabel Hull, *Absolute Destruction: Military Culture and the Practices of War in Imperial Germany* (Ithaca: Cornell University Press, 2013).

**5** I use the phrase in the sense that has become current since Dipesh Chakrabarty's groundbreaking study in historiography and political theory, *Provincializing Europe: Postcolonial*

which marks the breakdown of European or Eurocentric historiography, requires a new mode of writing altogether – one freed of established (but defunct) principles of interpretation and significance, the loss of which it has to account for. With the "genealogical" method, Nietzsche had already begun such a critical historiographical project. Kafka radicalizes the project by relinquishing the final ascetic remnant of "morality" that Nietzsche, insofar as he still believes he is doing "science," does not abandon – and writing fiction. It is in this sense, in undertaking the literary exploration of a state exposed to the faultiness and failure of its ostensibly universal political and epistemological principles, that *In the Penal Colony* is "postcolonial literature."

## Pain after Punishment

Nothing is more doubtful than the officer's statement in Kafka's *In the Penal Colony:* "Guilt is always beyond doubt" (*Die Schuld ist immer zweifellos*).[6] It was the already dubious character of guilt that had led Nietzsche, some thirty years before Kafka wrote the story of a research traveler (*Forschungsreisender*) who visits a penal colony and finds himself curiously involved in its fate, to conduct a research trip of his own to discover "the vast, distant and hidden land of morality."[7] What Nietzsche discovered in *On the Genealogy of Morality* (1887) was a land not unlike a penal colony, at least of the sort described by the officer in Kafka's novella.

At the center of Nietzsche's research is an increasingly complex and historically Christological machinery, "a whole, hidden machinery of salvation," that ascribed, or rather prescribed, guilt as the univocal significance of suffering.[8] The history of morality is, according to Nietzsche, an encounter, primarily in the mode of an evasion, with the debilitating experience of pain. It was not suf-

---

*Thought and Historical Difference* (Princeton: Princeton University Press, 2012). Chakrabarty takes the phrase from Hans-Georg Gadamer speaking of Heidegger in 1977, citing as an epigraph Gadamer's claim: "Europe... since 1914 has become provincialized,... only the natural sciences are able to call forth a quick international echo," *Provincializing Europe*, 3.
6 Franz Kafka, *In der Strafkolonie* (Leipzig: Kurt Wolff Verlag, 1919), reprinted in facsimile in *Historisch-Kritische Ausgabe sämtliche Handschriften, Drucke und Typoskripte* (HKA), ed. Roland Reuß and Peter Staengel (Basel: Stroemfeld/Roter Stern, 2009), 18. Subsequent citations in-text; translations mine. I have consulted and drawn freely on Stanley Corngold's translation, *In the Penal Colony*, in *Kafka's Selected Stories* (New York: Norton, 2006).
7 Friedrich Nietzsche, *On the Genealogy of Morality*, ed. Keith Ansell-Pearson, trans. Carol Diethe (Cambridge: Cambridge University Press, 1994), 8.
8 Nietzsche, *Genealogy*, 44.

fering itself, however, but rather "the meaninglessness of suffering" (*das Sinnlose des Leidens*) that was, or would be, Nietzsche argued, truly unbearable.[9] The whole machinery of guilt arrived, therefore, to guarantee that the bewildering sense of finitude experienced in pain was not all for nothing. It was there, in other words, to ensure that guilt, as the meaning of suffering, was never in doubt.

Guilt – as processed by the arcane "machinery of salvation" – is the interpretation of otherwise inexplicable pain as punishment for a forgotten and immeasurable debt to God. Once this logic of what Nietzsche calls the ascetic ideal was in place, it produced ever more elaborate mechanisms for suffering and the infliction of suffering and indeed the desire for more suffering; "they *thirsted* for pain," he observes, concluding: "The interpretation – without a doubt – brought new suffering with it, deeper, more internal, more poisonous suffering, suffering that gnawed away more intensely at life: it brought all suffering within the perspective of *guilt*...."[10] Bringing all pain under the perspective of guilt provides a certainty – "without a doubt" (*es ist kein Zweifel*) – to the vicissitudes of finite earthly existence. Indeed, it underwrites certainty as such.

Already by 1887, however, guilt is anything but beyond doubt. The dubious character of guilt is, indeed, the very condition of a project of the sort Nietzsche embarks on. Nietzsche's most shattering insight does not, however, concern the already fading theology of guilt but the epistemological grounds of modern science. The rational-scientific dismantling of the machinery of guilt destroys the surety on which science itself is based since the modern scientific will for truth proves to be a late, latent, and more extensive, if comparatively mild, off-shoot of the arduous and torturous ascetic mechanisms of the guilt-process. For Nietzsche the self-destruction of morality promises, at least in his more affirmative moments, in the 1887 preface to *The Gay Science* for example, a land of new-born innocence, an open field for a new kind of joyful research.[11] As, however, the cases of Kurtz and the Intended, and of Stevie and Winnie already show, indeed as the very problematic of "bureaucratic fanaticism" indicates, the promised liberation from morality had by no means produced a new innocence – a situation that would become all the more pronounced in the year 1914, when Kafka began writing *In the Penal Colony*.

---

**9** Nietzsche, *Genealogy*, 44, 120.
**10** Nietzsche, *Genealogy*, 105, 120.
**11** See the Preface to the Second Edition, in Friedrich Nietzsche, *The Gay Science*, ed. Bernard Williams, trans. Josefine Nauckhoff (Cambridge: Cambridge University Press, 2001).

Kafka's work in general can be read as a literary examination of the dubious grounds and desperate atmosphere of the post-moral landscape.[12] The default of guilt gave way not to a new innocence but to despair of the sort expressed in the indefinite, unattributable, inauthentic, and fundamentally uncertain sense that characterizes the world disclosed in, and through, Kafka's texts. In *The Trial* (written 1914–1915) the despair over guilt – and the economy of attribution, responsibility, and etiology it guarantees – finds a striking elaboration in gestures of shame. If, as Benjamin writes, shame is "Kafka's strongest gesture"[13] then the vague, barely discernible, but distinctly modern gestural landscape of *In the Penal Colony* might be the mildest. This I propose to call "embarrassment."[14]

Embarrassment could translate the German *Verlegenheit*, a word with which Kafka opens another of his stories from this period, *A Country Doctor* (written 1917).[15] Provisionally, one can characterize *Verlegenheit* as a displacement in, if

---

[12] That Kafka was an engaged reader of Nietzsche is attested to in Max Brod, *Über Franz Kafka* (Frankfurt am Main: Fischer, 1966), 45. The relevance of Nietzsche's *On the Genealogy of Morality* to *In the Penal Colony* has long been acknowledged, see for example Malcolm Pasley, "Introduction," in Franz Kafka, *Der Heizer, In der Strafkolonie, Der Bau*, ed. Malcolm Pasley (Cambridge: Cambridge University Press, 1966).

[13] Walter Benjamin, "Franz Kafka" (1934), trans. Harry Zohn, in Benjamin, *Selected Writings*, ed. Michael W. Jennings, Howard Eiland, and Gary Smith (Cambridge, MA: Harvard University Press, 1999), 2:808.

[14] "Embarrassment" – in any case a word of recent origin – has only since the nineteenth century been associated with the awkward feeling it now commonly denotes. Historically, since its introduction in the seventeenth century, it referred either to an impediment or obstruction or to finding oneself in perplexing or other difficult circumstances, for the most part involving debt. These meanings can be traced to the derivation of the word via the French (which is not without significance in *In the Penal Colony*) to the Spanish *embarazar*, which carries both the sense of obstruction and of being put in a difficult situation though of a rather different – and distinctly corporeal – sort: *embarazada* means pregnant. With the ambiguity characteristic of such words, in English it also carries an opposite – exorbitant – sense, as in the phrase, also from the French: "an embarrassment of riches."

[15] The first line reads: "Ich war in großer Verlegenheit…," Franz Kafka, *Ein Landarzt. Reprint der Erstausgabe im Kurt Wolff Verlag 1920* (Frankfurt am Main: Stroemfeld Verlag, 2006). Indeed, the "state of embarrassment" is by no means exclusive to *In the Penal Colony*. Karl is sent to America by his parents in "Der Heizer" (The Stoker, 1913), and also in the novel-fragment, *Der Verschollene* (The Man who Disappeared), which serves as a catalogue for embarrassing situations, Franz Kafka, *Drucke zu Lebzeiten*, ed. Wolf Kittler, Hans-Gerd Koch and Gerhard Neumann (Frankfurt am Main: Fischer 1996); *Der Verschollene*, ed. Jost Schillemeit (Frankfurt am Main: Fischer, 1983). "Schlag ans Hoftor" (Knock at the Courtyard Gate, written 1917) is thematically closest to *In the Penal Colony*, where the inadvertent and perhaps non-existent knock culminates in the narrator being dragged into a literally *peinliche* situation, complete with torture apparatus and no prospect of escape, Franz Kafka, *Nachgelassene Schriften und Fragmente I*, ed. Malcolm Pasley (Frankfurt am Main: Fischer, 1993).

not indeed a displacement of, space and time. It is a dislocation of one's situation, the sense "no longer to be in a position to...," in German, *nicht mehr in der Lage zu sein*. Arising from the lack or misapprehension of a *Gelegenheit* – an occasion or opportunity – *Verlegenheit* is always out of place, comes at the wrong moment, and is therefore thoroughly inopportune. Like the doctor in *A Country Doctor*, to be in *Verlegenheit* is to find oneself "put on the spot" but deprived of the self-possession and sense of standing that the situation seems to demand. It is a painful awareness of one's own inauthenticity and the inappropriateness of one's words and actions. To be in *Verlegenheit* is to suffer an awkward, if not agonizing, sense of impotence and disorientation that solicits in turn an even more excruciating tendency to over-compensate.

This no doubt indicates why embarrassment could also translate *Peinlichkeit*, a term that relates more explicitly to *In the Penal Colony* and specifically to a letter Kafka wrote to the publisher Kurt Wolff on October 11, 1916 concerning the novella's possible publication.[16] It is a response to Wolff's reservations about what he, rather awkwardly, calls the *peinlich* aspect of the story, probably intending the term in the somewhat outmoded technical or legal sense of the "painful," "torturous," or "punishing," and referring thereby to the juridical torture apparatus that the officer describes. Kafka, it seems to me, takes the term in its milder, more current sense, referring to the embarrassing character of the story itself. This he explains in the letter by suggesting that such embarrassment was definitive of the times – his time in particular (which may refer to his recent broken engagement to Felice Bauer) and to the time in general (likely a reference to the ongoing war in Europe).[17] I use "painful" in the translation of the passage below in order to maintain the ambiguity of "peinlich" but suggest that Kafka's use always plays on, or betrays the more colloquial sense of, painfully embarrassing:

> Your reservation about the painfulness [*des Peinlichen*] altogether accords with my own opinion, although I feel that way with regard to just about everything that I have written so far. Notice how little is free of such painfulness [*von diesem Peinlichen*] in some form

---

[16] Although it was first written in October 1914, *In the Penal Colony* would not ultimately, despite Kafka's considerable efforts, appear in print before late 1919.

[17] On the system of inquisitional justice, "das peinliche Gericht," in contrast with the modern institution of justice in the context of Kafka's novella, see Wolf Kittler, "In dubio pro reo. Kafkas Strafkolonie," in *Kafkas Institutionen*, ed. Arne Höcker and Oliver Simons (Bielefeld: Transcript, 2007). Among those who deal more extensively with the "painful" conditions of the epoch in a legal and socio-political context, see Walter Müller-Seidel, *Die Deportation des Menschen: Kafkas Erzählung "In der Strafkolonie" im europäischen Kontext* (Frankfurt am Main: Fischer, 1989); Ulrich Schmidt, "Von der 'Peinlichkeit' der Zeit. Kafkas Erzählung 'In der Strafkolonie,'" *Schiller-Jahrbuch* 28 (1984): 407–445.

or another! To clarify this last story, I will only add that not only is it painful [*peinlich*] but rather that our general and my particular time was and is likewise very painful [*peinlich*] and my particular time painful [*peinlich*] for even longer than the general.[18]

It is here, around the question of the meaning of the word *peinlich*, that the problem of guilt comes into focus. If guilt, in Nietzsche's account – and the officer in *In the Penal Colony* appears to suggest something similar – is the interpretation of pain as punishment, as if it were payback for a forgotten debt, then its sole consolation is that suffering is thereby rendered meaningful. And the intensification and articulation of this meaningful suffering – of this suffering for the sake of meaning – drives history, according to Nietzsche, as a veritable "guilt-history."[19] In the epoch Kafka describes in the letter, however, suffering is no longer taken seriously as punishment and therefore is no longer the guarantee of guilt. As a result it is embarrassing. Pain (*Pein*) as such is simply embarrassing (*peinlich*).[20] The inability to make sense of manifest suffering gives rise to a disorder which itself is no longer convincingly explained in terms of guilt.[21] The time rep-

---

**18** Franz Kafka, *Briefe 1914–1917*, ed. Hans-Gerd Koch (Frankfurt am Main: Fischer, 2005), 253. Translation mine. The original reads: "Ihr Aussetzen des Peinlichen trifft ganz mit meiner Meinung zusammen, die ich allerdings in dieser Art fast gegenüber allem habe, was bisher von mir vorliegt. Bemerken Sie, wie wenig in dieser oder jener Form von diesem Peinlichen frei ist! Zur Erklärung dieser letzten Erzählung füge ich nur hinzu, daß nicht nur sie peinlich ist, daß vielmehr unsere allgemeine und meine besondere Zeit gleichfalls sehr peinlich war und ist und meine besondere sogar noch länger peinlich als die allgemeine."
**19** See Werner Hamacher, "Guilt History: Benjamin's Sketch 'Capitalism as Religion,'" trans. Kirk Wetters, *diacritics* 32.3 (2002): 81–106.
**20** Margot Norris argues that Kafka's novella entertains the possibility of a sheer "animal" pain absolved of morality, *Beasts of the Modern Imagination: Darwin, Nietzsche, Kafka, Ernst and Lawrence* (Baltimore: Johns Hopkins University Press, 1985), chap. 5. To me it seems that one could not help but be embarrassed by such beastliness.
**21** The genealogy of morality, at least as Nietzsche presents it, could be abbreviated in the historical transformation of the valence of the term "peinlich." It derives from Church Latin, *poena* which referred to damnation (*Höllenstrafe*), and already drew on the association in the Latin between suffering and exchange (so central to Nietzsche's argument), an association which is even more explicit in the Greek *poiné*, which means first of all payment (*Zahlung*) and then punishment and revenge in the sense of pay-back. The mapping of the calculation of the debtor-creditor relationship into the Christian relation of damnation in a term that refers to the experience or infliction of pain strikingly coincides with Nietzsche's genealogy (which expressly proposes this sort of investigation of the meaning of words). Around 1800 – a crucial moment, as it happens, for all German embarrassment-terms (*verlegen, genant*) – the Grimms list four senses of the term. The first: "*körperliche pein bereitend, qualvoll, schmerzlich*" (inflicting corporal pain, torturous, painful). The second relating to the language of the court according to which "peinliches gericht" refers to the process of inquisitional justice in which torture is applied to produce a confession of guilt, or alternatively according to a more modern legal regime as punishment for a

resented by the year "1914" was (and is) an embarrassing time (*peinliche Zeit*) above all because the pain (*Pein*) to which it so extensively attests gives rise to an unresolvable, somehow humiliating, laughable kind of perplexity.

## Beyond Doubt – The Officer

*In the Penal Colony* appears to take place in a period of transition from an "old regime" characterized by postures of sovereignty and by the apparatus for the torturous inscription of guilt, and an emergent dispensation that appears to exhibit more liberal, scientifically objective, and humanitarian aspirations. The apparent liberalization of the regime is also its bureaucratization. This transition is figured in the encounter between the officer who claims to operate on the principle: "Guilt is always beyond doubt" (*Die Schuld ist immer zweifellos*, 18) and the travelling researcher who, presented with the archaic system of justice of which the officer is a proponent, becomes equally certain that "The injustice of the procedure and the inhumanity of the execution were beyond doubt [*zweifellos*]" (32). While, however, it is tempting to read the story as a progressive history in which former certainties give way to new certainties of a different, more rational and humane kind, in fact, the novella presents a distinctive temporality that is "always already" a *peinliche Zeit* – one that is, so to speak, constitutively traversed by doubt (*Zweifel*). Embarrassment, it turns out, is a kind of despair (*Verzweiflung*).

---

guilty party, *peinlich* as corporal and usually capital punishment or *peinlich* as deserving of such punishment. Third: with Luther the term is introduced in the internal sense that will come to characterize its principle contemporary significance: "*innerlich quälend und ängstigend, von innerlicher qual und unruhe erfüllt*" (causing inner angst or agony, to be filled with inner agony or disturbance). Finally, the term has also long been associated with exactitude and attentiveness in the sense of painstaking: "*...eine ängstliche, pedantische, übertriebene, bis ins einzelne und kleinste sich erstreckende genauigkeit...*" (an anxious, pedantic, exaggerated, exactness applied to the minutest detail), *Deutsches Wörterbuch von Jacob und Wilhelm Grimm* (electronic resource), 16 vols (Leipzig: Hirzel, 1854–1961). From the "peinliches Gericht" in which, as with the apparatus in *In the Penal Colony*, the accused learns the meaning of his suffering and comes to know his guilt from his wounds, through the moment brought into focus by the Reformation of the internalization of this process, to the *Peinlichkeit* that comes to define the procedure, indeed the ethic, of the modern sciences – the mutations and variations in the sense of the term relate and articulate the entire field that Nietzsche uncovers in the *Genealogy*. What follows however, when finally pain is to be cut off from the nexus of debt/guilt that had defined it, is the indeterminate feeling of the *Peinlichen* in the prevailing contemporary sense to which I take Kafka to be referring in the letter to Wolff: embarrassment.

The officer is from the start in a state of embarrassment. His embarrassment finds its expression in a discourse that frenetically insists on guilt – most emphatically in the principle, "Guilt is always beyond doubt" – but precisely by such insistence betrays that there *is* doubt and so no guilt. Despite, or because of, his passionate insistence on the contrary, there proves to be no solid ground on which to stand, nor any categorical principle to which to appeal. The image of a regime built on the firm grounds of guilt is itself a product of embarrassment. The officer's not altogether coherent account is something like a screen memory of the former regime which, however, everywhere undermines what it would affirm. When describing the culmination of the in those days ostensibly popular, public spectacle of justice under the old commander, the officer recalls: "Many people had stopped watching and lay with their eyes closed in the sand; everyone knew: Now justice is being done!" (37). This ambiguity, or rather the *Verlegenheit* of this extraordinary gesture – so close to burying their heads in the sand – underlies and so undermines the officer's discourse.

The story he tells is as faulty as the machine he is constantly trying to repair. As the traveler observes: "If the wheel had not squeaked, it would have been glorious [*herrlich*]" (26); so too with the officer's speech. But this is a sign of the times: something grates or screeches about everything ostensibly *herrlich* – great, glorious, or sovereign – and this on account of something faulty about its own machinery. The officer's efforts to restitute, or even simply to evoke, a sense of the *herrlich*, which he painfully feels to be missing or to have been missed, only exacerbates his increasingly fanatical and ultimately altogether embarrassing behavior. "What times, my comrade!" he cries, carried away by his recollections, "The officer had evidently forgotten who was standing before him; he had embraced the traveler and laid his head on his shoulder. The traveler was greatly embarrassed [*in grosser Verlegenheit*]; he looked impatiently past the officer" (38). Pulling himself together, the officer proceeds: "I wasn't trying to move [*rühren*] you [...] I know, it's impossible to make those times comprehensible [*begreiflich*] today" (39). Former times come across only as a groping emotion. They cannot be made *begreiflich* insofar as the term refers to the kind of comprehension that is associated with the rigorous communication of firmly-grasped concepts (*Begriffe*). Nonetheless, the impression of the past, as the officer seeks to convey it, is preeminently that of its supposed conceptual fastness (*Begreiflichkeit*), which is felt – in *Verlegenheit* – to have become somehow disfigured in the present day. For nothing in *In the Penal Colony* is quite *begreiflich*. The whole interaction between officer and traveler takes the form of misapprehensions of every order culminating in the awkward embrace. The officer's efforts to make the apparatus *begreiflich* to the traveler – and we should not forget that the machine itself is supposedly devoted to the production of a certain unequivocal understanding – only embarrass the traveler, an embarrassment the officer fails to grasp. "Had the officer finally grasped this?" (*Begriff es schon der Offizier?*) the narrator asks at one point, "No, he still had not grasped it" (*Nein, er begriff es noch nicht*, 44).

The possibility of *Begreiflichkeit* – of substantiality, clarity, understanding, definition – is guilt. The inarticulacy of mere suffering, the indefiniteness of mere emotion, achieve surety and directedness – rightness – in guilt which, taken as a "principle" (*Grundsatz*), provides the grounds for conceptual articulation and sure-handed interaction. In contrast, nothing in embarrassment is certain. It seeks certainty and security in the evasiveness that betrays its own uncertainty and insecurity – and in this way gives itself away. The ungraspable (*unbegreiflich*) character of the penal colony is that of a gestural world in which the functional implication of hand and mouth, suggested in the German term, *Begriff*, is demonically disrupted and uncoupled. The officer's discourse – his words as well as the gestures that accompany them – is full of awkward groping, hand holding, fist clutching, hugging, and fidgeting that seem to bear no relation to the matter at hand. This is, however, the heart of the matter. Anyone seeking conceptual clarity in *In the Penal Colony* is left grasping at straws.

The officer's discourse, so easily read on the level of statement, would be better characterized as one of excuse. His "explanation" is really an apology for the apparatus and is accompanied from the start by excuses that, like the repairs he is engaged in, betray its already defunct character: "[The officer] sought to safeguard himself against all contingencies by saying: 'Naturally malfunctions do occur [...] But if malfunctions do occur, they will only be very minor, and they will be resolved immediately'" (7–8). Here, the excuse does not mean the removal, nor the delegation, nor even a tacit acknowledgment of guilt, but is testament to the withdrawal of its ground, the definitive discrediting of its indubitability.

The officer's many and overdetermined excuses betray the fact that he is always also making excuses for the mortifying fact that he is just a bureaucrat – a petty bureaucrat. He is a bureaucrat and perhaps an altogether exemplary one insofar as he desperately tries to make out that his office has real, indeed transcendent, significance of the sort that he claims the machinery is there to inscribe. To draw on the distinction popularized from the work of Max Weber, the officer's passionate speeches are but sorry attempts on his part to inflate his atrophied and empty career (*Beruf*) to the status of enthusiasm that once attended a proper calling (*Berufung*). His eccentric and altogether excessive behavior, which makes him so insufferably officious and embarrassing, is but the expression of the embarrassment about just being a bureaucrat.

Thus while he attempts to recall the greatness and glory of former times, when he claims to have had a decisive function, he constantly betrays what he so desires to conceal: the petty character of his office. In contrast with those times when he was free, as he claims, to be wanton in his use of funds, there is now a very limited budget for the upkeep of the machine and the usual, infuriating, administrative checks and balances: "if I send for a new strap, the torn one is demanded as proof, and it takes ten days for the new one to get here, and then it's of an inferior sort and not much good. But how I'm supposed to operate the machine in the meantime without a strap – that's something no one is concerned about" (31). Despite his best efforts to pose as a self-respecting officer in the style of the old regime, his preoccupations with the machine increasingly betray the behavior characteristic of an officer of the bureaucratic type from which he so strenuously seeks to distinguish himself. The officer is already the kind of bureaucrat he worries about becoming.

The epitome of what the officer describes as the indifferent and laughable "consultations" (*Beratungen*) of the new government concern the development of the harbor – "mostly harbor construction, always harbor construction!" (47). The harbor, site of the final scene of the published novella, comes to replace the "machinery of salvation," to use Nietzsche's phrase, as the order of the day. While the inflection of the harbor over and against the penal apparatus can be read as indicating the transition from a penal colony and its political-theological apparatus to a colony oriented to more "horizontal" concerns such as developing trade infrastructure and entering the global economy, a shift also marked by a deflation of the category of guilt (*Schuld*) to that of mere debt (*Schuld*), it cannot be ruled out that the whole affair might be just a rather petty case of "office politics," a kind of bureaucratic in-fighting about the assignment of influence and resources.

The excuse-character of the officer's discourse culminates in the incrimination of the new commander, "It's all the fault [*Schuld*] of the commander!" (33). The "guilt" that there is no longer guilt properly speaking, only the indeterminate and uninspiring feeling of guiltiness characteristic of what I am calling "embarrassment," lies with the new commander, whose regime, on account of its reticence with regard to the machinery of guilt, puts the officer in charge in a state of embarrassment. The colony, according to the officer at least, is suffering a crisis of authority. For what the new commander is really guilty of is not being an authentic commander in the old style that the disillusioned officer admires. The new commander lacks the sovereign definitiveness and resolution that the officer longs for – and pretends that he (still) has. Insofar as he fails to take responsibility and assume command, he is an embarrassment to any self-respecting officer of a so-called penal colony. This crisis of authority, about which the officer complains, is nothing other than the bureaucratization of authority. The old commander whom the officer claims to represent is dead and gone and the new one is not of the same type, which is simply to say, there is no new commander, only, to use a word that the officer would not have used, a "governor," which names an essentially bureaucratic function.

The preponderance of what is called, following Weber's typology, "bureaucratic authority" expresses in fact a crisis of authority, a crisis brought about by bureaucratization.[22] For bureaucracy undermines all instances of authority but its own. Insofar, however, as it understands and legitimates its own authority as delegated, it produces a situation in which it exercises an authority authorized by an instance that it ultimately undermines. The bureaucratization of authority ends in embarrassment. Thus, despite himself, but as a function of his office, the officer is the agent for the discrediting of the authority he claims to represent – as will be made unambiguously clear when he, in order to prove the justness of its operation, brings about the complete destruction of the machine and of himself.

For all his talk, the uncertainty regarding the new commander is profoundly unsettling to the officer, who seems anxious about the status and security of his own post: "'Will the commandant attend the execution?' 'It is not certain [*gewiss*],' said the officer, painfully moved [*peinlich berührt*] by the direct question, and his amiable expression grew contorted" (21). The new commander is a sore point – and it is unclear whether the officer's embarrassment is because of the unlikelihood of his coming or the danger that he might actually do so. The latter seems however to be the more pressing preoccupation as the officer apologetically continues: "That's precisely why we have to hurry. As sorry as I am, I will even have to curtail my explanations. [...] So for now, only the essentials" (42).

Everything must be done before the decisive intervention of the new commander that the officer seems to fear. The distinctive power of the new regime – the regime of bureaucratic authority – seems, however, to consist in its inability or unwillingness to take such decisive measures, which puts the officer into even more of a state. It soon becomes apparent that the officer's real concern lies not with the new commander himself, but rather with his "ladies" (*Damen*). To judge from the officer, these "ladies" (of whom there is reason to suppose they are anything but *ladies*) are the power behind the new dispensation. While the commander seems to be little more than a voice – "his ladies call it a thunderous voice" (42) – the "public spectacle" (*Schaustellung*, 47) by

---

**22** See Max Weber, *Economy and Society: An Outline of Interpretive Sociology*, ed. Guenther Roth and Claus Wittich (Berkeley: University of California Press, 1978).

which the officer characterizes the proceedings of the new regime is dominated, or so it seems to him, by the gaze of these "ladies."

Thus the officer's uncertain and "painfully moved" expression when asked about the new commander may well be on account of the "ladies," before whom he is anxious about embarrassing himself. *Before* in the double sense: in front of and in advance, for, in the officer's account, the "ladies" are always imminent, watching and obstructing, on the horizon of his discourse. Whence the frantic urgency that in fact always already betrays the embarrassment he anticipates. As he tells the traveler, no matter how categorically he may speak before the assembly, his words will be taken up and perverted and the "ladies" will furthermore bar any attempts at clarification: "You are on the point of interjecting that you never said what he proclaims [...] – but it is too late; you never get onto the balcony, which is already full of ladies; you want to draw attention to yourself; you want to shout; but a lady's hand covers your mouth – " (42–43). With the one hand the "ladies" obstruct, with the other they expose. The new dispensation, "the new mild direction" (*die neue milde Richtung*)," as the officer refers to it, is ruled with a "lady's hand" (34). Nothing is more threatening to the officer's embarrassing fantasy of self-assertion and possession. Hence his peculiar warning when he imagines the traveler reporting on the execution: "And now you step up to the railing. Place your hands for everyone to see, otherwise the ladies will grab them and play with your fingers" (48).

The officer proves to be engaged in an increasingly fanatical attempt, before the "ladies," to assert himself, to stand upright, to be his own man, to grasp his situation. As his insistent talk of principles, certainty, and guilt also shows, he thus emerges as the paradigm of the discourse of "authenticity" with its sovereign, masculinist, and heroic resonances that became prominent around the time of the First World War. In the figure of the officer – and it is precisely in this respect that he is altogether paradigmatic – such discourse, far from being "authentic," only betrays the embarrassment that he is trying to hide.

The awkward opening exchange between the traveler and the officer is already touched by the "mild new direction." The traveler, noticing two "ladies' handkerchiefs" stuffed under the officer's collar, and feeling obliged to say something, remarks on the inappropriateness of the officer's European-style uniform in the tropical heat: "'These uniforms are surely too heavy for the tropics,' said the traveler [...]. 'Certainly [*Gewiss*],'" the officer said [...] 'but they represent our homeland, and we don't want to lose touch with our homeland. – But now look at this machine'" (7). While they only offer a mild relief, the ladies' handkerchiefs precisely thereby undermine the uniform in its proper, weighty significance. As the traveler's remark shows, the homeland was already lost, the moment the sweltering officer stuck the handkerchiefs under his collar and showed, despite his protestations to the contrary, that the only certain thing is the inappropriateness in the tropical colony of such significant displays of conventional meaning. Indeed, the casual use of the utterance "certainly" (*Gewiss*) throughout the story always attests to the only certainty in the colony, one that the officer, and the traveler too, would strenuously deny: that nothing in the penal colony is beyond doubt.

The disconcerting power of the "new mild direction" appears to be a function of its mildness and lack of direction. Indeed it seems to operate by indirection, which is why the enmity, that the officer claims exists, never becomes an open hostility but remains only "almost hostile" (33). The decision on the friend-enemy is never made or is always deferred, which only makes the nature of relations within the colony the more intensely "political."[23] The influence exercised by the new regime is most effective when it withdraws altogether, discharging its own representatives and executors. The officer, claiming to be the "only representative" of the old regime, insists that the colony is full of "followers" (*Anhänger*) of the old commander who, however, with the demise of his authority, are embarrassed to appear: "[the followers] have crept into hiding, there are still plenty of them, but no one admits to it" (35). The new regime seems to draw its strength precisely from this withdrawal. The "potency" of the new command, if that is the right word, consists in the withdrawal of all followers – including its own.

The new regime avoids open confrontation. As the officer remarks, "they are cowards and send you, a foreigner, out in advance" (35 – 36). It is perhaps not so much a question of cowardice as of embarrassment. In any case, as the officer continues, the intervention of the foreigner is supposed to be decisive: "Although he has enough power to proceed against me, he does not yet dare to do so; instead he will subject me to your judgment, the judgment of a distinguished foreigner" (40). Here the story articulates a decisive moment in the history of the bureaucratization of authority: for the point is that *this* judgment – a judgment of "knowledge" rather than one of "power" – emerges as the more powerful form of judgment. Rather than asserting its own authority, the new regime draws on the expert and cosmopolitan authority attributed to a respected foreigner, this being in fact exemplary of the indirect and indecisive manner by which the regime governs. The new regime, in the style of many a modern post-

---

[23] See Carl Schmitt, *The Concept of the Political* (1932), trans. George Schwab (Chicago: University of Chicago Press, 2010).

colony, relies on the purported objectivity of foreign experts who seem to exercise decisive authority.[24]

Without meaning to, the research traveler thus finds himself to be an exemplary and influential agent of the new dispensation. The officer imagines the new commander's declaration in the light of even the traveler's most "innocent remarks" (*unschuldige Bemerkung*):

> "'A great researcher from the West, with the mission of examining the legal procedures in all the countries of the world, has just declared that our old traditional procedure is inhumane. Given this judgment by so distinguished a personage, it is naturally no longer possible for me to tolerate this procedure. And so, from today on, I decree' – and so forth." (42)

The traveler is a "technocrat" *avant la lettre* and an all the more effective one to the extent that he continues to think of himself in ostensibly objective and apolitical terms as just an observer – "he was traveling with the sole purpose of observing" (32). Even before it becomes clear that everything hangs on his judgment as a scientist and a foreigner, the traveler reflects: "It is always a sensitive matter to interfere decisively in other people's affairs. He was neither a citizen of the penal colony nor a citizen of the country to which it belonged" (32). Although it was strictly none of his business, he speculates his intervention might prove decisive precisely on account of his foreignness and impartiality. He did, after all, come "with recommendations from high authorities" (32–33). Thus the research traveler, representative of the rationalized humanitarian dispensation, exhibits a "mild" modality of bureaucratic *Schwärmerei*, the consequence of which, however, may prove no less painful than the fanaticism of the officer.

## Beyond Doubt – The Research Traveler

Despite the awkwardness of the situation, when he is eventually challenged by the officer to take sides, the research traveler knows what he has to do. The right judgment is certain, and he is convinced that he has the right, and, furthermore, is right, to express it: "The answer that he had to give had been for the traveler from the outset beyond doubt" (50). If the embarrassment of the officer is betrayed by impatient excuses, the traveler's embarrassment expresses itself in hesitation and delay. Despite claiming to have been certain "from the outset," he is never quite sure when decisively to intervene and leaves it late – indeed by his own standards too late, since the condemned man's execution would al-

---

[24] See Achille Mbembe, *On the Postcolony* (Berkeley: University of California Press, 2001).

ready have begun were it not for the disrepair of the machine. Like the officer's discourse, the traveler's apparently lucid reflections thus betray a doubt, one that will soon be revealed in a wholesale disintegration of the capacity for objective judgment on which he prides himself as a researcher and on which his right to condemn the officer's regime lies. This crisis is foreshadowed by all his justifications, or excuses, for the deferral of intervention, which in turn serve to defer intervention.

His unacknowledged doubts arise at the troubling sight of the condemned man and the soldier, a sight that leads him to hesitate, one last time, before doing the right thing: "Finally, however, he said, as he had to: 'No'" (50). The condemned man in particular makes an impression on him despite, or because, of appearing to be "without understanding" (29). Although the condemned man never explicitly attempts to communicate with him, and it is certain that, like the soldier guarding him, he does not understand French, the language in which the officer and traveler are communicating, it seems nonetheless as though his vague gestures, uncomprehending gaze, and gaping mouth are somehow addressed to the traveler in a profoundly unsettling way. Described as "a dull-witted, wide-muzzled man" (5), the condemned is an embarrassing figure above all on account of his mouth, which is invariably open – but does not disclose.

Apart from some alleged whispering to the soldier, the condemned is characterized entirely in mime. The only words that he is reported to have said are those for which he was to be punished – threatening to eat his captain, after he was caught sleeping rather than guarding: "Now, instead of getting up and begging forgiveness, the man seized his master by the legs, shook him, and cried, 'Throw the whip away, or I'll eat you alive'" (19). The guard-dog reverts to wolfdom. It is not, however, his alleged carnivorous wildness but rather his open-mouthed doggishness that the traveler finds unsettling.[25] What the traveler interprets as a lack of understanding, the officer takes to be a failure to acknowledge guilt. And such guiltlessness is precisely grounds for punishment insofar as the machine provides instruction in the meaning of guilt as the basis of all understanding: "Understanding dawns even on the dumbest" (28). The machine involves nothing less than the internalization of the wolfish bite to produce the "bite of conscience,"[26] thus generating interiority and turning the sheer gaping

---

[25] From the beginning the "doggish" character of the condemned man is emphasized: "In fact, the condemned man looked so doggishly submissive that it seemed you could let him run around freely on the slopes and would only have to whistle at the start of the execution for him to come" (6).

[26] On the "*morsus conscientiae*," see Friedrich Nietzsche, *On the Genealogy of Morality*, 55–56.

mouth, which threatens to devour, into the site of an articulate, regulated and unambiguous disclosure of authentic meaning. Under the new dispensation, the open mouth just seems doggish – and for that reason all the more embarrassing. He who has nothing to conceal, who is not "guiltily" racked by doubts and anxieties, lets his mouth hang open. As an empty cavity, the condemned man's mouth stands for the nothing that guilt was supposed to obscure and which now returns to focus as the very "truth" embarrassment is embarrassed about.

The embarrassment to open one's mouth – let alone read in public or publish in print – is characteristic of the *peinliche Zeit* Kafka described in the 1916 letter to his publisher (*Verleger*), a certain *Wolff*, who had, albeit rather doggishly, raised reservations about the novella's publication. Indeed, in the very next sentence of the letter, Kafka describes his own writing as a lip-biting, non-opening, embarrassed disclosure of what would be concealed: "God knows how much further I would have gone along this road had I written more or better, had my circumstances and my condition permitted me, teeth biting lip, to write as I longed to."[27] Lip-biting betrays embarrassment in dis-closing the mouth, which is embarrassing because one cannot keep one's mouth shut – not without thereby drawing attention to it.

It is precisely such a symptom of *Verlegenheit* that the traveler exhibits, biting his lip, as he watches the officer *do the right thing*, so to speak. Following the traveler's damning judgment of the apparatus, the officer places a paper that he claims reads "Be just!" under the writer (*Zeichner*) and submits himself to the machine: "The traveler bit his lip and said nothing. He knew what would happen, true, but he had no right to stop the officer in any way" (58). For the first time, in this moment of pure bureaucratic fanaticism, the traveler identifies with the officer precisely on account of the rightness of his conduct, "for the officer was now acting quite correctly; the traveler would not have acted any differently in his place" (58). Not so the condemned man however. When he sees the officer trade places with him, "A broad, noiseless laugh appeared on his face and never left it again" (59). As the machine sets to work on the officer, the traveler cannot but be distracted by the condemned man, who appears to be mimicking for the soldier the explanation he had seen the officer giving the traveler:

> The traveler looked over at the soldier and the condemned man. The condemned man was the livelier of the two; everything about the machine interested him; now he was bending down, now he was stretching up: his index finger was continually pointed to show the soldier some-

---

**27** Kafka, *Letters to Friends, Family, and Editors*, trans. Richard Winston and Clara Winston (New York: Schocken Books, 1977), 127; Kafka, *Briefe 1914–1917*, 253–254.

thing. The traveler was embarrassed [Dem Reisenden war es peinlich]. He was determined to stay here to the end, but he could not have borne for long the sight of the two. (61)

"Peinlich" refers not to the expected scene of torture but to the sight of the condemned man dumbly rehearsing, even ritualizing, the officer's explanation. As if the officer's excuses were not embarrassing enough, this guiltless repetition, the playing with a senseless pain, which is to say, a pain with no relation to guilt, gives rise in the traveler to a *Peinlichkeit* he can barely stand. In the midst of this embarrassment, the epochal event comes to pass: the definitive self-destruction of the institution of punishment on the island. In the penal colony without punishment, its "freed" occupants, freed that is from guilt, present a new kind of state – a state of embarrassment.

## Postcolony

It is likely that Kafka was referring to this point in the story when on December 2, 1914 he records being "not entirely dissatisfied" with a reading of the text to a private circle of friends, "except for its glaring and ineradicable faults [*Fehler*]."[28] Still in 1918, he would, in a letter to Wolff, point to a "deeper flaw [*Mangel*]" attested to by the botched final pages.[29] The faultiness or failure here might be characterized in relation to the difficulty of finding a positive characterization for the dispensation that succeeds the earlier regime, once it has been rendered completely defunct. While the officer's embarrassing speech

---

**28** Franz Kafka, *The Diaries of Franz Kafka, 1910–1923*, trans. Joseph Kresh and Martin Greenberg with the cooperation of Hannah Arendt (New York: Schocken, 1988), 318; Kafka, *Tagebücher*, ed. Hans-Gerd Koch, Michael Müller und Malcolm Pasley (Frankfurt am Main: Fischer, 1990), 703. Kafka read a version of the story to Brod on November 20, 1914 and again on December 2 with Franz Werfel and Otto Pick also present. He would go on, as indicated later in the letter to Wolff, to read the text in Munich on November 10, 1916, an event which Felice Bauer was to attend, but did not. Kafka had to submit a manuscript of the reading to the censors in Munich, which worried him as he wrote to Felice on November 3, 1916: "I am still nervous about it, and to tell the truth, however harmless [*unschuldig*] the nature of it is, I cannot imagine that permission will be granted," *Letters to Felice*, ed. Eric Heller and Jürgen Born, trans. James Stern and Elisabeth Duckworth (New York: Schocken, 1973), 533; *Briefe 1914–1917*, 272. Ultimately, he could proceed with the reading so long as it was advertised as a "Tropische Münchhausiade."
**29** "Perhaps there is some misunderstanding concerning the 'Penal Colony.' I have never been entirely wholehearted in asking for it to be published. Two or three of the final pages are botched, and their presence points to some deeper flaw; there is a worm somewhere which hollows out the story [...]," *Letters to Friends*, 136; *Briefe 1914–1917*, 312.

could operate against the model of a conceptuality (*Begreiflichkeit*) it turned out to lack, and thereby convey a particular, if highly questionable, kind of account, with the complete demise of the guilt-regime and the system of writing it was taken to represent, the story itself breaks down. It is by no means clear what kind of writing can account for the aftermath of this event, which is nothing less than the "end of history," according to the political-theological, but also the penal-colonial, historiography of "guilt-history." In short, the problem is what kind of writing might articulate the phenomenological as well as the political constitution of the postcolony and its distinctive history. This – the question of "postcolonial literature" – is the conundrum of Kafka's ending.

For the story "ends" in embarrassment. Awkwardly, *In the Penal Colony*, does not end with the end of the apparatus of writing and punishment that seemed to be at the center of its account. On the contrary, the end is missed or missing. The traveler's one resolution – a resolution he keeps despite himself, by default as it were – is to stay to the end, the end, that is, of the officer's execution. This he misses while distracted by the scene of the soldier and the condemned man, the sight of which "he could not have borne for long" (61). When, subsequently, "almost against his will" (65), he sees the officer's face, it presents none of the transfiguration the officer had described:

> It was as it had been in life; no sign of the promised deliverance could be detected; what all the others had found in the machine the officer did not find; his lips were firmly pressed together; his eyes were open, and had the expression of life, their look was full of calm and conviction, the forehead was pierced by the point of the great metal spike. (65)

The officer is not dead – he is mortified.[30] Even death loses its significance to become an embarrassment. If the regime of guilt was built on the determinate

---

**30** "Mortification," appearing in English from the French around the fourteenth century, was used primarily in a religious ascetic sense – the deadening of the body to the world by self-inflicted suffering. During the sixteenth and seventeenth centuries, the usage took on a more neutral physical (insensible) or medical sense (of a part of the body – necrosis, gangrene), also becoming increasingly deprecatory or ironic (abstemious). This en-lightening semantic shift produced the "mild" sense prevalent in current usage, immobilizing embarrassment, which emerged also from the French in the seventeenth century and entered common usage, along with other such embarrassment terms, in the eighteenth century. It is striking that German not only lacks a cognate but any translation that shares parallel connotations. The word with the most similar history is perhaps "Peinlichkeit." Certainly "mortifying" would be an apt candidate to translate "peinlich" in Kafka's 1916 letter to Wolff, especially for a writer who regularly describes the writing process in more or less mortifying terms – on August 6, 1914, for example, in his diary: "it is not death, alas, but the eternal torments of dying," *Diaries*, 302; *Tagebücher*, 546.

and ultimately terminal significance of suffering, if punishment was always ultimately the death penalty, then *Peinlichkeit* emerges as the uncoupling of the unity of suffering and meaning and the loss therefore of a definitive significance of death. Certainly this is not to say that there is no more death; rather mortification is dying of a different kind, one that is defined not as an "end" but by the failure to end. Nothing is more common, nor more painfully insignificant, than *to die of embarrassment*, which one can also do in German: *vor Peinlichkeit sterben*.[31]

In anticipating a decisive ending, traveler and reader alike are sorely disappointed, a condition that recalls Kafka's letter to Wolff: "To clarify this last story, I will only add that not only is it painful [*peinlich*] but rather that our general and my particular time was and is likewise very painful [*peinlich*] and my particular time painful [*peinlich*] for even longer than the general."[32] Embarrassment *was and is* because one is not embarrassed at a particular point *in* time, but rather one is *in* embarrassment as a distinct temporality. Embarrassment has no sense of the chronological account of time authorized by the dispensation of guilt. Nor does it disclose an ecstatic temporality. Rather, embarrassment upsets as a disjunctive temporality of chronic, feverish impotence. It does not disclose a situation but puts one in a state.

The state of embarrassment, as Kafka's letter further indicates, solicits a crisis of judgment, for it is characterized by a sense of incommensurability between the particular and the general that renders it impossible to subsume particulars to the generality of concepts.[33] Having passed judgment on the apparatus, the traveler now finds himself in a crisis of judgment. Or rather this critical moment makes the crisis of judgment underlying the whole of *In the Penal Colony* explicit. While he had only been embarrassed by the officer's arbitrary and archaic principle regarding the indubitable character of guilt, the self-assured traveler had maintained, "The injustice of the procedure and the inhumanity of the execution were beyond doubt" (46). Now, however, he is cast into a truly perplexing state as it seems, in line with Nietzsche's analysis, that the very certainty that

---

[31] Heidegger's unfolding of the question of death is reminiscent of the officer's *peinliche* discourse – and not just because he makes the authentic being-toward-death into a matter of assuming one's own guilt (*Schuld*). See especially the first three chapters of the second division of *Being and Time*, trans. Joan Stambaugh (Albany: SUNY, 2010).

[32] Kafka, *Briefe 1914–1917*, 253.

[33] Kant states that it is an "embarrassment [*Verlegenheit*] about a principle..." that is the occasion for the critique of judgment in the Preface to the *Critique of the Power of Judgment*, ed. Paul Guyer, trans. Paul Guyer and Eric Matthews (Cambridge: Cambridge University Press, 2000), 57. AA, 5:169.

underwrites the scientist's judgment depends on the inhumane system that he, rightly, condemned. In short: by the conscientious application of his judgment, the research traveler has bureaucratically done away with his own authority. For his power of judgment owed its force and authority to the jurisdiction he condemned. He now finds himself in the unregulated and ungovernable sort of space that the officer, by means of his penal machinery and the principle of guilt, had fanatically sought to suppress. The state of embarrassment is a bureaucratic nightmare.

In 1914 the attempt to convey the postcolonial state ended in failure, as Kafka described it, and he appears to have abandoned the story after a humiliating public reading in Munich in November 1916, where the ending (now lost) came in for particular criticism.[34] In August 1917, however, on three consecutive days preceding his diagnosis with tuberculosis, Kafka returned to the novella, noting down in his diary a series of possible concluding fragments. They are, of course, not really *concluding* fragments. Rather, in a manner that recalls Kierkegaard's definition of the task of the "essential author" in a time of religious and political confusion – "In order to find the conclusion, it is first and foremost necessary to perceive very vividly that it is lacking and thereby in turn very vividly to miss it"[35] – Kafka's fragments render vivid the inconclusiveness of conclusion at the end of the penal colony.

The fragments present the attempt to write in a manner distinct from the one for which the penal apparatus stood; to employ a different form of signification from the penal model of judgment/execution. They are experiments, in other words, with a "postcolonial" mode of expression, absolved of the fanatical fantasy of indubitable guilt, apodictic judgment, definitive ending, and "authentic" death on which the officer – to his mortification – still relied. Or, put another way: they present the attempt by the traveler to come to terms with his situation once even the certainty that had sanctioned his condemnation of the machine on just, humanitarian grounds proves itself to have presupposed the operation of

---

**34** The *Münchner Neueste Nachrichten*, for instance, complained that the story ought not to "ebb away so interminably slowly" (*so endlos langsam verebben*) following the "grotesque" death of the officer, while the *Münchener Zeitung* similarly stated, "the story is too long and not gripping enough to be read aloud" (*zum Vorlesen ist die Geschichte zu lang und zu wenig fasselnd*), cited in *Franz Kafka: Kritik und Rezeption zu seinen Lebzeiten, 1912–1914*, ed. Jürgen Born with assistance from Herbert Mühlfeit and Friedemann Spicker (Frankfurt am Main: Fischer, 1979), 120–122. Translations mine.
**35** Søren Kierkegaard, *The Book on Adler*, ed. and trans. Edna Hong and Howard Hong (Princeton: Princeton University Press, 1998), 8. See also Chapter One above.

just such an apparatus. To his own mortification, it turns out that he cannot bear to be without the assurance promised by the former regime.

The very end of the last of these fragments, dated August 9, 1917, expresses the traveler's vague sense of guiltiness that the whole course of events seems to have been obscurely but indubitably his fault. Exhausted, to the point of no longer being able to stand upright, he longs for his ship to appear to deliver him from the island – "that he would have preferred to everything."[36] Before departing once and for all, in a voice loud enough to be heard by the captain and sailors, he anticipates pronouncing a damning verdict to the officer (who reappears living before him), on the penal system, and the execution of the condemned man, which the confused traveler seems to believe has taken place. But the fantasy of publicly passing judgment ends in a humiliating spectacle. While the initial desire for the ship was in the subjunctive, the text has, with the actual arrival of the boat, shifted to the indicative. The officer (speaking before the suddenly attentive audience of captain and sailors) responds by showing the condemned man to be unharmed and then accusing the traveler of having already passed a judgment, a death sentence no less, upon the officer himself:

> "No," the officer said, "a mistake on your part; I was executed, as you commanded." The captain and the sailors now listened even more attentively. And all saw how the officer now passed his hand across his brow to reveal a crooked spike protruding from his shattered forehead.[37]

The notion of an abstract judgment turning into a concrete execution, a condemnation in principle turning into an execution in fact, is of course at the center, or really at the end, of Kafka's *The Judgment* and is thus one of the founding tropes of Kafka's writing. That Kafka in these closing fragments of *In the Penal Colony* might be reflecting on the transformation of his own writing practice is indicated furthermore by the "literalization" of metaphor, central to the first of the fragments (from August 7, 1917), in which the traveler loses his command of language.[38] The fragment begins with the traveler too tired "to still give commands

---

36 Kafka, *Tagebücher*, 826. Translations of the 1917 fragments are mine.
37 Kafka, *Tagebücher*, 826–827.
38 On "monstrous literalization" as both a rhetorical and political device, see Stanley Corngold, "Allotria and Excreta in 'In the Penal Colony,'" *Modernism/modernity* 8.2 (2001): 281–293; also Corngold's analysis, discussed in more detail in Chapter Seven, of Kafka's writing practice as a critique of metaphor by means of a "metamorphosis of metaphor," in "The Structure of Kafka's *Metamorphosis*," in *The Commentators' Despair: The Interpretation of Kafka's Metamorphosis* (London: Kennikat Press, 1973). On Kafka's critical self-reflection on his own writing, see Mark

here or even to do anything," and ends with him behaving like the dog he claims he would be: "With his hand on his heart he said, 'I am a cur [*Hundsfott*] if I allow that to happen.' But then he took his own words literally and began to run around on all fours."³⁹ The oath "mis-executes" the force of the performance, bringing about a literal execution of its metaphorical intention.⁴⁰ And it is not just any metaphor, but a dog-metaphor, which in *In the Penal Colony* and elsewhere in Kafka's work connotes that which lives at the edge of (human) language, that which precedes and exceeds the jurisdiction of language, and as such comes to stand for the most famous and, paradoxically, successful of Kafka's "failed" endings: "'Like a dog!' he said. It seemed as if his shame would live on after him."⁴¹ In *The Trial*, "like a dog" is the figure for the shame, which, as it seems to the dying Joseph K., would outlive him. Shame is that which appears to exceed the execution, that which cannot be assimilated to the judgment of guilt.⁴² The 1917 penal colony fragment in which he literally behaves like a dog presents, in contrast, the mode in which one outlives guilt – in embarrassment. While shame appears to have to do with the failure to be properly guilty, even at one's own death/execution, the failure or default of guilt produces an embarrassing crisis of judgment that culminates in mortification.

The most revealing aspect of the 1917 fragments is the transformation they imply in the traveler's attitude to the condemned man and the soldier – and thus to his humanitarian pretensions.⁴³ If, while the officer demonstrated the apparatus, these figures, especially the condemned man, had caused him some embarrassment on account of their doggishness, now in the complete absence of the failing authority of the officer, his embarrassment turns into animosity. In the second of the fragments, from August 8, 1917, the traveler finds himself unable to command his language sufficiently to give orders – orders that it is doubtful would in any case carry any force – and so resorts to hand gestures, stone-throwing and ultimately to all-out fisticuffs: "he dismissed the soldier and the condemned man with a gesture of his hand; they hesitated, he threw a stone

---

Anderson, "The Ornaments of Writing: 'In the Penal Colony,'" in *Kafka's Clothes: Ornament and Aestheticism in the Habsburg fin de siècle* (Oxford: Clarendon Press, 1992).
39 *Tagebücher*, 822.
40 These terms are from J. L. Austin's speech-act theory, *How to Do Things with Words: Second Edition*, ed. J. O. Urmson and Marina Sbisà (Cambridge, MA: Harvard University Press, 1975), 18.
41 Franz Kafka, *The Trial*, trans. Mike Mitchell (Oxford: Oxford University Press, 2009), 165.
42 For a recent discussion of guilt and shame with particular attention to *The Trial* see Lutz Ellrich, "Diesseits der Scham. Notizen zu Spiel und Kampf bei Plessner und Kafka," in *Textverkehr*, ed. Claudia Liebrand and Franziska Schößler (Würzburg: Königshausen & Neumann, 2004); Oliver Simons, "Schuld und Scham. Kafkas episches Theater," in *Kafkas Institutionen*.
43 On the questionable ethics and epistemology of the traveler, see Kittler, "In dubio pro reo."

at them, and when they still deliberated, he ran up to them and struck them with his fists."[44] This descent into violence is immediately followed by an expression of his perplexity: "'What?' the traveler suddenly said. Had something been forgotten? A decisive word? A firm grasp? A reach of the hand?'"[45] The traveler has lost his composure and can find neither the right word, nor the fitting gesture. He finds himself undergoing a transformation into a different and monstrous man, not unlike, but certainly less principled than, the officer whose gruesome apparatus he had just condemned: "Who can penetrate the confusion? Damned, miasmal tropical air, what are you doing to me? I don't know what is happening. My power of judgment [*Urteilskraft*] has been left back at home in the north."[46] Suddenly he fantasizes in an all-too-characteristic colonial fever about excessive displays of violence that would, so to speak, put the grinning figures of the soldier and the condemned in their place and in any case get them off his mind: "But who will set it right? Where is the man who will set it right? Where is the good old miller back home in the north who would stick these two grinning fellows between his millstones?"[47] How quickly the humanity and scientific objectivity of the research traveler can turn to gratuitous violence once the conditions of his authority have been undermined!

In the postcolony, the traveler is without authority – whence his frustration and anxiety. There is one instance, however, in which one of his commands appears to take effect, although this effect likely owes rather to the feebleness or failure of his speech act. It is at the beginning of the third fragment from August 9, 1917 that ends with the attempted departure by ship: "The traveler made a vague movement of his hand, abandoned his efforts, again thrust the two men away from the corpse and pointed to the colony where they were to go at once. Their gurgling laughter indicated their gradual [*allmählich*] understanding of the command; [...]."[48] The felicitous, or in any case grinning, uptake of the traveler's command, if that is what his indeterminate gestures really amount to, appears to repeat the process of judgment described by the officer in

---

44 Kafka, *Tagebücher*, 823.
45 Kafka, *Tagebücher*, 823.
46 Kafka, *Tagebücher*, 823.
47 Kafka, *Tagebücher*, 823. The text goes on to a scene involving a "great Madame," which recalls the figure of Brunhilde in *Der Verschollene* (1911–14/1927), and which Norris, for example, considers anomalous to the world of *In the Penal Colony* in her reading of the sado-masochism in the text, *Beasts*, 114. In contrast, Zilcosky reads the scene as integral to the "colonial masochism" that provides the counterpoint to the imperial sadism with which the main part of the story is made up, *Kafka's Travels*, 120–121.
48 Kafka, *Tagebücher*, 823.

which, upon the sixth hour of torturous inscription, understanding gradually dawns on the condemned. Whereas, however, under the machine the condemned, according to the officer, never misses the opportunity to lap up porridge at the beginning of the process but stops this consumption at precisely the sixth hour, spitting the remainder of the food out and devoting himself to the ascetic labor of the decipherment of guilt, here the understanding achieved is gradual, *allmählich*, which, given the *Mahl* in it, may more literally be translated as "piecemeal," or perhaps "bit by bit." While, in the officer's account, meaning in the form of guilt begins where the meal ends, understanding taking the place of eating, here the gurgling laughter presents the awkward intersection of eating and articulation, of consumption and expression, the over-lapping of mealing and meaning. The painstaking deciphering of judgment which promises to end in a disclosure of univocal meaning, a transfiguration based on a metaphorical or indeed metaphysical imputation of meaning to suffering, is here displaced – not without chagrin – by a time-consuming and rather unsatisfying piecemealing of partial and ambivalent bits of meaning. The always ambiguous and never fully determinate communication achieved by this speaking/eating game discloses the postcolonial form of life.

Despite the disorganization and disorientation and all the effects of the "miasmal tropical air,"[49] the traveler finds, to his dismay, that a curious communal bond seems to have formed between the three of them: "the condemned man pressed his face, which had been repeatedly smeared with grease, against the traveler's hand, the soldier slapped the traveler on the shoulder with his right hand – in his left hand he waved his gun – all three now belonged together."[50] With this ambivalent combination of camaraderie and aggression, a community asserts itself.[51] The fragment proceeds: "The explorer had violently to ward off the feeling coming over him that in this case a complete order [*eine vollkommene Ordnung*] had been established."[52] The fragments of 1917 break off with the image of the emergent political community as a state of embarrassment. The vi-

---

49 Kafka, *Tagebücher*, 823.
50 Kafka, *Tagebücher*, 823.
51 On Kafka's own awkward relation to community, expressed paradigmatically perhaps in the phrase in his diary dated January 8, 1914, "What have I in common with Jews? I have hardly anything in common with myself and should stand very quietly in a corner, content that I can breathe," *Diaries*, 252; *Tagebücher* 622, see Vivian Liska, *When Kafka Says We: Uncommon Communities in German-Jewish Literature* (Bloomington: Indiana University Press, 2009).
52 Kafka, *Tagebücher*, 825.

olence with which the traveler seeks to repel the encroaching *sensus communis*,[53] is precisely what binds him to it. For belonging to the state of embarrassment, is the longing to be apart and depart. Mortifyingly there is no departing. The ship-scene that follows (and so "concludes" the last of the fragments) plays out this failure to depart – to close, to pass judgment, to make an end of it, to properly die. One is instead stranded on the point of departure. The prolongation of the ending, of the time at the end from which there is no transport or transcendence, is excruciating – indeed, it almost cannot be borne. To recall Nietzsche's discussion of pain, what is truly unbearable is not suffering as such, but suffering devoid of definite significance – of what the officer called "guilt." The distinctive kind of pain that arises with the failure of the torturous machinery of guilt is captured in the figure of the officer with the spike in his forehead, which signifies nothing but the inscrutable agony of *Peinlichkeit*. The penal colony diary fragments are thus only commentaries on the transfixed image of the stricken officer with which the story had broken off. They repeat its in-conclusion: the sense of embarrassment without end is mortification. Freed from the regime of guilt and its death-sentence, the state that remains is an interminable dying of embarrassment.

## The Harbor

On November 11, 1918, the date of the Armistice – the significance of which may have precipitated his resolve – Kafka committed himself to publishing *In the Penal Colony* and sent the final proof back to Wolff. He abandoned the attempt to produce a coherent ending, even of the distended sort attempted in the fragments, and opted instead to insert a large break into the text, punctuated, as he suggested, by three asterisks.[54] He then cuts to a scene in a quite different space: a colonial harbor. In this form, the appended section in the port is not a continuation of the main story but something like its repetition, insofar as both have to do with the embarrassment arising from the persistent "pastness" of the former commander and everything for which he is taken to stand. Even under the new dispensation, the postcolony cannot shake off its colonial history.

Upon entering the colony, the traveler is attracted to the teahouse, where "he felt the power of earlier times" (66). It turns out that the former commander, who

---

[53] Kant, *Critique of the Power of Judgment*, §18; for the political significance of judgments of taste, see Hannah Arendt, *Lectures on Kant's Political Philosophy*, ed. Ronald Beiner (Chicago: University of Chicago Press, 1982).
[54] See Kafka, *Briefe 1914–1917*, 51.

was denied a place in the cemetery, is buried there. The locals have to move their tables aside to show the "foreigner" (67) the gravestone. On the one hand, the former commander is displaced and disparaged, not only buried in but hidden under the tables of the teahouse. On the other hand, his memory is ubiquitous insofar as it is recalled as what is concealed in that most public house. The open secret of the former commander's memorial is thus a persistent reminder of what is lost, namely, guilt. If approaching the teahouse the traveler had been given "the impression of a historical memory" (66), then this impression is that, so to speak, guilt is history. What remains, to judge from the scene that follows, is the pervasive sense of guiltiness that I have been calling embarrassment.

Disdaining the "poor, abused people" (67) who are hanging about, the traveler insists on reading the tombstone as if seeking to turn his inarticulate impression into something more significant. But although the engraving on the tombstone is perfectly legible, its transparent signification does not provide the kind of meaning – the "closure" – the traveler expects. It contains a prophecy: "There exists a prophecy, that after a certain number of years the commander will rise again and lead his followers [*Anhänger*] from this house to reconquer the colony" (67–68). Having read the inscription, the traveler gets up and sees, to his chagrin, "the men standing around him smiling, as if they had read the inscription along with him, found it ridiculous [*lächerlich*], and were inviting him to share their opinion" (68). Under the *peinlich* apparatus the condemned were supposed to gain understanding (of their guilt) by learning to read. As the officer had stated: "Understanding dawns even on the dumbest" (28). Here it is the understanding achieved through reading itself that seems foolish. And it turns out that everyone can read. The post-guilt epoch is not characterized so much by a loss of meaning as by its proliferation, betrayed by the apparent universality of literacy. While the traveler had found the condemned man embarrassing because of the lack of understanding he imputed to the man's open mouth, here he is embarrassed by these grinning figures because they understand, or rather because they, like he, can read but do not know what to make of it. Surely it cannot be taken in earnest. For fear of looking fools themselves, all demand acknowledgement of the foolishness of the grave script. Out of embarrassment, none of those hanging around the teahouse want to be mistaken for a "follower" or "hanger-on" (*Anhänger*) of the old commander, as if they were "believing and waiting," as the gravestone directs, and grinning at their own laughable, but nonetheless disappointed, expectations.

The research traveler is not prepared to acknowledge the awkward smiles of his fellows and so to acknowledge his belonging to this ironic community of postcolonial readers. Instead, pretending not to notice, he responds with a most painful gesture of embarrassment, handing out some small change before making a dash for it. But can one get away from the state of embarrassment? The last line is by no means conclusive. It presents the traveler on the point of departure, fending off the soldier and the condemned man who had pursued him to the port: "They probably wanted to force the traveler at the last minute to take them with him" (69). The traveler threatens them with a knotted rope, as they, like dogs, are about to leap into his retreating boat. Suspended in a desperate posture of departure, the three figures present the still image of the state of embarrassment.

# 7 Society of Security: Kafka's *The Metamorphosis*

At the beginning of the nineteenth century, Schiller's *The Criminal out of Lost Honor* showed how the brutalizing and exclusionary attentions of the law made a wolf of the criminal. The new administration advocated in Schiller's text promised a more inclusive model for the management of human life and social relations. This alliance of liberalism and bureaucracy – of liberalization on the condition of bureaucratization – came to define what Foucault would later call the "society of security." At the end of the long nineteenth century the proliferation of the dispositive of security, which was above all concerned with securing human life, was occasion for considerable anxiety. The scope of life outside of bureaucratic oversight and beyond constraining biopolitical determinations was felt, ever more urgently and impotently, to be threatened or lost. In Kafka's *The Metamorphosis* (1912), the rejection of the humanizing concerns that inform the norms and institutions of social security is presented as a transformation into vermin. The underside, for there is no longer an outside, of bureaucratization is verminization. Vermin is the abject and unnatural residue of the bureaucratization of modern life. Yet it also offers in Kafka's text the monstrous, or rather *schwärmerisch*, figure of an ungoverned life.

## Security and its Discontents

Literally in the background of *The Metamorphosis* is the institution of care. The prospect from Gregor's bedroom window affords a view of "a section of the endless, grayish-black building opposite – it was a hospital – with its regular windows starkly piercing the façade."[1] It is as if the outside were an endless care-institution – one, however, that finds its concentrated reflection in the family home opposite. The cozy world of the home, and the intimate family relations it shelters, proves to be structured on the by no means particularly affectionate model of care-giver and care-receiver. It is, furthermore, connected to other net-

---

[1] Franz Kafka, *Die Verwandlung* (Leipzig: Kurt Wolff Verlag, 1915), reprinted in facsimile in *Historisch-Kritische Ausgabe sämtliche Handschriften, Drucke und Typoskripte* (HKA), ed. Roland Reuß and Peter Staengel (Basel: Stroemfeld/Roter Stern, 2003), 19. Subsequent citations in-text; translations mine. I have once again consulted and drawn freely on Stanley Corngold's translation in Franz Kafka, *The Metamorphosis: Translation, Backgrounds and Contexts, Criticism*, ed. and trans. Stanley Corngold (New York: W.W. Norton, 1996).

works of care that bind it to society as a whole and are represented in the story by the *Prokurist* – an authorized representative, literally a representative carer or pro-carer – who seems able, at his discretion, to freely enter and exit the family home. Before the transformation, Gregor, assuming the role and responsibilities of a healthy, average, family man, takes it upon himself to look after his aging father, his invalid mother, and his sister, who is still a child, as well as to take on the family debt inherited from his father's bankrupt business venture – a debt that binds the family ever more inextricably into broader circuits of social concern. After the transformation, the family is obliged to reorganize itself and redistribute the burden of care across the household – and the story becomes, from the point of view of the family at least, about various iterations and experiments in care-giving and solicitude but also – in the interests of financial security and of securing social standing – about neglect, domestic violence, and abuse. Thus the whole economy that provides the backdrop to *The Metamorphosis* is a kind of managed care or, drawing on Foucault, what might be called a "dispositive of security."[2]

The society of security Foucault discusses corresponds to a historical moment in which the machinery of transcendence – be it theological or sovereign – has, to recall a famous exchange about Kafka between Benjamin and Scholem, lost its force or significance or both.[3] It is characterized by a diffusion of authority and the emergence of an epistemology of uncertainty that gives rise to a new kind of governance and a re-evaluation of human life. In the context of accelerating urbanization and industrialization, the increasingly irrelevant questions of guilt and individual responsibility are gradually superseded by the statistical regulation of populations organized around the concept of risk. Henceforth the individual is responsible for behaving "normally" and is considered responsible insofar as they are "normal." While security adapts juridical and disciplinary measures, it does so with an eye to dynamically regulating given conditions according to distributions of "the normal" rather than imposing definitive normative statutes. The dispositive of security responds to a condition of structural,

---

[2] See especially the first lectures in Michel Foucault, *Security, Territory, Population: Lectures at the Collège de France, 1977 – 78*, ed. Michel Senellart, François Ewald, and Alessandro Fontana (New York: Palgrave Macmillan, 2007).

[3] See the exchange between July 17 and September 9, 1934, reprinted in *Benjamin über Kafka: Texte, Briefzeugnisse, Aufzeichnungen*, ed. Hermann Schweppenhäuser (Frankfurt am Main: Suhrkamp, 1981), 63 – 92; the exchange is important to Giorgio Agamben's conception of modernity, marking the point, in the terms developed here, at which sovereign politics turns into security politics, see Giorgio Agamben, *Homo Sacer: Sovereign Power and Bare Life*, trans. Daniel Heller-Roazen (Stanford University Press, 1998), 49 – 62.

epistemological, and political insecurity which it seeks to manage according to a statistical calculus of risk.

As such, security was certainly a preoccupation of Kafka's work – his office work, that is, in the insurance industry.[4] As recent scholarship has shown, it was equally a preoccupation of his literary work.[5] Kafka started his career with a trans-national private insurance firm, *Assicurazioni Generali*, and from 1908 he worked for the parastatal Workmen's Accident Insurance Institute for the Kingdom of Bohemia in Prague. While he appears, despite his indefatigable complaining about the job, to have been a conscientious agent of security in his official capacity, his writing is expressive of a profound ambivalence towards the emergent society of security. *The Metamorphosis* is in this respect a kind of *Institutionennovelle*,[6] one that is however not structured around an "unheard of event" (*unerhörte Begebenheit*), to cite Goethe's memorable definition of the novella, but rather around an unhappy and accidental event, "a great misfortune" (*ein großes Unglück*), for which there could be no oversight (16). It is, in short, an uninsurable event.

If the dispositive of security defines the field of the normal according to a distribution of the probable, then the great misfortune is not considered fabulous nor supernatural but abnormal. It is not simply an accident but an embarrassment that reflects badly on those to whom it happens. There is, to be sure, no insurance policy against the eventuality that one's son or brother, employee or bread-winner, might suddenly transform, as Gregor does one morning, into a "monstrous vermin" (*ungeheueres Ungeziefer*, 3). And, for those whose home is

---

4 On the relevance of the emergent discourse of insurance as paradigmatic of the dispositive of security, see François Ewald, *L'état providence* (Paris: Grasset, 1986).
5 See the work of Benno Wagner, especially his unpublished *Habilitationsschrift*, "Der Unversicherbare. Kafkas Protokolle" (University of Siegen, 1998), much of which was published in English in the book he wrote with Stanley Corngold, *Franz Kafka: The Ghosts in the Machine* (Evanston: Northwestern University Press, 2011). For Kafka's "office writings" as well as a selection of contemporary material on the insurance industry, see *Amtliche Schriften*, ed. Klaus Hermsdorf and Benno Wagner (Frankfurt am Main: Fischer, 2004), and in English, *Franz Kafka: The Office Writings*, ed. Stanley Corngold, Jack Greenberg, and Benno Wagner, trans. Eric Patton and Ruth Hein (Princeton: Princeton University Press, 2009).
6 On the *Institutionenroman*, see Rüdiger Campe, "Kafkas Institutionenroman. Der Proceß, Das Schloß," in *Gesetz. Ironie*, ed. Rüdiger Campe and Michael Niehaus (Heidelberg: Synchron Wissenschaftsverlag, 2004). See also Joseph Vogl, "Lebende Anstalt," in *Für Alle und Keinen. Lektüre, Schrift und Leben bei Nietzsche und Kafka*, ed. Friedrich Balke, Joseph Vogl, and Benno Wagner (Zurich: Diaphanes, 2008); as well as the essays in Arne Höcker and Oliver Simons, eds., *Kafkas Institutionen* (Bielefeld: Transcript, 2007); and Ritchie Robertson, "Kafka, Goffman, and the Total Institution," in *Kafka for the Twenty-First Century*, ed. Stanley Corngold and Ruth Gross (Rochester: Camden House, 2011).

so infested, there can hardly be a more embarrassing situation. For the monstrous vermin presents a paradigmatic political-zoological problem for a society of security. Its significance is indicated by the emergence, in the period in which Kafka was writing, of such hybrid disciplines as "applied entomology," which, with military and colonial as well as domestic, urban and agricultural objectives, brought together techno-scientific knowledge with administrative measures connected with hygiene and public health to specifically address vermin and other pests as a security concern.[7] Needless to say, such developments were accompanied by the rise of an age-old but increasingly virulent derogatory rhetoric – strongly inflected by social Darwinism, eugenics and other quasi-scientific discourse – marshaling the supposed analogy between pests and parasites in the animal kingdom and those to be found in the human polity.[8]

As Kafka would have known from the Grimms' dictionary, *Ungeziefer* refers etymologically to an unclean animal that is not fit for sacrifice. As a juridical term, *Ungeziefer*, like the English "vermin," refers to a living being that has no inherent worth as an independent existence, and cannot, like domestic animals or even "wild" game, count as the lawful property of a person or corporation. Regardless of where they are found and what they may be up to – and it is characteristic of vermin to appear in the most repulsive and obnoxious way precisely where they are least wanted – they can, and in certain situations must, be killed with impunity. Vermin also carries the suggestion, inflected by the *Un-* of *Ungeziefer*, of an aberration of nature, a naturally unnatural being at odds with, if not a destabilizing threat to, the laws of nature and natural law alike.

Within the modern dispositive of security, the *Ungeziefer* emerges as a particular irritation: while it cannot be said to take care of itself and cannot be cared for, it insinuates itself as a pest into institutions and organizations to upset the very possibility of care and carefulness. The appearance of an *Ungeziefer* is taken as the sign or symptom of an endemic carelessness or a constitutive oversight. Where vermin appear the system of security is rendered unsure, and excessively so – an infestation is imputed. As such, the only prudent policy with regard to vermin, when they can no longer be simply ignored or kept out

---

[7] See Markus Jansen, *Das Wissen vom Menschen: Franz Kafka und die Biopolitik* (Würzburg: Königshausen & Neumann, 2012); Sarah Jansen, *"Schädlinge": Geschichte eines wissenschaftlichen und politischen Konstrukts 1840–1920* (Frankfurt am Main: Campus Verlag, 2003).

[8] On the specifically anti-Semitic resonance of such rhetoric, see Sander Gilman, *Franz Kafka, the Jewish Patient* (New York: Routledge, 1995); also Eric Santner, "Kafka's *Metamorphosis* and the Writing of Abjection," in *The Metamorphosis*, ed. Stanley Corngold; and Simon Ryan, "Franz Kafka's *Die Verwandlung*: Transformation, Metaphor, and the Perils of Assimilation," *Seminar: A Journal of Germanic Studies* 43.1 (February 2007): 1–18.

of view, is "to take care of" them in the terminal sense. They must be, as one tellingly says in German, "de-cared" – *entsorgt* – eliminated from the field of concern altogether.

It is the servant who ultimately takes care of Gregor – or what is left of him. She laughingly announces, as if reporting "a great good fortune" (*ein großes Glück*) to the family, "Look, you don't have to worry [*keine Sorgen machen*] about getting rid of the stuff next door. It's already been taken care of [*Es ist schon in Ordnung*]" (71). The family appears to be more irritated than relieved by the news that they need have no further worries. After all, they are anxious to put the monstrous fate of Gregor Samsa behind them by carefully avoiding the subject. There remains therefore one more thing to be taken care of: "She'll be fired tonight," states Herr Samsa after the servant has left (72). On account of her habit of indiscretion as much as for the dirty work that she so brazenly carries out, the servant is the last remnant and reminder of the vermin that has plagued the household since Gregor's transformation. Once she is "let go," the unhappy episode in the history of the family can be forgotten once and for all.

With the forgetting or de-caring of their great misfortune, the Samsas can indulge once again in an idyll of security. The epilogue of *The Metamorphosis* – a closing that Kafka later described as "unreadable" – presents a vision of a world overdetermined by sight or rather oversight.[9] The family takes the day off and heads out of town into the "open air on the outskirts of the city" (*ins Freie vor die Stadt*). En route they reflect in the light of the sun's comforting rays on the reassuring prospects (*die Aussichten für die Zukunft*) of their respective employments and on "the greatest immediate improvement in their situation," a plan to move from their apartment to something smaller but better situated. Finally, the parents exchange knowing glances (*durch Blicke sich verständigend*) as they realize how their daughter has blossomed "into a good-looking, shapely girl," for whom they should look for a suitable husband. This sequence of meaningful looks, outlooks, plans and projects on the part of the newly restituted family unit culminates when, as if confirming their dreams and intentions (*wie eine Bestätigung ihrer neuen Träume und guten Absichten*), upon reaching their destination (*Ziel*), "their daughter got up first and stretched her young body" (72–73).

The journey into the "open air" is not therefore an escape from the habitual expectations and exigencies of the everyday business of life. On the contrary, the much-needed break from work presents the opportunity to more vigorously plug themselves back into the system of the circulation of labor, of property, of bod-

---

9 Kafka, *Tagebücher*, 624.

ies – they do after all take "the electric" out of town and by no means go off the grid. The freedom of the Samsas, freedom as the Samsas understand it, is the freedom of circulation that includes, as an immanent aspect of its integrating logic, the possibility, occasionally, to take some "free time" from home and work and travel into the "open air" where they can comfortably reflect on the feasibility of their modest, entirely realistic and realizable, dreams and intentions. Far from being carefree, this idyll of security is composed of foreseeable cares that together constitute the petty bourgeois vision of the good life.

When Gregor Samsa wakes up one morning "from unsettling dreams" (3), he wakes up, I want to suggest, from *this* nightmare of his family's reasonable dreams and realizable intentions. And the whole transformation is a function of this awakening, in which Gregor creeps out of the institutions of security in which he finds himself inscribed. In this respect, the servant's disposal or *Entsorgung* of Gregor's decrepit body is not only a relief to the family, but also his liberation or *Ent-sorgung* from the attentions and exigencies of care. For it cannot be ruled out that Gregor too escapes albeit via the sewer or the refuse heap or perhaps by being simply chucked out of the window into the open air – *ins Freie*.

## Authenticity

Far from being, as the etymology of the word suggests, carefree or careless (*securitas*), modern security is taken up with the regulation of manageable cares based on the principle of the manageability of all cares.[10] If Gregor's transformation – so embarrassing for the family – was brought about because he became restless about the normative exigencies of security that saturate everyday life, he was by no means alone. On the contrary, a concern about security – about the perceived amnesia, apathy, and anesthesia of security – and above all the concern that security had forgotten or lost sight of *what one ought really to care about* was very much a preoccupation of the epoch. The security complex of modern life was reproached for lacking or obfuscating "great cares."

A generation earlier, Nietzsche had diagnosed the historical condition in which "the highest values devaluate themselves" as "nihilism." Far from being the disappearance of values or value-thinking, nihilism was rather the devaluing of those values that had hitherto posed as transcendent. "The aim [*Ziel*] is missing," Nietzsche continues in the famous fragment published posthumously by

---

[10] On the history of "security," see John Hamilton, *Security: Politics, Humanity, and the Philology of Care* (Princeton: Princeton University Press, 2013).

his sister, "'Why?' [*Wozu?*] finds no answer."[11] To readers of Nietzsche in the first part of the twentieth century, the emergence of the dispositive of security seemed to be a virulent form of nihilism – the nihilism that discredits the highest values in favor of an economy of manageable ones. Security ran the risk of obscuring *real* cares by reducing all cares to the level of insurable concerns. It presented the danger of an indifferent, functionalized, reified mode of existence in what was seen to be the increasingly regulated and uneventful milieu of everyday life.

One typical response, of the sort that Kafka thematized with the figure of the officer in *In the Penal Colony*, was to become the proponent of a discourse of authenticity – that is to say, the vision of a proper selfhood, and sometimes a proper community, based on a commitment, beyond any considerations of security, to authentic cares. The tendency, in other words, was to move in the direction of care – of more significant cares or of the intensity of care itself. In contrast, I want to suggest, Kafka explored in *The Metamorphosis* an unforeseeable alternative: *why care?*

Just over a decade after the publication of *The Metamorphosis*, Heidegger provided the following answer: because we are the kind of beings whose very being is caring. It is altogether revealing of the historical overdetermination of and preoccupation with care/security that in *Being and Time* (1927), *Sorge* – usually translated as "care" – is taken to be the meaning of being of Dasein. In his famous analysis of the inauthenticity of everyday life under the governance of the "one" (*das Man*), Heidegger describes a generic mode of existence taken up, as it is in the Samsa household, with the management of manageable concerns: "Everydayness takes Dasein as something at hand, that is taken care of [*besorgt*], that is, regulated and calculated [*verwaltet und verrechnet*]."[12] Forgetting the care that really defines him, Dasein is for the most part but a "fallen" functionary, or patient, of the security dispositive of the everyday.

The possibility of authenticity is the disclosure of a fundamental insecurity – "anxiety." When everyday concerns withdraw or lose their significance, it can be shown that Dasein still cares – about nothing. Heidegger takes this to be the fundamental care-structure (*Sorgestruktur*) of Dasein. It is not the case, therefore, that there are other more significant cares beyond everyday "security concerns";

---

**11** Friedrich Nietzsche, *The Will to Power*, trans. Walter Kaufmann (New York: Random House, 1968), 9. See Nietzsche, *Der Wille zur Macht 1884/88; Versuch einer Umwerthung aller Werthe*, ed. Elisabeth Förster-Nietzsche (Leipzig: Naumann, 1906), 11; compare Nietzsche, *Sämtliche Werke: Kritische Studienausgabe* (KSA), ed. Giorgio Colli und Mazzino Montinari, 15 vols. (Munich: Deutscher Taschenbuch Verlag, 1988), 12:350.
**12** Martin Heidegger, *Being and Time*, trans. Joan Stambaugh (Albany: SUNY, 2010), 277; compare *Sein und Zeit* (Tübinger: Niemeyer, 1967), 289.

rather Dasein itself takes the form of an unaccountable, unanswerable, and fundamentally insecurable *Sorge*. Authenticity, and here Heidegger's analysis presents the form of authenticity in general, is ultimately the conviction that caring matters (more than anything one cares about).

In *The Metamorphosis*, Kafka responds to the concern about the reification of everyday life by taking a different tack, one that goes beyond the human and certainly beyond what Heidegger would call Dasein, in order to pose the question of what it would mean not to care at all. Read in this way, *The Metamorphosis* presents an analysis of the possibilities of a singular but thoroughly inauthentic mode of being which I propose to call, using a term attributed, perhaps improperly, to Kafka speaking of *The Metamorphosis*, "indiscretion" (*Indiskretion*). The term is from Gustav Janouch's book, *Conversations with Kafka* (1951), which has become controversial for the questionable authenticity of the alleged conversations its author purports to recall.[13] In an early conversation, Janouch, for whom *his* Doctor Kafka ("mein Doktor Kafka") was always first and foremost the writer of *The Metamorphosis*, recalls asking whether *Samsa* was a cryptogram for *Kafka*, to which Kafka supposedly replied: "Samsa is not Kafka without remainder. *The Metamorphosis* is not a confession, although it is – in a certain sense – an indiscretion."[14] Without doubt, many are Janouch's indiscretions with regard to the dubious "conversations" – on more than one occasion, he even records himself apologizing to Kafka for the shamelessness of his insistent and prying questioning. In this context, however, far from presenting a shortcoming, I consider the questionable authenticity of Kafka's remark about "indiscretion" all the more felicitous. For it belongs to the very problem – one intimately to do with questions of propriety regarding Kafka's writings as well as writing about Kafka – that I take *The Metamorphosis* to be engaged in elaborating.

To outline the course of the argument that follows: Indiscretion, in accordance with the everyday usage of the term, can be defined as the inadvertent, or simply careless, transgression of the latent norms, limits, and inhibitions that regulate the cares that consume a normal life. Indeed, it is out of such indiscreet transgressions that these norms are first made – excruciatingly, or embarrassingly – explicit. Gregor's transformation challenges the habitual attitude of security

---

[13] Gustav Janouch, *Gespräche mit Kafka: Aufzeichnungen und Erinnerungen* (Frankfurt am Main: Fischer, 1951). For a scathing assessment of Janouch's book, and the authenticity of the alleged conversations it records, critical in particular of the second, expanded edition of 1968, see Eduard Goldstücker, "Kafkas Eckermann?" in *Franz Kafka: Themen und Probleme*, ed. Claude David (Göttingen: Vandenhoeck & Ruprecht, 1980).

[14] "Samsa ist nicht restlos Kafka. *Die Verwandlung* ist kein Bekenntnis, obwohl es – im gewissen Sinne – eine Indiskretion ist," Janouch, *Gespräche*, 29. Translation mine.

insofar as it is preoccupied with the self-conscious discipline of the body, with the cultivation of presence of mind, with the privileging of the attentions of consciousness, with discretion in expression and articulateness in speech, and, finally, with a guilty sort of conscientiousness. *The Metamorphosis* thus presents a reduction of care – as an ostensibly defining human capacity – to its supposedly most primitive or primordial features. Indiscreetly probing the limits of the human, the story indicates disparaged or forgotten aspects of the self and of the world that suffer abjection in order that a supposedly decent human existence might be secured.

But Gregor's indiscretion not only turns him into a monstrous vermin but also into a peculiar kind of *Schwärmer*. The danger of security, as Gregor runs the risk of demonstrating, consists not in a failure to properly account for authentically human cares but, on the contrary, in the tendency to reduce the world to the horizon of human concerns, as if it were only there for the sake of human life and only meaningful on account of human life. At such moments when he abandons his cares altogether, most notably in his apparent death, a different sphere of "peculiar" (*eigentümlich*) cares relating to the self and the world is disclosed. While in *The Metamorphosis*, the biopolitical concerns of security are always ultimately about a kind of sustenance – about feeding, eating, fulfillment, or substantial satisfaction – the other "cares" that emanate from beyond or outside of the dispositive of security impose themselves obscurely as questions of taste. This is why, as Gregor's confusion attests, the peculiar predilections of taste are always in danger of being mistaken for the exigencies of fulfillment, and why the realm of taste tends for the most part to be forgotten by the more pressing concern with consumption. Recalling Michael Kohlhaas and his related perplexities, one can say: taste concerns the givenness of the *vergeblich*. It discloses a peculiar horizon of concerns that are not coterminous with the properties and proprieties of being insofar as they are secured by and for human beings. Gregor's fallen and altogether abject *Schwärmerei* tends towards this insecurable, inhuman, and apparently futile realm.

## Who cares? / *Wen juckts?*

*The Metamorphosis* can be read as the story of what happens when an ordinary man wakes up one morning and finds, to his bemusement, that he does not care anymore – and has therefore turned into a monstrous vermin. To be sure, the transformation does not correspond to a responsible decision on Gregor's part – the decision not to care would still be a mode of caring – rather it takes the involuntary form familiar from the everyday, horrifying, but typically transient experience of indiscretion. If indiscretion is the eruption of monstrosity into the everyday – a humiliating and embarrassing experience which one is habitually careful to repress – then a chronic case of indiscretion turns the ordinary man into an *ungeheueres Ungeziefer.*

Kafka presents the disintegration of the care-structure that defined Gregor's life up until the transformation as a series of indiscretions or, rather, as a continually unfolding indiscretion that persistently exacerbates his monstrosity in a manner that at once frees and alienates him from the networks of care to which he had previously been bound. A passage shortly after his waking provides an exemplary analysis of this indiscretion:

> "Oh God," he thought, "what a gruelling job [*Beruf*] I've picked! Day in, day out – on the road. The stresses of doing business [*geschäftlichen Aufregungen*] are much worse than the actual business back home [*eigentlichen Geschäft zu Hause*], and, besides, I've got the plague of traveling, worrying about changing trains [*die Sorgen um die Zuganschlüsse*], eating miserable food at all hours, an ever-changing, never lasting, contact with people that never becomes truly intimate [*ein immer wechselnder, nie andauernder, nie herzlich werdender menschlicher Verkehr*]. To the devil with it all!" He felt a slight itching [*ein leichtes Jucken*] up on top of his belly; shoved himself slowly on his back closer to the bedpost, so as to be able to lift his head better; found the itchy spot covered with small white dots, which he had no idea what to make of [*die er nicht zu beurteilen verstand*]; and wanted to touch the spot with one of his legs but immediately pulled it back, for the contact sent a cold shiver through him. (4)

The paragraph is cut into two distinct but synchronous parts: an apparently conscious set of reflections that amount to a catalogue of complaints concerning the exhaustingly banal exigencies of his professional life, accompanied by a more or less unconscious exploration of his newly unfamiliar body. The disjunction between the discourse of his thoughts and the articulation of his body indicates the very lapse in which the indiscretion asserts itself.

As he fusses about the agitations or arousals (*Aufregungen*) of professional life, in contrast to what he refers to rather curiously as one's own proper business at home (*eigentliche Geschäft zu Hause*), and as he regrets the absence of intimate or heartfelt (*herzlich*) human intercourse (*menschlicher Verkehr*), he is at the same time attending to the itchy spot (*juckende Stelle*) on his lower body.[15] Whether the white dots there are supposed to refer to part of his body or a residue on his body, and whether in that case they are to be taken for a

---

[15] Concerns about a business trip that Kafka himself had to take plagued him during the writing of *Die Verwandlung*, see John Zilcosky, "'Samsa war Reisender': Trains, Trauma, and the Unreadable Body," in *Kafka for the Twenty-First Century*, ed. Stanley Corngold and Ruth Gross (Rochester Camden House, 2011).

not terribly discreet reference to masturbation or to some kind of more or less involuntary evacuation or discharge, whether they are not perhaps a kind of rash or infection – symptoms of the "plague of traveling"? – the point is that Gregor seems not to be quite sure himself.[16] He is not in a position to judge (*nicht zu beurteilen verstand*) the sticky situation in which he finds himself – and finds himself incompetent (not to mention incontinent) to determine the contours and orifices of his own body.

*The Metamorphosis* thus provides an instance of Kafka's ongoing literary interrogation of judgment going back to *The Judgment* (1912). The itchy spot presents a site of displacement or incapacitation of his power of judgment. A suitable word for this condition would be *Verlegenheit*. It is true that Gregor wakes up to find himself in an embarrassing situation but, bemused though he may be, he is not embarrassed at all. The appropriate term for this lapse in judgment, that is not even, or not yet, in a position to judge its own confusion, is indiscretion. Accordingly, with regard to the current indiscretion – and indeed to the entire transformation – the in other respects fastidiously apologetic and careful Gregor feels in no way obliged to find or make up an excuse. On the contrary: "To the devil with it all!" he thinks, recoiling with a shudder from touching the spot.

Gregor's transformation emerges when his power of judgment – his apparatus of secure discernments – is cut off from the body, which appears therefore cut up into an insect-like, yet nevertheless indiscreet, articulation of its own. While it is the body that first of all orients and positions the power of judgment, judgment is most immediately responsible for taking care of the body by judiciously regulating its articulations in accordance with prevailing conventions and practices. It is, in short, concerned with the normalization of the body, which is thus integrated into a system of security that is generally concerned with regulating its openings and closings, its ins and outs, its articulations and disclosures, with uniform coverings and discreet gesticulations, and all in all, with organizing the processes that constitute corporeality as an integral component of a broader system of circulation. Indiscretion upsets such careful arrangements. The power of judgment loses its bearing in a body that suddenly discloses itself in an unrecognizable and uncontrollable fashion. In thus failing to take care of

---

[16] For a reading of this irritation as a case of autoeroticism, in the context of turn of the century discourses about onanism, see Frank Möbus, *Sünden-Fälle: die Geschlechtlichkeit in Erzählungen Franz Kafkas* (Göttingen: Wallstein, 1994), 66–67. Foucault's lectures on the *Abnormal* focus on the related figures, all of whom seem indeed to converge in the figure of Gregor Samsa, "the great monster," "the incorrigible," and "the little masturbator," see *Michel Foucault, Abnormal: Lectures at the Collège de France, 1974–1975*, ed. Valerio Marchetti and Antonella Salomoni, trans. Graham Burchell (New York: Picador, 2003).

the self, indiscretion makes legible, on account of its monstrous transgression, the "political anatomy" of the human body – specifically, the average male body – as the achievement of a kind of self-governance.[17] The *Ungeziefer* is the figure that emerges out of the incommensurability between the out-of-joint judgment and the unregulated body.

It is a structural characteristic of an unfolding indiscretion that one does not recognize it until it is too late. The lapse of judgment is also a disconcerting lapse in time. It is not by chance that Gregor's initial agitation had to do with sleeping through his alarm and missing his early train. He is late for work and the efforts to make good on this unfortunate start to the day only exacerbate the disjunction between his body and his thoughts. Crucial is that Gregor cannot hope to regain his presence of mind without regaining command of his body. It is as if the articulation of his judgment can only operate with perspicuity when it is oriented in space and time by the regular articulations of his body. It is not, therefore, that Gregor fails to become fully conscious of his transformation but that the transformation itself is a disruption and deformation of his conscious, as of his bodily, composure. Indiscretion opens up a fault in the mapping that coordinates the anatomy of the consciousness and that of the body as a field of discrete – and discreet – articulations.

A fault it may be, but it is no-one's fault.[18] It is for this reason that Gregor's transformation appears – and first of all to Gregor himself – as if he were the victim of an alien power: "What's happened to me?" he asks upon waking (3). Indiscretion throws the distinction between passivity/activity as orchestrated by an attentive, conscious agency into question. Indeed, indiscretion brings about an action or a disclosure that is not directed by an agency, be it individual or institutional. If indiscretion is a failure to do what "one does," in Heidegger's sense, then it is equally a failure to recognize the unfolding indiscretion as one's own. And if indiscretion therefore exhibits a certain freedom from the "governance of the one," such transgressive activity does not assert itself as authentic self-expression, but quite the contrary, it betrays an unassimilable, irresponsible, uninhibited, "monstrous" exhibition of the self. Gregor has, as one says, let himself go.

Thus while Gregor remains – or thinks he remains – concerned about *Sorge*, even if only in the form of worrying about train connections (*die Sorgen um die*

---

[17] On the concept of "political anatomy," see Michel Foucault, *Discipline and Punish: The Birth of the Prison* (1975), trans. Alan Sheridan (New York: Vintage, 1995).

[18] On the ambiguity of action and accident in the text, departing from Freud's reflections on apparent accidents as instances of self-punishment, see Walter Sokel, "Kafka's 'Metamorphosis': Rebellion and Punishment," *Monatshefte* 48.4 (1956): 203–214.

*Zuganschlüsse*), he at the same time attends to an inexplicable itch, *Jucken*. In everyday language, *Jucken* is used metaphorically to refer to particularly pressing or irritating cares or concerns suggested in such colloquial phrases as: *Wen juckts? – Who cares?* Kafka's much-discussed literary practice of the literalization of metaphor applies here to the indiscrete uncoupling of *sorgen* and *jucken*. If *Sorgen* are metaphors, they are not just concerns about transport connections but modes of transport themselves. They generate correspondences between mind and world and provide the form and motility of intentionality. In the transformation, they are shown, however, to be no more than bodily agitations and irritations transported or transferred into a mental, metaphysical or metaphorical sphere. Caring is just itching – discreetly. By a process of metaphorization, it conceals, covers-over, and forgets the itching to which it responds.

But, as Stanley Corngold has shown, to speak in earnest of a literalization of metaphor is misleading insofar as it suggests a self-evident "return" to a non-metaphorical language.[19] Quite the contrary, as Kafka's practice demonstrates, the rendering "proper" of improper language produces an indecipherable monstrosity. Accordingly, the itchy spot, taken literally, does not expose the "body proper" and certainly not the "natural body" freed from the discernments of care. Rather it exposes the monstrous deformity of a "nature" that is not assimilable to, and is in fact the by-product of, the exigencies of security. In waking late to *Sorgen* that no longer move his body and to an itch that no longer corresponds to his *Sorgen*, Gregor finds himself transformed into an itching, scratching, crawling *Ungeziefer*.

The "aberrant literalization," to use Corngold's phrase, that characterizes Kafka's so-called mature writings since *The Judgment* is thus, in the case of *The Metamorphosis* at least, a practice of indiscretion. When Kafka speaks to Janouch of "indiscretion" regarding the story – if he ever did so in these words – he is not necessarily inviting a metaphorical, nor a referential, nor indeed any kind of encrypted reading (into which indiscreet friends and critics can forever delve more deeply and intrusively) but may well be providing a full disclosure. Indiscretion is not then the revelation of a hidden or repressed truth – it is nothing so profound. It is simply: disclosure without security. Indiscretion is the completely superficial failure to regulate the normal mechanisms of disclosure – and

---

**19** Stanley Corngold, "Kafka's *The Metamorphosis:* The Metamorphosis of Metaphor," in *Franz Kafka: The Necessity of Form* (Ithaca: Cornell University Press, 1988), and more recently, "Thirteen Ways of Looking at a Vermin," in *Ghosts in the Machine*, co-written with Benno Wagner (Evanston: Northwestern University Press, 2011). For an early and influential discussion of Kafka's practice of literalization, see Günther Anders, *Kafka, pro & contra: die Prozess-Unterlagen* (Munich: Beck, 1951), 40–41.

so to show discretion in matters relating to expression, communication, publication, etc. As such, however, it performs a certain – sometimes scandalizing, sometimes comical, sometimes off-putting – truth-function, for it discloses the inconspicuous norms and normative operations that latently structure what "one" cares about and how.

## Indiscreet Disclosures

Bodies, which are forever described with attention to their intakes and outbursts, their coverings and uncoverings (Gregor's mother even takes to sewing fine underwear to earn a living), are organized in much the same way in Kafka's text as other institutional spaces. If attention is, quite indiscreetly, paid to the way the contours, articulations, openings, secrets and secretions of the human body are regulated, the family home is likewise described with regard to the layout and articulation of walls, windows and doors. These in turn, as countless allusions to the textile business, the train system, banks, and so on suggest, participate in broader similarly regulated systems of circulation. It is significant in this respect that, apart from the servants, whose place, function, and corporeality within the household is of constant concern throughout the story, the representative figures who are called on for help are the locksmith (*Schloßer*) and the doctor (*Arzt*). It does Gregor good – although he very likely misinterprets this good feeling – when the two are called on by his worried parents with such certainty and assurance: "The assurance and security [*Zuversicht und Sicherheit*] with which the first measures had been taken did him good. He felt integrated into human society once again and hoped for marvelous and amazing feats from both the doctor and the locksmith, without really distinguishing sharply between them" (17). The semblance of security makes Gregor think that he has returned to the human fold. One has, however, to question his judgment on this as on all issues, given that in his indiscreet state he cannot exactly distinguish the doctor from the locksmith. In terms of the economy of the body and of the home as it is presented in *The Metamorphosis*, the doctor and the locksmith are representatives of the institutionalization of security insofar as they facilitate and maintain the openings and closings of human intercourse and circulation. They are each, in their own domain, in charge of the mechanics of significant dis-closings. It is little wonder, therefore, that Gregor in his current state would confuse them.

Indeed, it is precisely to prepare for a significant disclosure – one that will be put off until he literally opens the door with his mouth – that Gregor, expecting the imminent arrival of doctor and locksmith, clears his throat, "taking pains, of course, to do so in a very muffled manner, since this noise, too, might sound different from human coughing, a thing he no longer trusted himself to decide" (17). That he can no longer trust himself to decide between human and non-human sounds has to do with the alarming transformation of his voice that he notices when he first attempts to respond to the concerned call of his mother:

> Gregor was shocked to hear his voice answering. It was indeed unmistakably his own voice from before, but, as if from below, an irrepressible, painful chirping [*Piepsen*] intruded into it, which left the clarity of his words intact only for a moment really, before so badly garbling them as they reverberated that one could not be sure whether one had heard right. (6)

The chirping brings about a literal in-discretion of distinct words such that they can no longer be made out with certainty. When subsequently, prompted by his "manager," the *Prokurist*, Gregor gives a long and rambling excuse that amounts not so much to an explanation for his late start as to an apology for his very existence, the speech that he "hastily blurted out [*hastig ausstieß*], hardly knowing what he was saying," is altogether incomprehensible (15).

The *Prokurist* observes – and his word counts in such matters – "That was the voice of an animal [*Tierstimme*]" (16). Gregor's voice is not, however, a natural voice as if it returned to a primitive tonality prior to the discipline and regulation productive of human discourse. Rather, it is as if, like his *Ungeziefer* appearance, his voice betrays the disfigured, repressed, and painfully silenced residue of the humanizing process, which Gregor can no longer hold in check. For this reason it would be a mistake to contradict the judgment of the *Prokurist* and hear in Gregor's chirping an authentic mode of expression. Instead the "insistent distressed chirping" that can no longer be repressed is, as it were, the underside of the everyday language of concern that Heidegger calls *Gerede*. As the condition of its articulacy, *Gerede* requires a certain *Verschwiegenheit* (another word for *discretion*) – a disciplinary system of "silencing." And it is these silenced sounds, constrictions, closures, and other such "discretions" that habitually accompany everyday discourse, which "chirp up" from time to time in indiscretion. When the *Verschwiegenheit* that is the secret condition of discursive disclosure is broken or blurted out, it signifies nothing; it is the very repulsion of significance. This is why Gregor's incomprehensible excuse so irritates but also worries the *Prokurist*, who asks: "'Did you understand a word?' [...] 'He isn't trying to make fools of us, is he?'" (16).

The *Prokurist* is, as his title suggests, responsible for taking care of the business of someone else, in this case the company boss (*Chef*), whom he represents. He is also, however, as soon becomes apparent, the representative of institutional caring in general. The *Prokurist* operates according to a kind of hermeneutics of suspicion: every irregular closure or disclosure is a security concern, one that is in principle accessible to the probing techniques of procuration. Thus the *Prokurist*, in his official capacity, feels authorized to reveal in public the most private and intimate concerns and to behave, even in ostensibly private spaces, in the most presumptuous and intrusive manner. Accusing Gregor of causing them "serious, unnecessary worry" (*schwere, unnötige Sorgen*), the *Prokurist* immediately presumes to speak in the name of Gregor's parents as well as of his boss (14).

There is nothing that is not admissible to the prying eyes of the lowliest security representative. Gregor's father unhesitatingly tells him to open the door for the *Prokurist*, adding, "He will surely be so kind as to excuse the disorder of the room" (12). And it is his family members who are distressed, to Gregor's bemusement, when he refuses, "In the room on the left there was an embarrassed silence [*eine peinliche Stille*]; in the room on the right his sister began to sob" (13). It is this institutional prying and presumption – posing as concerned intervention in order to freely traverse the normal boundaries of discretion – that is the source of the perverse, "pornographic,"[20] pleasure associated with the excessive conscientiousness and obsessive attentiveness, in short, with the officiousness of such officials. Officiousness is officially sanctioned indiscretion.

The *Prokurist* cannot bear disclosure that is not an admission – an admission (be it spatial or moral) that therefore facilitates and legitimizes the interventions of procuration. This is why he is anxious that the "repulsion" of Gregor's animal voice may be a mockery of his officious position. When Gregor, having heedlessly (*besinnungslos*) bitten down on the key opens the door and reveals himself, the *Prokurist* does indeed make a fool of himself: "he heard the manager burst out with a loud 'Oh!' – it sounded like a rush of wind – and now he could see him, standing closest to the door, his hand pressed over his open mouth, slowly backing away, as if repulsed by an invisible, unrelenting force" (18). As John Hamilton has observed, the moment marks the dissolution of the *Prokurist* in his representative caring function – he is driven to abandon his office out of an altogether "unofficial" concern for his personal security.[21] But the repulsive appearance of the monstrous *Ungeziefer* causes first of all a crisis of procuration

---

[20] On the pornographic character of the law in Kafka and the eroticism that motivates its representatives, see the chapter, "Eros, Macht und Gesetz: Der Verkehr der Behörden," in Rainer Stach, *Kafkas erotischer Mythos: Eine ästhetische Konstruktion des Weiblichen* (Frankfurt am Main: Fischer, 1987).

[21] See John Hamilton, "Procuratores: On the Limits of Caring for Another," *Telos* 170 (Spring 2015): 15. Hamilton reads the *Prokurist* as representative of the logic of delegation, specifically the delegation of care for the sake of security, characteristic of liberalism.

itself. Not the abandoning of his office but the desertion of his officiousness contributes to his comical loss of composure in a pantomime of discrete bodily gestures as he flees the Samsa home. Upon reaching the foyer, he stretches his right hand towards the stairs "as if nothing less than an unearthly deliverance were awaiting him there," and shortly is described "ridiculously holding onto the banisters with both hands," while his still open mouth lets out sounds – "Oh!" "Argh!" – involuntary expirations of an unmonitored body (21–22).

Gregor's indiscretion is the undoing of procuration. With the *Prokurist* expelled, it will now be a matter of finding an accommodation – although "there could never, of course, be any question of a complete adjustment" (*von vollständiger Gewöhnung*) – within the home and family for dealing with the great misfortune (31). Henceforth, to prevent further disclosures, a rather crude security measure is resorted to: Gregor is locked up in the home. The expulsion of the *Prokurist* does not mark the moment of liberation from the relations of care, but on the contrary marks the beginning of a period of incarceration and of neglect by those, his family, who care the most – as they are supposed to.

Neglect, as it is exhibited by the family over the coming months, is not a failure to care, but a deficient mode of caring. And even if, given the circumstances, this neglect is understandable – "Who in this overworked and exhausted family had time to worry about Gregor any more than was absolutely necessary?" (52) – before long it turns into positive abuse. While the family fulfills its obligations not to *not* care, Gregor remains entrapped in the household of concern. He is stuck in a kind of limbo that oscillates between a confused plan to return to the human world of everyday concerns, which he retains as a vague memory of the past, and the *schwärmerisch* prospect of abandoning his human cares and so perhaps his humanity once and for all.

## Carefree Comportments

The drama of the expulsion of the representative of security from the Samsa home is accompanied by Gregor's apparently conscientious efforts to detain the *Prokurist*, a gesture of solicitude that comes across as threatening and so only hastens his departure. In the attempt to execute this plan, Gregor is distracted by the sight of a more immediate, but in fact merely incidental and irrelevant, concern. His screaming mother knocks over a pot of coffee: "'Mother, Mother,' said Gregor softly and looked up at her. For a minute the manager [*Prokurist*] had completely slipped his mind; on the other hand he could not refrain [*versagen*] from snapping his jaws several times in the air at the sight of the spilling coffee" (22–23). The scene that started with the disfiguration of Gregor's voice ends with the snapping of his jaws. The snapping, which may well be Gregor's failure to say what he thinks he is saying softly, namely, *Mother*, in fact repulses his mother as well as the *Prokurist*, who, not without reason, interpret it as aggression. Insofar as it concerns the privileged site of opening and closing – the mouth – the whole scenario of indiscretion has been one not simply of *versagen* in the habitual sense of "failure" or "omission," but rather of a specific failure or infelicity in the relation of speaking and acting, *ver-sagen*. It thus describes the lapse or disjunction between the still-concerned, albeit thoroughly inapt, discourse of his thoughts and the indiscreet gestures of his body. It is not just that Gregor fails to speak and so misspeaks. He cannot fail to mis-speak – as displayed in the culminating expression of snapping. The indiscretion thus consists in the failure of not being able to fail to misspeak (*nicht versagen können*) by, for example, cultivating a discreet and inexpressive silence (*Verschwiegenheit*). One by one, all of the inhibitions that are the condition of his human capabilities begin to fail – a failure, however, that perhaps occasions an alternative mode of disclosure that is revelatory of a different kind of openness or "Freie."[22]

This ambivalence is captured in his attitude to the window of his bedroom/cell. Early in his incarceration he takes to looking out the window, "evidently in some sort of remembrance of the feeling of freedom he used to have from looking out the window" (36). The view is no longer liberating because the world of his circumspection has undergone a contraction figured first of all in his reduced capacity of sight. He can no longer make out the hospital across the street, "and if he had not been positive that he was living in Charlotte Street – a quiet but still very much a city street – he might have believed that he was looking out of his window into a desert where the gray sky and the gray earth were indistinguishably fused" (36). This desertification of the urban world, culminating in the indifference of the earth and the sky, is doubtless a function of incarceration. The freedom idealized by security that he had once associated with looking out and with outlooks – the freedom of "das Freie vor die Stadt" – is closed off. But the reduction of his field of concern and the shutting out of the prospect of the freedom it promises makes room for experimentation in a new kind of relation to body, time, and space – adapting to a new comportment (*Haltung*) that seems to be, at times at least, both carefree and careless.

---

[22] On "the open" – so central to Rilke's poem and subsequently important to Heidegger and then Agamben – see Leland de la Durantaye's critical overview, "The Suspended Substantive: On Animals and Men in Giorgio Agamben's *The Open*," *diacritics* 33.2 (2003); an historical contextualization of Rilke's preoccupation with "the open" is also the point of departure for Eric Santner, *On Creaturely Life: Rilke, Benjamin, Sebald* (Chicago: University of Chicago Press, 2006).

Earlier, before Gregor's sight began to fail, he congratulated himself on having the foresight (*Voraussicht*) to try to prevent the *Prokurist* from leaving the house. But, in accordance with the logic of *Versagung*, his concern with winning over the *Prokurist* slips up, "groping for support [*nach einem Halt suchend*], Gregor immediately fell down with a little cry onto his many little legs" (22). His human concern is accompanied by a failing attempt to assume a human comportment that sees him fall flat on the floor. If the next moment he suddenly, and for the first time that morning, enjoys "a feeling of physical wellbeing" (22) and finds himself in command of his body, this is only because he has altogether abandoned or lost sight of the attempt to stand upright, to be understood, and to consciously, not to mention conscientiously, take care of himself. This lack of attention or *Rücksichtlosigkeit*, to use another term for indiscretion, is the condition of his newfound comportment. He has adopted a horizontal and for that reason altogether indiscreet and inhuman posture such that while he interprets – mistakenly of course – his good feeling as a good sign, "that final recovery from all his sufferings was imminent," his mother in that very moment, "as he lay on the floor rocking with restrained motion" (*schaukelnd vor verhaltener Bewegung*), is horrified by this newly adopted, and all the more monstrous, stance (22). Reflexively, she leaps away – spilling the coffee.

The *Ungeziefer* that cannot be depicted, nor even indicated from a distance,[23] invites comparison with the photograph Gregor notices when he shows himself for the first time, framed by the door of his room: "On the wall directly opposite hung a photograph of Gregor from his army days, in a lieutenant's uniform, his hand on his sword, a carefree smile on his lips [*sorglos lächelnd*], demanding respect for his bearing [*Haltung*] and his uniform" (19). Abandoning the ideal image of upright conduct suggested in his posed smile and imposing uniformed posture, Gregor's transformation – his naked monstrous body, abject crawling comportment, and especially the absence of his hands, for he has only legs and feelers – presents a thoroughgoing inversion, or rather perversion of the image.[24] The attitude of indiscretion thus presents a kind of "counter-conduct,"

---

[23] "Das Insekt selbst kann nicht gezeichnet werden. Es kann aber nicht einmal von der Ferne aus gezeigt werden" (The insect cannot be drawn. It cannot even be indicated from a distance), letter to Kurt Wolff Verlag regarding the proposed cover illustration for *The Metamorphosis* (October 25, 1915), Kafka, *Briefe 1914–1917*, ed. Hans-Gerd Koch (Frankfurt am Main: Fischer, 2005), 145.

[24] For a discussion of the crisis of masculinity – Gregor as a man who fails to be manly – in relation to this photograph, see Elizabeth Boa, "Creepy-Crawlies: Gilman's *The Yellow Paper* and Kafka's *The Metamorphosis*," *Paragraph* 13.1 (1990): 19–29; also Santner, "Kafka's *Metamorphosis* and the Writing of Abjection."

one that indeed upsets the very idea of conduct insofar as the term suggests a composed form of self-governance.[25]

The overturning of the ideal of upright and respectable comportment reaches its high-point when, out of distraction regarding his incarceration, Gregor starts crawling all over the walls of the room and especially the ceiling. He takes to hanging there upside-down: "He especially liked hanging from the ceiling; it was completely different from lying on the floor; one could breathe more freely; a faint swinging sensation went through the body" (39). An oscillation – *Schwingen* – without aim or orientation, light and subtly playful, he breathes freer, liberated from the "spirit of gravity," to quote Nietzsche's Zarathustra, whose polemical vocabulary seems, almost naively, to be recited here.[26] By no means instances of transcendence, and certainly not flights of fancy, these moments of suspension are instead described as bringing about an "almost happy absent-mindedness" (*fast glücklichen Zerstreutheit*, 39).[27] In the midst of his incarceration, such moments suggest a kind of freedom that is not outside the enclosed room, not modeled on an open space in which one would be free to move, but is instead an opening up in the mode of a fortuitous disaggregation of "human room" (*Menschenzimmer*), to use a strange term from the first page of *The Metamorphosis*, the "somewhat too small" (*etwas zu klein*) space that one habitually occupies as the functional condition of one's own composure (3). The almost happy loss of composure of the self is thus a near liberation from the transcendental constraints of human room.

Gregor lets himself go – literally – to the point indeed that from time to time "it could happen to his own surprise that he let go and plopped onto the floor" (23). Such falls no longer do any harm, however, for "he now naturally [*natürlich*] had control of his body in a very different way [*ganz anders in der Gewalt*]" (39). This change in the governance of his body is a function of the loss of his conscious powers of concentration. And while it is by no means a "natural" relation to his body that he now enjoys, the change is a matter of course – *natürlich* – insofar as he is accustoming himself to a physical and gestural embodiment and corresponding environment that comes at the cost of human determinations – *Sorgen* – to which he had hitherto been subject. In such a state of almost happy distraction, it is no longer clear whether one can say that Gregor is really

---

[25] See Foucault, *Security, Territory, Population*, 191–226.
[26] See Friedrich Nietzsche, *Thus Spoke Zarathustra: A Book for All and None*, ed. Adrian Del Caro and Robert B. Pippin, trans. Adrian Del Caro (Cambridge: Cambridge University Press, 2006), 153–156.
[27] On the question of distraction in Kafka, see Paul North, *The Problem of Distraction* (Stanford: Stanford University Press, 2012), 74–108.

*there* – his Dasein has more or less dissolved into a pleasurable sense of embodied non-composure. Until he is called back to himself by the natural power of gravity, he is nearly free of cares. For the most part, however, Gregor does not so easily let go of his former cares – more often than losing himself in near happy distraction, he gets worked up into a sorrowful confusion.

## Bites of Conscience: Guilt, Debt, Hunger

"'Life' is a 'business' [*Geschäft*], whether or not it covers the costs," writes Heidegger in *Being and Time*, ironically ventriloquizing the idle talk of *das Man*.[28] In his analysis, "one" obscures the existential possibility of authenticity disclosed in guilt by converting such concerns into the improper business of debt. Kafka, in contrast, "literalizes" this fallen discourse; in *The Metamorphosis*, all talk of *Schuld* turns out to be just debt. Ironically – but perhaps precisely in this respect he is all the more typical – Gregor committed himself to a "life as business" out of a conscientious determination to transcend it, a longing for authenticity, that has only entangled him more deeply in the world of debt.

Since his father's bankruptcy, Gregor's overriding concern had been with taking care of the family and paying off the family debt. This well-intentioned gesture on Gregor's part – assuming the family debt as though it were (his) guilt – soon, on account of its inherently business-like character, instrumentalizes the warm and authentic familial relations he had sought thereby to affirm and to preserve. After the initial excitement at his promotion and rise in income, the family habituates itself to the new state of affairs: "the money was received with thanks and given with pleasure, but no special feeling of warmth went with it anymore" (33). Gregor's aspiration for an authentic relationship beyond the cold calculations of business is also his motivation for capitalizing on the "innocent" (*unschuldig*) remarks of his sister about attending the conservatory. He looks forward to announcing at Christmas his intention to pay for her studies, "regardless of the great expense involved" (33). But even such generous gestures are poisoned by the ineradicable implication of debt from which there seems to be no escape. Debt has imposed itself as the common denominator of all cares.[29]

---

**28** Heidegger, *Being and Time*, 267.
**29** François Ewald, considerably less ambivalent about the emergence of security than Foucault, nevertheless describes with some reservation the new understanding of "the social" around 1900, as, essentially, social security, based on a fundamental relation of indebtedness that touches every aspect of one's being, *L'état providence*, 693. Drawing on Ewald and Foucault, Benno Wagner relates the preoccupation with *Schuld* in Kafka's world to the emergent under-

His only hope, therefore, a hope he has still not given up the morning he wakes to find himself turned into an *Ungeziefer*, is to finally pay off the family debt and thereby "make the big break" (*den großen Schnitt*) that would free him from his employer and traveling job – as well as from his obligation to his family, or perhaps simply from his family altogether (4).

Of course, on the morning of his transformation, he fails to notice that a decisive cut has already been made – as indicated by the in-sect like articulation of his body. It is equally indicated by the dissolution of conscience as demonstrated in Gregor's seemingly guilty reflections when the *Prokurist* arrives. Complaining of being condemned (*verurteilt*) to serve a firm where the slightest lapse (*Versäumnis*) is seized on with the greatest suspicion, he asks himself:

> Were all employees louts without exception, wasn't there a single true, dedicated human being [*treuer, ergebener Mensch*] among them who, when he had not fully utilized a few hours of the morning for the firm, was driven mad by pangs of conscience [*Gewissensbissen*] and was virtually unable to get out of bed? (11)

The thought proceeds, or rather unravels, against its own line of inquiry to self-condemnation. Confirming, if not introjecting, the suspicions of the *Prokurist* that were supposed to be called into question, Gregor concludes that all employees are indeed guilty good-for-nothings insofar as they are debilitated by the gnawing of their own consciences.

As he is dimly aware, his judgment is compromised by the transformation he is trying to make sense of and the whole consideration is patently nonsensical. The bites of what Gregor takes or mistakes to be his conscience prevent him from going to work and thus bring about, paradoxically, an abandonment of his obligations, investments, and everyday concerns. The incisive interventions of his conscience thus indiscreetly betray what they ought to affirm, that *there is nothing to be guilty about* – and so Gregor finds himself, painfully and confusingly, without grounds to get up in the morning. If the conscience is supposed to govern the machinery of caring in the self, if it is the last instance and insistence of care in the self, then, like the machine in *In the Penal Colony*, it seems to disintegrate by its own rigorous and ultimately ridiculous execution. No longer sure of itself, the "conscience" eats away at the formerly conscientious Gregor – and accordingly the "true, dedicated human being" turns into a monstrous *Ungeziefer*.

---

standing of the society of security as a distribution of mutual obligation understood not on the liberal model of responsibility but on the insurance model of risk, "Kafkas Phantastisches Büro," in *Kontinent Kafka: Mosse-Lectures an der Humboldt-Universität zu Berlin*, ed. Klaus Scherpe and Elisabeth Wagner (Berlin: Vorwerk 8, 2006), 106.

Waking up unsatisfied, Gregor is eaten away by a hunger for something more fulfilling – or perhaps he is just hungry. The gnawing crisis of conscience introduces among other things an uncertainty about its proper place and function, which expresses itself as an ambivalence as to whether it is a metaphysical or a natural phenomenon, to do with the transcendence of the spirit or the concretion of the body, and ultimately, then, whether it is really a conscience or just a bite. The confusion between conscientious *Sorgen* and mere hunger is a persistent preoccupation throughout the story. On the one hand, Gregor thinks he is concerned about the family, recollecting good times and anxious about what the future might bring. But on the other, and often at the very same time, he is vaguely aware that this might all just be, as it were, the hunger speaking – or rather biting and snapping. The first night, for instance, he spends, "partly in a sleepy trance, from which hunger pangs kept waking him with a start, partly in worries [*Sorgen*] and vague hopes" (28). Later, he plans to "take charge of the family's affairs again," but the next moment finds himself in no mood "to worry about his family" (*sich um seine Familie zu sorgen*) and fantasizing about raiding the pantry like the vermin he has become, even though "he could not imagine anything that would pique his appetite" (53).

Gregor's hunger is a perversion of care and his ostensible *Sorgen* are perversions of hunger. It is not the case, therefore, that in the form of hunger Gregor returns to his biological or primal instincts and so to a state of pre-care or "primitive" care. Rather his indiscreet behavior is always a decadent and indecent hybrid of the biological and the cultural, indeed, it is a by-product of the biopolitical techniques that no longer recognize the traditional distinction between nature and culture. Gregor's vermin-like-behavior might be better classified therefore as the exhibition of unnatural instincts and inhuman concerns.

## Family Ties

The scene in which Gregor's tenuous family membership reaches a crisis point revolves around an ambivalent gesture that relates to food and guilt, falling and feeding, communion and excommunication. The father, upon receiving a curt and, to Gregor's mind, misleading report from Grete – that the mother had fallen unconscious (*Ohnmächtig*) and that Gregor had broken out (*ausgebrochen*) – takes matters in hand, throwing apples at his *Ungeziefer*-son. In defending his family from the immediate threat, it is as if the father is at the same time attempting to reassert the genealogical tradition of guilt that Gregor's transformation had thrown into question. To be sure, one could insist that the father is just throwing apples – literal apples – that happen to be at hand. But there is no such thing as a literal apple. This is not to say that the apple serves as a metaphor of guilt – specifically, for the origin and genealogical transmissibility of guilt. Rather, it is no-longer-a-metaphor of guilt and communicates, therefore, the historical sense, which pervades Kafka's work, of the failure of the tradition of guilt. That guilt has lost its force and determinate significance, a fact that eminently threatens the institution of fatherhood which in the story is in a laughable state of dissolution, drives the father to resort to such violent but also meaningless projections. For while the apples strike Gregor causing a "startling, unbelievable pain," he does not get the message, that is to say, he does not feel guilty, but is left rather "in a complete confusion of all his senses" (49).

This morsel combat is not just a family fight or a war of the generations but one in which the very existence of the family as the inter- and trans-generational institution par excellence is at stake. To the extent that the securing of authentic commonality and continuity over time is based on the genetic transfer of a conscientious bond or debt, be it biological or metaphysical, Gregor's transformation poses an existential threat to the very fabric of the family. Ultimately, however, the "modern family" proves to be bound neither by the inheritance of blood nor of guilt. Instead the distinctive character of familial care is recalled by the sight of the wound inflicted by an apple embedded in Gregor's back:

> Gregor's serious wound [...] seemed to have reminded even his father that Gregor was a member of the family, in spite of his present pathetic and repulsive shape, who could not be treated as an enemy; that, on the contrary, it was the commandment of family duty to swallow their disgust and endure, nothing but endure [*zu dulden, nichts als zu dulden*]. (49)

If Gregor, for a time at least, returns to the family fold on account of his wound, this is perhaps because it acts as a reminder of the distinctive modality of care that constitutes the family bond. Far from being based on an originary or organic debt of solicitude, the family bond consists first and foremost in "enduring" or "putting up with" one another and doing so without reprieve – *nichts als zu dulden*. *Dulden* is family-care, defined as a particular relation of suffering over time. Specifically, it involves the obligation to exercise, continuously, a kind of patient lassitude with regard to family-members irrespective of their irresponsible, unlicensed, or distasteful behavior.

*Dulden* emerges as a security concept. Indeed, it may be construed as its very principle, insofar as security, as Foucault describes it, is oriented towards economically regulating the broadest possible spectrum of concerns – the entire population in its milieu – in a manner that aspires to treat the greatest feasible deviance and diversity. *Dulden* presents therefore the minimal, but also the almost inescapable, mode of the caring that constitutes social and political life.

The family presents in a society of security, as it does in *The Metamorphosis*, the exemplary care institution but also the institution of last resort – the depository responsible for discretely dealing with the most egregious cases of abnormality.

Even *dulden*, however, has its limits. At a certain point the repulsion (*Widerwillen*) becomes such that it can no longer be swallowed (*hinunterschlucken*) – although it remains unclear whether this is a physiological reflex or a moral one. Of course, it was first of all Gregor who could no longer keep down his *Widerwillen* – neither the solicitous platitudes of secure family life nor the food of his incarceration – indiscreetly disgorging all. And this failure to show not positive solicitude (*sorgen*) but patient lassitude (*dulden*), transforms him into something truly monstrous – *infamiliaris* – a positive threat to the family in the form of a verminous figure that is more than the family itself can stomach. The wound inflicted by his father with the apple presents, for a time at least, the repulsion he has to swallow as the traumatic condition of family life.

The family, for its part, puts up with it all because that is what is expected. When the sister finally declares that Gregor cannot be borne any longer, her decision is cast in terms of the anticipation of reproach:

> "I won't pronounce the name of my brother in front of this monster [*Untier*], and so all I say is: we have to try to get rid of it. We've done everything humanly possible [*das Menschenmögliche*] to take care of it and to put up with it [*es zu pflegen und zu dulden*]; I don't think anyone can blame us in the least." (63–64)

The decision on the limits of family membership is also, given the foundational role of the institution of the family, at once a decision on the limits of humanity, indicated here by the use of, or refusal to use, Gregor's name. And it emerges tellingly in the sister's declaration as a function of the degree of care or rather patience (*Geduld*) shown – they have done the humanly possible.

The limit of humanity is presented as a function of the human capacity for care, which in turn seems to be measured against the vague norms of opinion or potential reproach. Any entity – even a brother – that neither shows nor suffers human care, exhibiting neither solicitude nor patient endurance (*Geduld*), and cannot, therefore, be so much as put up with, loses in turn the right to be spoken of by name and must, for the security of the family and society alike, be gotten rid of. Thus the question of humanity is not structured according to the distinction human/animal but according to a functional model that determines belonging to the human milieu on a sliding scale of the human-animal. The scale is graded in terms of the feasibility and effectiveness of, and at the very least with a measure of patience or passivity (*Geduld*) with regard to, the interventions of care. The question of humanity has thus become a functional and administrative question of "humanitarianism."

What cannot be put up with is the *Untier*, to use Grete's carefully selected term. Like the *Ungeziefer*, it refers to a being that has no proper place – neither by rights nor by nature – in the human milieu. Insofar as it remains indifferent to any humanly possible measure, it marks, as Grete insists, the legitimate, indeed, necessary end of all humane efforts. Thus, while it may be tempting to read *The Metamorphosis* metaphysically, as a story of the Fall, or naturalistically, in the light of contemporary developments in the natural sciences, as an account of "the descent of man," the story is rather an elaboration of a peculiar figure: an inhuman animal and unnatural human that appears at the site of the confusion of these two interpretations of the original or primordial meaning of care.

The *Ungeziefer* is the political-zoological exponent of an aspect of nature (including so-called human nature) that is treated as unnatural insofar as it cannot be secured by human measures. It is not, however, for this reason representative of an innocent or untouched nature outside of and inaccessible to human exploitation. On the contrary, it emerges deformed and debilitated in the midst of the human milieu as what has been rejected and rendered useless, the *Unrat* (refuse) of the economy of human concerns. For a time at least, Gregor's room is turned into a refuse heap – a site within the household in which *Ungeziefer* and *Unrat* could be put up with, insofar as they were kept out of sight, as unpleasant but necessary and never altogether eliminable by-products of everyday life. Thus there is after all nothing natural about an *Ungeziefer*. It is rather what is left of nature as a result of the interventions and aspirations of security, once the humanly possible has been exhausted.

## Dying from Taste

When his sister finally denounces him for his endemic indiscretions, refusing to call Gregor by his name, she terminates a movement that has characterized his transformation from the start – the thoroughgoing loss of all that, in everyday discourse at least, one takes to define one's own self. Calling him an *Untier* consigns him to a seemingly generic category of disparagement, ostracism, and abjection. But the privative and exclusionary terms, *Untier* or *Ungeziefer*, do not refer to general categories or common nouns. *Ungeziefer* are not simply excluded from the taxonomy, legal as well as natural-scientific, of living beings; they are rather characterized by the trouble they cause the normative process of classification, for *Ungeziefer* have no common properties apart from the rather indiscreet habit of indulging in improprieties. In other words, while the term *Ungezie-*

*fer* is certainly a derogatory de-nomination, exercised by the paternalistic, procuratorial, or familial powers that be, the criterion for the *Ungeziefer*'s ostracized status is its singularity – or rather, its peculiarity. Thus, the so-called *Ungeziefer* is not a merely passive determination. It is in a certain sense defined or scrawled in its "own" improper terms – or, rather, according to its peculiar taste. In short: Gregor as *Ungeziefer* cannot just be written off as "vermin," he also presents an inscrutable kind of *Schwärmer*.

When his sister brings him an assortment of food the morning after his transformation, she thoughtfully, as he sees it, puts out both fresh and rotten stuffs for him to sample, including some cheese that two days before he had declared unpalatable (*ungenießbar*). In his transformed state he is immediately drawn to the cheese and sucking greedily on it, in one of the few really comic moments in the later part of the story, Gregor wonders: "Have my senses become less refined?" (29). He has indeed lost the discrimination for what passes in his milieu for good taste – and as a result his emergent sensibilities, which he cannot fail to express, come across as distasteful, if not altogether repulsive, to those around him with whom he had hitherto shared a *sensus communis*.

Some indication of Gregor's questionable taste is given by the picture on the wall of his room, which is the first thing – perhaps on account of that indeterminable itching – that he looks at upon waking before looking out of the window opposite. It depicts a woman in furs that Gregor had recently cut out of an illustrated magazine and put "in a pretty gilt frame" (3). Gregor had made the frame himself. His mother remarks on this pastime – fretwork – to the *Prokurist*, stating that it is his only distraction from worrying about work. She adds, drawing attention to the frame, "You'll be amazed how pretty it is" (12).

In this regard, the picture is functionally related to the window opposite insofar as both seem to represent a certain relation to freedom as Gregor conceived of it from within the confines of his somewhat too small *Menschenzimmer*. They are representations of freedom that complement but also, in being closed off from the space in which he actually lives, aggravate his captivation by everyday concerns. Gregor is free to indulge his risqué if perhaps clichéd tastes in photography as long as he keeps it all contained in a pretty gilt frame in his room. The frame is the operator of the acceptability of the picture, regardless of its source or content. And one is inclined to suggest that Gregor's transformation begins when all the fretting that goes into his fretwork is abandoned in favor of indiscreet itching and scratching – and an altogether more fluid relation to the frames and compartments that had defined his former, human, life.

After the transformation, when his sister proposes moving all the furniture out of his room so that he may crawl about more freely, his mother worries that it will be a sign that they have given up altogether on a return to normality.

And indeed, Gregor is suddenly recalled by his mother's concerned words to his human cares, wondering: "Had he really wanted to have his warm room, comfortably fitted with furniture that had always been in the family, changed into a cave, in which, of course, he would be able to crawl around unhampered in all directions but at the cost of simultaneously, rapidly, and totally forgetting his human past?" (41). In the midst of this de-humanizing emptying out of his *Menschenzimmer* – "They were clearing out his room; depriving him of everything that he loved" (43) – Gregor, worked up into a state of distraction, breaks out from his hiding place ostensibly to try and save something that he cares about from his human past:

> He really didn't know what to salvage first, then he saw hanging conspicuously on the wall, which was otherwise bare already, the picture of the lady all dressed in furs, hurriedly crawled up on it and pressed himself against the glass, which gave a good surface to stick to and soothed his hot belly. At least no one would take this picture away now that Gregor completely covered it. He turned his head toward the living-room door to watch the women when they returned. (44)

Mark Anderson describes this scene as a record of a great and groundbreaking piece of avant-garde performance art.[30] Indeed, there can be no more fitting statement of Gregor's transformation in taste and judgment than this indiscreet aesthetic, or anti-aesthetic, performance. He does not save an aspect of his old *Menschenzimmer*, nor does he allow it to be turned into a hollow cave; instead he intervenes, by his singular exhibitionism, to turn it into an exhibition space for a new scandalizing experiment in taste. For, as he takes pleasure in pressing his hot belly against the cool glass, he does not simply cover the picture of the woman in furs, with which he enters into a sticky promiscuity, but also, and more disturbingly, the "pretty" frame.

When his mother returns to the room, she "caught sight of the gigantic brown blotch on the flowered wallpaper" (45) – a monstrous demonstration, unframed, de-framing, that ruins not just the decor but the cozy (*gemütlich*) coher-

---

[30] Mark Anderson, *Kafka's Clothes: Ornament and Aestheticism in the Habsburg fin de siècle* (Oxford: Clarendon Press, 1992), 123. Anderson is perhaps mistaken to make performance characteristic only of contemporary art. Kafka was familiar with avant-garde artistic practices that could be construed as early iterations of performance art. Notable in this regard is Else Lasker-Schüler, for whom Kafka professed a profound antipathy, letter to Felice, February 12/13, 1913, in *Briefe 1913–14*, ed. Hans-Gerd Koch (Frankfurt am Main: Fischer, 2001), 88. On her 1913 visit to Prague, where she performed a reading as well as staging a poetic spectacle in the streets that attracted the attention of the police, see Hartmut Binder, "Else Lasker-Schüler in Prag," *Wirkendes Wort* 3 (1994): 405–438.

ence and compartmentalization on which his mother's delicate sensibilities rely. She really cannot handle the provocation and, before she even becomes fully conscious that the hideous brown blotch is Gregor, she cries out and faints. As Anderson observes, "No avant-garde artist of the modern period could ask for a more satisfying public response"[31] – although it is perhaps not strictly satisfaction but something less fulfilling that Gregor's exhibitionism is after.

It is significant, also in the light of her final gesture of renunciation, that his sister, on this occasion, addresses him directly for the first and last time calling him by name: "'You, Gregor!' cried his sister with raised fist and piercing eyes. These were the first words she had addressed directly to him since his transformation" (45). To be sure, it may just be an "anthropomorphizing" reflex. But it may be that she, who seems to have known him best, sees in this performance something that does not refer to Gregor's "normal" behavior but is nonetheless singularly, and to her unpleasantly, Gregor-like. As if this most inhuman and indeed *Ungeziefer*-like action betrayed something peculiarly characteristic about Gregor, something indeed that his transformation may be working out or working towards – and that is ultimately more than Grete can put up with – namely, his *Eigentümlichkeit*.

In an unpublished, quasi-autobiographical fragment written in 1916, Kafka opens what amounts to his most programmatic reflection on the problem of singularity and the institutional normalization of modern life with the statement: "Each human is peculiar [*eigentümlich*] and, by virtue of his peculiarity, called to act, but he must develop a taste for his peculiarity."[32] Since a really singular taste, by definition, cannot but appear anomalous to any commonly held standards of judgment, it will always appear indiscreet. Thus being indiscreet, which is always in bad taste, may indicate not a lack of discrimination but the insistence of a singular or, as I would suggest, in order to capture the abnormality implied in Kafka's use of the term *eigentümlich*, a *peculiar* one. Indiscretion could thus present a tactic – to be sure a thoroughly tactless one – to break with the standard inhibitions of the normal and generate an exhibition space in which one might be free to develop a taste for one's peculiarity. In so indiscreetly concerning himself with his *Eigentümlichkeit*, Gregor engages in the abject and unedifying kind of *Schwärmerei* to which, in response to the stifling normative exigencies of institutional life, the figures I call "peculiar fanatics" are driven.

---

31 Anderson, *Kafka's Clothes*, 123.
32 "Jeder Mensch ist eigentümlich und kraft seiner Eigentümlichkeit berufen zu wirken, er muß aber an seiner Eigentümlichkeit Geschmack finden," Kafka, *Nachgelassene Schriften II*, ed. Jost Schillemeit (Frankfurt am Main: Fischer, 1992), 7. Translation mine.

In the face of the uniformity associated with the governance of "the one," *The Metamorphosis* would present a risky attempt to distinguish not an authentic self, but rather something peculiar to the self.[33] If authenticity, *Eigentlichkeit*, is ultimately a caring about and coming to grips with one's whole existence (*Ganzsein*) as a distinct and meaningful project, *Eigentümlichkeit*, in contrast, would be an attention to the peculiarities that are distinctive or characteristic of the self but that precisely cannot be appropriated as one's own proper self. *Eigentümlichkeit* does not, and cannot, properly exist. It enjoys the tenuous modality of existence of something that would once have been called an aesthetic phenomenon – a feeling, an itch, an irritation, perhaps, even a "*Sorge*" – that cannot, however, be seized upon as a property, grasped as a concept, or made into a principle or a task like an idea.

It is significant therefore that while Gregor's senses lose their facility to make secure human discriminations, they seem to acquire new and indeed highly discriminating sensitivities and none more so than his, literal, sense of taste. Indeed, his taste proves to be a more powerful concern than his hunger. For during his incarceration, he soon loses interest in eating although he persists in some more or less disgusting behavior, taking to playing with his food in his mouth and then spitting it out rather than swallowing it: "Gregor now hardly ate anything anymore. Only when he happened to pass by the food laid out for him would he take a bite into his mouth just for fun, hold it in for hours, and then mostly spit it out again" (56). But he never loses interest in seeking new sources of nourishment or perhaps simply tasting new, albeit repulsive, things.[34]

Late in the text, the family having by now taken in lodgers, Gregor is taunted by the sound of the men eating: "'I'm hungry enough,' Gregor said to himself, sorrowfully [*sorgenvoll*], 'but not for these things. Look how these lodgers [*Zimmerherren*] are gorging themselves, and I'm dying!'" (58). To be sure, he is feeling sorry for himself and deploring his neglect by the family, which no longer takes any interest in his meals while exaggerating their polite attentions to the *Zimmerherren*. But he is *sorgenvoll* above all because his appetites seem to have nothing more in common with the sort of sustenance upon which the *Zimmerher-*

---

**33** Reflections on *Eigentum* and *Eigentlichkeit* have to address Kafka's note in the Zürau papers from 1917, "Das Wort 'sein' bedeutet im Deutschen beides: Dasein und Ihm-gehören" (The word "being" in German means both: to exist and to belong to him), *Nachgelassene Schriften II*, 56 (Oktavheft G). See Joseph Vogl, *Ort der Gewalt: Kafkas literarische Ethik* (Zurich: Diaphanes, 2010), 103–130; Paul North, *The Yield: Kafka's Atheological Reformation* (Stanford: Stanford University Press, 2015), chap. 1.

**34** On Kafka's peculiar taste for the disgusting, see Winfried Menninghaus, *Ekel: Theorie und Geschichte einer starken Empfindung* (Frankfurt am Main: Suhrkamp, 2002), 333–484.

*ren* subsist. Indeed, under the prevailing circumstances, his taste seems to have parted ways with the possibilities of subsistence altogether. He is dying on account of his taste. But he does not for that reason make death into the point of his taste or try to make a point with his death. Gregor is not on a hunger strike – and nor does he hope by his death to attain some higher satisfaction or indeed to leave a distinctive memorable mark. In short, although his taste has rendered him suicidal, his is not a taste for death and he does not interpret it as one.

In Heidegger's analysis, one cares ultimately about death, it is only a matter of how. The fallen everyday attitude is that of an evasive turning away. And an authentic being-toward-death (*Sein zum Tode*) is therefore about achieving an understanding that, in contrast, resolutely comes to terms with one's finitude – and owns it. In *The Metamorphosis*, it is as if this fallen attitude has not gone far enough. It is not enough to fly from or cover up death but to forget about it altogether (if it is possible to *live* in such a way, which is by no means clear). There are *other* cares qualitatively, that is temporally, distinct from the exigencies that structure our own finitude as being-towards-death. Taste takes pleasure in such distinction. What makes one distinctive is not one's being-towards-death but one's *Eigentümlichkeit*, which belongs to a different order of cares altogether. Gregor – and this defines his fallen *Schwärmerei* – is not in the manner of being-toward-death; he is dying-from-taste.

The society of security leaves no room and has no time for the cultivation of such singular tastes – and with good reason. When the *Zimmerherren* invite Grete into the living room to play the violin, they do so, expecting to hear "beautiful or entertaining violin-playing" (60). It turns out that she plays pretty badly – or perhaps she is just too avant-garde. In either case the *Zimmerherren* are ill at ease and retreat awkwardly to the window. Gregor meanwhile finds the music that the *Zimmerherren* consider not fit for their consumption very much to his taste – "And yet his sister was playing so beautifully" (60).

If Grete's playing inadvertently grates on the nerves of the average *Zimmerherr*, the entrance of the repulsive Gregor is also something of a performance and, at least at first, a more entertaining one. Drawn by the violin, he barely notes the absence in himself of the quality he formerly most prized, his solicitude for the family: "It hardly surprised him that lately he was showing so little consideration [*so wenig Rücksicht*] for the others; once such consideration had been his greatest pride" (60). Despite his disgusting and bedraggled state, he shamelessly (*keine Scheu*) makes his way onto the immaculate floor of the living room. With this show of abject indiscretion, Gregor carefully approaches his sister, wondering: "Was he an animal, that music could move him so? He felt as if the way to the unknown nourishment he longed for were coming to light" (61).

Once again and for the last time, Gregor confuses his peculiar taste with a desire for fulfillment, confuses "existential" with "aesthetic" *Sorge*. Instead of allowing himself to be gripped by the music, in the manner that he associates with the captivation of an animal, he intends to appropriate it as his own. Thus he approaches his sister resolved (*entschlossen*) to get her to retreat with him into his room, on the grounds that only he can or will properly appreciate her playing. Exploiting his monstrous appearance (*Schreckgestalt*), for the first time useful (*nützlich*) to him, he will defend the room on all sides, "he would be at all the doors of his room at the same time and hiss and spit at the aggressors" (61), while turning it into a kind of secure play-room – and by no means just for playing with the limits of taste in music. The convoluted plan includes telling his sister that he had intended at Christmas to send her to the conservatory, at which point she would break down "into tears of emotion" and Gregor would "kiss her on the neck which, ever since she started going out to work, she kept bare, without a ribbon or collar" (61). The unfulfilled plans for the conservatory thus converge with the attempt to convert his room into a conservatory for the cultivation of her music, but also into a kind of wildlife conservation or bestiary.

The fantasy of establishing himself as a perverse sort of *Zimmerherr* – sovereign protector of an unbanded, uncollared, but all the more intimate and incestuous community of music-playing and appreciation as well as of spitting, growling, hissing, kissing – accompanies his progress into the living room. This advance in fact culminates in the *Zimmerherren* indignantly terminating their contract – in doing so, the leader amongst them spits "curtly and decisively" onto the floor (63) – leaving the family once again in a state of humiliation and despair. Although she can by no means fathom his obscene intentions, the sister understandably concludes – in terms that are typical of the anxious and inflationary talk about vermin and people called vermin – that the *Ungeziefer* has predatory, if not outright imperious, intentions: "But as things are, this animal persecutes us, drives the lodgers away, obviously wants to take over the whole apartment and for us to sleep in the gutter" (65). They must, she therefore insists, be rid of him.

Gregor, having returned painfully slowly to his incarceration, could not agree more: "His conviction that he would have to disappear was, if possible, even firmer than his sister's" (67). Abandoned by his sister, the last bond of human care that had held him in check falls away, and he is free to abandon, or finds himself to be quite simply without, his life's concerns and his concern about life. This ascetic attitude is also a purely aesthetic one and may indeed be a kind of performance – one that makes him look dead or decrepit. For does Gregor really die or does he just stop being-there? Perhaps, insofar as he no longer cares about death, he has only lost the ability, which Heidegger reserves for Dasein, to die. And perhaps, therefore, he is instead, in the manner of certain insects, simply playing dead – not in order to preserve his life but in order to evade the attentions of security.

Finding himself back in his room, he discovers he can no longer move, although he feels in fact "relatively comfortable" (67). The pain throughout his body – including the apple-wound – seems to him to be fading away as he enters a "state of empty and peaceful reflection," thinking back on his family, "with deep emotion and love" (67). Letting go of his *Sorgen*, his last moments of life or of lived experience coincide with a brightening disclosed outside the window: "He still lived to see [*erlebte noch*] that outside the window everything was beginning to grow light. Then, without intending it [*ohne seinen Willen*], his head sank down to the floor, and from his nostrils streamed his last weak breath" (67). Contrasting this scene to the Romantic tradition, Corngold observes: "The light in which Gregor dies is said explicitly to emanate from outside the window and not from a source within the subject."[35] In the terms I have developed here, this external light does not owe its existence to, nor is it there for the sake of, a caring human being. It is not the flood-lighting of the security dispositive, but a "natural" light that, emanating from elsewhere, does not, so to speak, care about the human. The lighting outside the window that coincides with the expiration of his living intentions indicates the other realm of peculiar "cares" to which Gregor's taste obscurely attests. This is the sphere of "transcendence" to which Gregor's peculiar *Schwärmerei* has inadvertently tended – *das Freie*.

---

35 Corngold, "The Metamorphosis of Metaphor," 79.

**Figure 1:** Raphael, *Saint Michel terrassant le démon, dit le Petit Saint Michel*, 1503–1505 oil on wood, 31 x 26 cm, Musée du Louvre, Paris © Photo SCALA, Florence

# Epilogue: Guerrilla / Gardener
# Coetzee's *Life & Times of Michael K*

## Of Monstrous Pictures of Michaels

In 1801, following his "Kant crisis," Kleist traveled to Paris. En route he seems to have undergone a conversion to the visual arts upon seeing Raphael's *Sistine Madonna* (1512) in Dresden. The extent to which these two events, his despair about scientific knowledge in Berlin and the revelation of painting in Dresden where he discovered "a whole new world full of beauty" that allowed him to forget "the sad field of science," spurred Kleist's own turn to artistic production remains subject to debate.[1] What is certain is that any initial inclination in the direction of the arts was dampened by his experience of what he perceived to be the corruption of culture in Paris – even, or especially, at the Louvre where stolen Italian masterpieces like Raphael's *Transfiguration* could be informally viewed in exchange for a small bribe.[2] It is in this state of mind that, on his return from Paris, Kleist entertained the idea of tending a garden in the Swiss countryside. While weighing whether to turn to writing as a means of earning a living, and under the influence of reading Rousseau and Voltaire, he wrote to his fiancée, Wilhelmine von Zenge, "a human being could do nothing more pleasing to the divine, than to till a field, to plant a tree, and to beget a child," a pastoral prospect that understandably only accelerated the breaking off of their ill-starred engagement.[3]

One of the few paintings hanging in the formal galleries of the Louvre that made an impression upon him was another Raphael. In a letter to Adolfine von Werdeck, he wrote, "Lastly among the few Raphaels that are installed is an archangel, of whom one can rightly say [*von dem man recht sagen kann*], that he storms down [*heranwettert*] to crush a devil."[4] There is some ambiguity about which of the two Raphaels that the Louvre possessed, both with the title *Saint Michel terrassant le démon*, Kleist might have seen, the smaller and earlier image known as *Le petit St Michel* (1503–1505) in which Michael carries a

---

[1] Letter to Wilhelmine von Zenge, Leipzig, May 21, 1801. Heinrich von Kleist, *Briefe 1, Sämtliche Werke, Brandenburger Ausgabe* [BKA], ed. Roland Reuß and Peter Staengle (Basel: Stroemfeld/Roter Stern, 1988–2010). Hereafter BKA IV/1. Translations mine.
[2] "It is a disgrace [*Unwürdig*], the way the stolen works of art are treated here," letter to Adolfine von Werdeck, Paris and Frankfurt am Main, November 1801, BKA IV/1.
[3] Letter to von Zenge, Paris, Oktober 10, 1801, BKA IV/1.
[4] Letter to von Werdeck, Paris and Frankfurt am Main, November 1801, BKA IV/1.

sword, or the larger, later work in which Michael carries a pike, *Le grand St Michel* (1518). If, however, Kleist had a painting in mind when, almost a decade later, he wrote *Michael Kohlhaas*, who, to recall, at the height of his armed rebellion named himself, "a deputy of Michael, the Archangel,"[5] one would be inclined to conclude that he saw the smaller version [Figure 1].

*Le petit St Michel* presents a desolate landscape. A large rock-formation on the right gives way to a valley on the left and a city or castle going up in flames on the horizon, the smoke arching around and across, overcasting the bluish sky that is still to be glimpsed in the upper right-hand corner. In the background are figures drawn from Dante's *Inferno*. Under the rocks on the right criminals struggle, helplessly entangled by snakes and pecked at by black birds. On the left, hypocrites in gilded lead-lined cloaks are paraded on the plain before the burning city. In the center St Michael stands poised, sword raised, with his foot on the neck of a small, dragon-like, feather-winged creature. With its eyes bulging and tongue hanging out of gaping jaws, it curls its tail around and flails with its hind legs against the Archangel's armored lower leg. Scattered around the foreground, a disparate group of disconsolate figures can be discerned creeping, crawling and scratching in an aimless fashion across the barren earth. Hybrids of birds, insects, and reptiles, with stunted wings, or dog-like expressions and panting mouths, they reveal the northern influence of Hieronymus Bosch on the young Raphael.

Kleist continues in the letter, "The highest task of art is to present *one* feeling [eine *Empfindung*] but with its full force, and that is why Raphael is also a favorite of mine."[6] Viewed in Napoleon's Louvre in 1801, and recalled even more vividly perhaps from the perspective of Berlin around 1808, the painting might have presented a feeling of righteousness – a *Rechtgefühl* – in much the same ambivalent way that it would be presented in *Michael Kohlhaas*.[7] For it is unclear whether this intervention for the sake of divine justice is the result of, or the cause of, the earthly devastation the painting depicts.

While one can rightly say (*recht sagen kann*) that Michael strikes with the violence of a storm or a lightning bolt (*heranwettert*) out of the sky, it is more difficult to say – and certainly it is not clear that one could establish the right to say – by and for the sake of what right he does so. In the Kantian terms that Kle-

---

[5] Heinrich von Kleist, *Michael Kohlhaas* (1808 and 1810), BKA II/1, 140. Translations mine.
[6] Letter to von Werdeck, Paris and Frankfurt am Main, November 1801, BKA IV/1.
[7] On the possible influence of this painting on *Michael Kohlhaas*, see Peter Horwarth, "Auf Den Spuren Teniers, Vouets und Raphaels in Kleists *Michael Kohlhaas*," *Seminar: A Journal of Germanic Studies* 5.2 (September 1969): 102–113; see also Gernot Müller, *Kleist und die bildende Kunst* (Tübingen: Francke, 1995).

ist draws on to describe the painting, what one can rightly say is that Michael arrives with a superhuman display of force in the name of a supersensible order to which, however, the viewer has no access. If it does not lie in the exalted person of Michael himself, there is no sphere of transcendence presented in the painting. The devastated landscape would be the vision of a figure who sees the need to establish in the world an order of the heavenly sort that he feels in his own breast. To someone recovering from a Kant crisis, *Le petit St Michel* would, in other words, appear to be a representation *avant la lettre* of *Schwärmerei*. And it comes as no surprise therefore that Kleist would refer to the figure of Michael – perhaps this very figure of Michael – in the presentation of the bureaucratic fanaticism of Michael Kohlhaas in the first part of the novella, departing from the assumptions of Fichte.

A century later the painting might well have presented a rather different feeling – if it made an impression at all. On visiting the Louvre in 1911, Kafka seems not to have been struck by Raphael in particular. Indeed, as unhappy in Paris as Kleist had been, Kafka seems hardly to have been in the mood for looking at paintings at all. Reportedly moping around with Max Brod, he noted down a short list of paintings amongst which neither of the Raphaels features.[8] Whether Kafka even saw either Raphael is uncertain. His attention might have wandered from *Le grand St Michel* to the empty space in the same room where the *Mona Lisa*, stolen two weeks before, had hung. It is hard to imagine, however, that the person who would go on to write *The Metamorphosis* would not have been drawn to the monstrous vermin-like figures strewn under Michael's feet had he come across *Le petit St Michel*.

Such figures, forgotten or ignored like Kohlhaas's hapless *Rappen*, would in Kafka's life and times solicit renewed attention. Are they fallen angels or have they simply fallen by the way side? Abandoned, apathetic, and earthbound with deformed limbs and stunted wings, they appear rather as the collateral damage of a war in which they have no stake. They find themselves, like many of the figures in Kafka, subject to a justice, or in this case a just war, which they neither recognize nor understand. Seen under this optic, which inflects the horizontal rather than the vertical axis of the painting, the dominant mood would be less one of righteousness than of bewilderment. As Michael metes out justice to the rebellious angel, the creatures strewn around seem to have no relation to what is happening. Insofar as they are neither rebelling nor receiving punishment, they are the only figures in the picture that seem,

---

**8** See Hartmut Binder, *Kafka in Paris: Historische Spaziergänge mit alten Photographien* (Munich: Langen Müller, 1999), 162–187.

for the moment at least, to evade, or not to heed, the overwhelming force of Michael's righteous project. Perhaps one could say that here they just live. In the midst of the apocalypse, they present the image of a life that pays no heed to the millenarian project to institute a just order on the earth.

Viewed towards the end of the twentieth century Rapheal's painting, presuming a picture hanging in the Louvre still has the power to provoke such significant associations, might be evocative of scenes from *Apocalypse Now* (1979). It is the landscape that increasingly draws attention in a time in which the human and ecological devastation the painting presents has become a real techno-political possibility. In this regard the painting could be added to, or even inaugurate, a series of *Dusklands*, to take the title of Coetzee's first novel (1974), that would sketch out the consequence, taken to the extreme, of attitudes to the earth through modernity. While *Apocalypse Now* adapted Conrad's *Heart of Darkness*, from the "new imperial" ruination of the Congo to the Cold War neo-colonial conflict in south-east Asia, in *Dusklands* Coetzee sought to explore the relation between the Vietnam War around 1970 and the early European colonization of Southern Africa around 1760.[9] Each of the two novellas in *Dusklands* is taken up with the account that a dwindling, or I would say, despairing, subjectivity gives of itself as it defiantly responds to its own incapacity with escalating and annihilating violence. Despite their divergence in time and space, the novellas formally and thematically correspond to one another as if drawn to the same desolate horizon – *Dusklands*.

The first of the novellas, "The Vietnam Project," is narrated in California by a man who, presented as a bureaucratic type and employed by a RAND-like corporation, is tasked with developing propaganda for the war. The narrator's own psychological disintegration, occasioned not only by the violent photographs of war he obsessively studies but equally by the related domestic anxieties about masculinity that are a function of his desk job, expresses itself in his report which becomes a Swift-like satire on the psychology of ideological warfare and culminates in the proposal to lay-waste to the earth. No longer appealing to the likes of St Michael, the exalted narrator from his glass-cubicle in the California office declares himself the proxy of the B-52:

> When we attack the enemy via a pair of map co-ordinates we lay ourselves open to mathematical problems we cannot solve. But if we cannot solve them we can eliminate them, by attacking the co-ordinates themselves – all the co-ordinates! For years now we have attacked the earth, explicitly in the defoliation of crops and jungle, implicitly in aleatoric shelling and bombing. Let us, in the act of ascending consciousness mentioned above,

---

9 J. M. Coetzee, *Dusklands* [1974] (New York: Penguin Books, 1996).

admit the meaning of our acts. We discount 1999 aleatoric missiles out of every 2000 we fire; yet every one of them lands somewhere, is heard by human ears, wears down hope in the human heart. A missile is truly wasted only when we dismiss it and are known by our foes to dismiss it.[10]

In a display of *Schwärmerei* that makes even his manager (named Coetzee) uncomfortable, the narrator proceeds, "Let us show the enemy that he stands naked in a dying landscape."[11] It is in this direction – the production of bare life in the midst of a devastated landscape – that the literary history of fanatical bureaucrats traced in the present study has tended.

## Translating Michael Kohlhaas

It is, however, to *Life & Times of Michael K*, the novel Coetzee began in 1979 and published in 1983, that I would like to turn for a closing reflection on the literary preoccupation with the peculiar perplexities of political life under fanatical bureaucratization.[12] Set in South Africa in an indeterminate future of civil war, *Life & Times* features a picture of a different kind. Cut out of an old newspaper by Michael K, it is the photograph of the captured "Khamieskroon killer" and of his weapon, probably a Kalashnikov. This picture presents, it seems to me, the last or the latest of the line of the historical iterations of Raphael's *Petit Michel*. In the published text of *Life & Times*, this image is all that remains of a preoccupation that can be traced back in Coetzee's notes and drafts to the very conception of the book project, namely, to "somehow draw upon the passion + urgency of Michael Kohlhaas to write about... by writing what amounts to an interpretation (an interpretive translation) of MK" (JMC's ellipsis).[13]

Coetzee turned to *Michael Kohlhaas* as he began his fourth novel in 1979 at a moment when the cycle of popular unrest and violent suppression by the apartheid state was intensifying, accompanied by a resurgence in the paramilitary op-

---

10 Coetzee, *Dusklands*, 29.
11 Coetzee, *Dusklands*, 29.
12 J. M. Coetzee, *Life & Times of Michael K* (London: Secker & Warburg, 1983). Citations in-text.
13 J. M. Coetzee Papers, *Life & Times of Michael K* (Fiction, 1983), gray casebound notebook, includes notes on other subjects, 1972–1982, Container 33.5: 6. This epilogue draws on papers housed in the Coetzee Papers at the Harry Ransom Center at the University of Texas at Austin. Between 1979 and the publication of *Life & Times of Michael K* in 1983, Coetzee worked on seven hand-written versions of the book, as well as two annotated typed versions, and kept a notebook of ideas, commentaries, and plans for revision.

position to the regime by the force known as "MK."[14] Writing under such oppressive conditions in South Africa, which in certain respects paralleled those to which Kleist was responding with his partisan text,[15] *Michael Kohlhaas* seemed to present a literary and political "passion + urgency" that Coetzee felt the times called for, but that his own writing lacked. Kleist's partisan novella was meant to lend Coetzee's protagonist the resources to generate his own means of political expression by assuming the posture of a "citizen of the world" (an interpretive translation of one of Kohlhaas's declarations), beyond the law of nations as well as the norms of historical and literary representation. Michael K was to represent a cosmopolitan politics freed from the immediate exigencies, injustices, and discriminations that constituted everyday political life, especially in apartheid South Africa.

From the start, Coetzee's attempt to draw on Kleist was complicated by persisting reservations about the political and aesthetic implications of a literary project in the South African context, especially one that so self-consciously drew on European literary history. The work of "interpretive translation" thus involved interrogating the ideological investments of literary and historical representation. This is explicit in the short, third version that Coetzee drafted in December 1980, which focuses on a nine-year-old boy who, in order to take care of his indigent grandmother in Cape Town during an imagined period of civil war, engages in petty crime and prostitution. His grandmother reads to him from a translation of *Michael Kohlhaas* whom the child imagines to be his father. The boy would speak, Coetzee wrote in his notebook, "of God and the angels and of Michael," and in a subsequent entry he considered the possibility that "the child is an angel" who follows the "adventures of M.K. fervently."[16]

In this draft Coetzee describes Michael K entering the apartment where his grandmother worked as a domestic servant after it had been ransacked during a riot – a scenario that is kept in the published version of the text. A colonial picture of oxen straining to pull a wagon across a mountain pass hangs on the wall: "It is a scene which seems to have had a great meaning for people

---

14 uMkhonto we Sizwe, MK for short, was the armed wing of the ANC founded in response to the Sharpeville massacre in 1960. Following the Soweto Uprising in 1976, and especially in the early 1980s, the MK carried out a series of attacks on targets orchestrated to demoralize and undermine the regime. For a survey of the South African political context of Coetzee's book, see Susan VanZanten Gallagher, *A Story of South Africa: J. M. Coetzee's Fiction in Context* (Cambridge, MA: Harvard University Press, 1991), 136–142.
15 See Wolf Kittler, *Die Geburt des Partisanen aus dem Geist der Poesie: Heinrich von Kleist und die Strategie der Befreiungskriege* (Freiburg: Rombach, 1987).
16 J. M. Coetzee Papers, Container 33.5: 10.

in the past, but whose meaning will now become forgotten." Nonetheless, the child K is "fond of the picture," for "it is painted without combativeness. [...] there is even a child, without pants, bending to examine something (a lizard?), oblivious of the struggle to cross the pass." K asks:

> Is that child me? [...] going about my childish business while the ox-wagon of history trundles on behind my back? Or is the ox-wagon an image of the cart in which I am to pull my grandmother to safety? Dare I not look at a picture without it threatening to take on a meaning in my life?[17]

The child who asks himself these things is clearly no longer the child of whom K speaks, who seems to indicate a way of life and order of experience other than the one governed by the narrative exigencies of historical time.[18] In the published version of *Life & Times*, the effort to present such a figure is complemented by the thematization, made explicit in Part II of the novel, of the way in which K nevertheless, in spite of himself, takes up significance in the system of political and literary representation – if nowhere else then in a book called *Life & Times of Michael K* – in which he does not care to take part.

In the earliest drafts of the novel an elderly K-figure is actually translating *Michael Kohlhaas*, while in the fourth and fifth versions, K, a grown man, discovers and reads a translation in the ransacked apartment of his grandmother, although it is missing the cover and the last pages so he is never able to find out whether Kohlhaas is executed. Given his own experience with institutions and authorities, K suspects so.[19] At one point, as he makes his way out of Cape Town, he recalls the story and reflects:

> It seemed a pity to K that people like Kohlhaas did not exist, or no longer existed (for he was prepared to locate him in a past, though he did not know where to locate the past). It did not seem possible to act purely as the embodiment of a passion – say the passion

---

**17** J. M. Coetzee Papers, "Versions 1–4," handwritten draft with revisions, 31 May 1980 – 14 January 1981, Container 7.1: #4, vn 3 (1), 9–10.
**18** For a reading that refers to this moment in Coetzee's drafts in a discussion of the "idiocy" of Michael K, as a figure ostensibly outside of politics and therefore of history, see John Bolin, "Modernism, Idiocy, and the Work of Culture: J. M. Coetzee's *Life & Times of Michael K*," *Modernism/modernity* 22.2 (2015): 343–64.
**19** "Nothing in his own experience led him to believe that one could challenge the might of princes and survive. Very likely the horse-dealer has been tricked and executed. At least, though, he hoped that the two horses that had been the cause of so much suffering had found a good home," J. M. Coetzee Papers, Container 7.2: #4, vn 5 (5).

for justice; it seemed that nowadays one was called on to be more ~~interesting~~ various, but also to act from motives whose ground was mundane.[20]

This remark regarding the historicity of political affect is also a commentary on the historical determinants of literary presentation. Yet Coetzee's problem in bringing a Kohlhaas-like figure into the contemporary South African political and literary landscape was already integral, I would argue, to Kleist's *Michael Kohlhaas* insofar as it responded to the oppressive Prussian context around 1800. For Kleist's partisan novella, as shown in Chapter Two, is also a bureaucracy novella. Kohlhaas is constrained by an inability to interpret his passion for justice as anything other than a passion for Right (*Recht-gefühl*), namely, for legal-bureaucratic rectification, the inadequacy of which is attested to by the "enthusiastic" intensity and fanatical escalation of his righteous frustration. And it is precisely because his passion cannot be recognized under the bureaucratic conditions in which he finds himself, that, in the first part of the novella, the most law-abiding (*rechtschaffensten*) of citizens turns to a terroristic outburst of para-military violence.

In my reading, *Michael Kohlhaas* – very like *Life & Times* – is the staging of the diminishing scope and intensity of political affect, or its expression, under historical and institutional conditions that incite, but also obstruct and censor, it. Only in the second part of the novella, which in Coetzee's draft K never gets to read, does it occur to Kohlhaas, once he has surrendered to the authorities, that his passion for justice may require non-legal means of expression. In contrast to his outright resistance in the first part which, even as an outlaw, he understands in fact to accord with the letter of the law, his peculiar passive resistance during his arrest in the second indicates an extra-legal, if ultimately unformulable, criterion for justice. This is demonstrated by his swallowing of the note, even if it coincides almost exactly, as staged in the scene on the scaffold, with the execution of the law.

Kleist's text thus presents a tension between a para-military "partisan" mobilization that seeks to articulate and attain justice in the terms of the order it openly opposes, and an immanent "peculiar" resistance that rejects the terms of the order to which it seems to submit but – by renouncing such articulacy – deprives itself of the kind of declarative self-determination that defines modern political movements and communities. Although, to judge from his notes, Coetzee, like his protagonist in the early drafts, seems only to have read (or to recall) the first part of Kleist's novella, it is this very antinomy of the passion for justice

---

[20] J. M. Coetzee Papers, Container 7.1: #4, vn 3 (1), 10.

legible in *Michael Kohlhaas* that is interpreted and translated in Coetzee's *Life & Times of Michael K.*[21]

By the time he published *Life & Times* in 1983 both the narrative and formal traces of *Michael Kohlhaas*, as well as the concerns about historical representation for which the colonial painting had functioned, are filtered out and concentrated in the newspaper photograph of the "Khamieskroon killer." The "Khamieskroon case" concerned a historical figure (not unlike the historical Hans Kohlhase) who, according to Coetzee's notebook, "went on a rampage shooting whites and was eventually tracked down."[22] No particulars are mentioned in the novel. Instead K comes upon the article among newspapers "so old that he remembered none of the events they told of," where the photograph of "a handcuffed man in a torn white shirt standing between two stiff policemen" under the headline "KHAMIESKROON KILLER TRACKED DOWN" arrests his attention: "the Khamieskroon killer looked at the camera with what seemed to K a smile of quiet achievement. [...] K stuck the page with the story on the refrigerator door; for days afterwards, when he looked up from his intermittent work [...], his eyes continued to meet those of the man from Khamieskroon, wherever that was" (22). The last hint of Kohlhaas can be found in this image, and in the Khamieskroon killer's "smile of quiet achievement."

In very late drafts, Michael K speaks obscurely in Part II of the Khamieskroon killer, suggesting that the man who, in the style of Kohlhaas, had declared to the authorities that he was a "citizen of the world," was his father.[23] By inventing such an ancestry, one only legible in the published version in the photograph and in his name, K was to present in his own terms the figure of a radical opponent to the powers that be. Ultimately, Coetzee would cut out altogether the references to K's attempt to invent a paternity (and a past) for himself. Whatever the attraction of Kohlhaas and the Khamieskroon killer for K, he does not ultimately join the literary or political lineages they represent. Such paternity, and perhaps the very notion of a paternal line, would risk falling into the cycle of historical violence that is the subject of the book's unacknowledged epigraph from Heraclitus: "War is the father of all and king of all. / Some he shows as gods, others

---

[21] Without reference to the archival material, Peter Horn reads *Life & Times of Michael K* as a parody of *Michael Kohlhaas*, which he considers to be about liberal bourgeois sentiments for justice and right. See Peter Horn, "Michael K: Pastiche, Parody or the Inversion of Michael Kohlhaas," *Current Writing: Text and Reception in Southern Africa* 17.2 (2005): 56–73.

[22] J. M. Coetzee Papers, Container 33.5: 14.

[23] The later references to the Khamieskroon killer and the "Citizen of the World" in Part II were overworked in every draft only to be finally struck out in J. M. Coetzee Papers, "Typed Late Draft with Corrections," 1983, Container 8.3: 132–133.

as men. / Some he makes slaves, and others free." Instead, Michael K joins the subterranean genealogy of "peculiar fanatics" that I have traced in this study.

## The Savor of the Earth

For all its South African particularity, or perhaps precisely because of the type of regime South Africa had become,[24] Coetzee's novel presents the paradigm of the modern state machinery that Foucault was beginning to investigate initially under the title "security" and subsequently "governmentality" in the late 1970s.[25] Characteristically concentrating on European history, Foucault traced a genealogy of governmentality back to a practice of paternalistic authority modeled on the Christological notion of "pastorship" adopted by the early Church. In contrast, Coetzee's "criticism of 'political reason'" (to cite the title of Foucault's 1979 Tanner Lectures) proceeds by means of a literary presentation of a peculiar figure in South Africa who finds himself caught up in a future state of emergency in which the security regime that protects a precarious post-colonial order is threatened and destabilized by on-going civil war.[26]

Part I of the published version of *Life & Times* tells the story of the thirty-year-old son of a domestic servant who attempts to flee the escalating violence

---

[24] In the midst of continual and escalating unrest since at least the mid-1970s, South Africa declared a State of Emergency from 1985 to 1990.

[25] See Michel Foucault, *Security, Territory, Population: Lectures at the Collège de France, 1977–78*, ed. Michel Senellart, François Ewald, and Alessandro Fontana (New York: Palgrave Macmillan, 2007) and *"Omnes et singulatim:* Towards a Criticism of 'Political Reason,'" in *The Tanner Lectures on Human Values*, ed. Sterling McMurrin (Salt Lake City: University of Utah Press, 1981).

[26] For recent readings of *Life & Times* in the context of contemporary theories of biopolitics, see Catherine Mills, "Life beyond Law: Biopolitics, Law and Futurity in Coetzee's 'Life and Times of Michael K,'" *Griffith Law Review* 15.1 (2006): 177–195; David Babcock, "Professional Subjectivity and the Attenuation of Character in J. M. Coetzee's *Life & Times of Michael K*," *PMLA* 127.4 (2012): 890–904; Daniele Monticelli, "From Dissensus to Inoperativity: The Strange Case of J. M. Coetzee's Michael K," *English Studies* 97.6 (2016): 618–637. Many such readings respond to Michael Hardt and Antonio Negri's influential remarks in *Empire* in which they argue that K, like Bartleby, presents a politics of refusal that, however, on account of the emptiness of absolute refusal and the isolation it involves ultimately fails to constitute a new social body, Hardt and Negri, *Empire* (Cambridge, MA: Harvard University Press, 2000), 203–204. Late in the notebook, Coetzee makes reference to "the great internment" that was at the center of Foucault's early work on the history of madness but his familiarity with Foucault seems to be filtered through the essay by Shoshana Felman, "Madness and Philosophy or Literature's Reason," *Yale French Studies* 52 (1975): 206–228; J. M. Coetzee Papers, Container 33.5: 69.

and hardship of Cape Town and take his ailing mother back to the rural home in which she had grown up. Failing to get the requisite passes to travel, K sets off on foot pushing his mother in a makeshift cart. They suffer the deprivations and humiliations of travelling through a territory where they are vulnerable to the threat of marauders and common criminals and discriminated against, not to mention taken advantage of, by the corrupt representatives of the crumbling regime. On the way, his mother dies in a hospital in Stellenbosch; however K, carrying her ashes, continues the journey.

For a novel that took its "original inspiration" from *Michael Kohlhaas*, and indeed for one written in the South Africa of the 1980s, the question that imposes itself is why K, despite the interminable provocations of the state's various institutions, never takes up arms, like Kohlhaas, and joins the guerillas in the mountains.[27] Instead, he attempts to cultivate an inconspicuous patch of earth in the middle of the Karoo and to evade, rather than oppose outright, the predations of the regime, which increasingly seeks to control, contain, and round up into camps the unpropertied and dispossessed. Translated into South Africa at the end of the twentieth century, the "passion + urgency" of *Michael Kohlhaas* gives rise to a figure whose passion for justice is impassively expressed in his "gardening."

Most striking of the differences between Michael Kohlhaas and Michael K is their articulacy. Kohlhaas exhibits a charismatic capacity for expression during the uprising. K, in contrast, has a hare lip that impedes his speech. It is on account of this natural (and easily "corrected," 99), defect that he is placed in an institution for "variously afflicted and unfortunate children" (4) and brought up under conditions in which the disfigurement accentuates his dependence and disability. While being subjected to the institutional demands of authorities and their mouthpieces, he is also alienated by his impairment from the language he can only imperfectly pronounce and deprived of the possibility of finding his own voice. In contrast with Kohlhaas's initial faith in the legal discourse on which he eloquently draws, K's experience of his disability as it is treated institutionally makes him aware of the ways in which discourse can perpetuate and obscure structural injustices.

At the beginning of the novel, K is entirely a creature of such institutions. In the course of his journey, especially once he is freed of his obligation to his needy and neglectful mother, he gradually dispenses with the institutional exi-

---

[27] During the drafting, this was often Coetzee's question too: "The book started off with Kleist behind it. Is Michael K– ever going to take to the hills and start shooting?" J. M. Coetzee Papers, Container 33.5: 42.

gencies with which he has been instilled and becomes, in his own words, "a different kind of man" (93). This transformation is presented as the reduction of the voices that call upon K as well as of the claims and imperatives that saturate the South African landscape to the point where he can access, albeit only for a time, a different relation to the "earth."

What is at stake in what he calls "gardening" then is the attempt to find a way in which, to cite K's words in the closing line of the book, "one can live" (250), in a manner that is not qualified by the security apparatus, which, according to the biopolitical imperative famously formulated by Foucault, operates: "to 'make' live and to 'let' die."[28] It is no coincidence, therefore, that the most remarkable aspect of his transformation is his loss of interest in food as a source of sustenance, which he associates with institutional life. Like Gregor, and also Kafka's hunger artist, to whom he is more often related, K turns into, or turns out to be, a man of peculiar tastes.[29] But where the sphere of Gregor's taste remains indeterminate, in *Life & Times* taste takes on apparent concretion, its locus being the earth: "When food comes out of this earth, he tells himself, I will recover my appetite, for it will have savour" (139).

It is his encounter with the desolate landscape of the Karoo that first discloses the possibility of a livable life: "I could live here forever, he thought, or till I die. Nothing would happen, every day would be the same as the day before, there would be nothing to say" (63–64).[30] Its appeal is first of all its silence:

> He could understand that people should have retreated here and fenced themselves in with miles and miles of silence; he could understand that they should have wanted to bequeath the privilege of so much silence to their children and grandchildren in perpetuity (though by what right he was not sure). (64)

The silence of the Karoo is brought into relief by the indistinct voice he hears, or thinks he hears, as he approaches the town of Prince Albert: "As he descended the hillside towards the town, he began to be aware of a man's voice rising up to

---

[28] Michel Foucault, *"Society Must Be Defended": Lectures at the Collège de France 1975–1976*, ed. Mauro Bertani and Alessandro Fontana, trans. David Macey (New York: Picador, 2003), 241.

[29] On Kafka in *Life & Times*, see Peter Horn, "Kafka in der Karoo: John M. Coetzee's *Life & Times of Michael K*," in *Interkulturelle Erforschung der österreichischen Literatur*, ed. Herbert Arlt and Alexandr W. Belobratow (St. Ingbert: Röhrig Universitätsverlag, 2000); Patricia Merivale, "Audible Palimpsests: Coetzee's Kafka," in *Critical Perspectives on J. M. Coetzee*, ed. Graham Huggan and Stephen Watson (London: Palgrave, 1996).

[30] On Coetzee's attachment to the Karoo, see David Attwell, *J. M. Coetzee and the Life of Writing: Face to Face with Time* (New York: Viking, 2015), 40–54.

meet him in an even and unending monologue without visible origin. Puzzled, he stopped to listen. Is this the voice of Prince Albert?" (67).

This is the voice, I am inclined to suggest, of a certain Eurocentric-colonial tradition, or rather the voice that transmits that tradition. Much like the painting the young K sees in the earlier version of the novel, it seems "to have had a great meaning for people in the past, but whose meaning will now become forgotten."[31] While in the early draft of Coetzee's book K reads *Michael Kohlhaas*, "bored," like many readers, "by all the talk of counts, chancellors and protectors,"[32] in the published novel K's aversion to any kind of reading is explicitly a function of the postcolonial situation: "He had never liked books, and he found nothing to engage him here in stories of military men or women with names like Lavinia" (22–23).

But this world, so uninteresting to K in books, in fact traverses the South African landscape through the European names of towns, streets, parks, buildings, and public works. At one point, when he is taken to hospital in Prince Albert, he hears the voice again and asks the nurse: "'Tell me, I have always wanted to know, who is Prince Albert?' [...] She paid no attention. 'And who is Prince Alfred? Isn't there a Prince Alfred too?'" (97). As it happens Prince Albert, the Prince Consort to Queen Victoria, and his son Prince Alfred, were Saxon aristocrats, their line enduring, in name at least, in the country of South Africa. But K, the latter day Kohlhaas, is no longer provoked or threatened by this discourse. He is leaving this history behind – and abandoning the interpellations of institutions and authorities that had determined him since childhood.

At the Visagie farm, which he takes to be the childhood residence of his mother, no voice announces itself – not even his own. K, whose disfigured mouth has made pronunciation difficult and social intercourse awkward, finds himself released from the obligation to train his faculties to articulate sounds that he could never master and were not his: "He coughed, and gave a little hoot like an owl [...]. He thought: Here I can make any sound I like" (77). When contemplating what to do with his mother's ashes,

> He closed his eyes and concentrated, hoping that a voice would speak reassuring him that what he was doing was right – his mother's voice, if she still had a voice, or a voice belonging to no one in particular, or even his own voice as it sometimes spoke telling him what to do. But no voice came. (80)

---

[31] J. M. Coetzee Papers, Container 7.1: #4, vn 3 (1), 9.
[32] J. M. Coetzee Papers, Container 7.3: #4, vn 6 (2), 15.

It is around this time that K begins his "life as a cultivator" (81). His repudiation of the injunctions of the voice is accompanied by a search for a "silent" space, free from the property claims by which the land has been taken over. While still en route to Prince Albert and making his way across the vast seemingly deserted landscape punctuated by intermittent fences, he wonders, "whether there were not forgotten corners and angles and corridors between the fences, land that belonged to no one yet" (64). K embarks on his gardening as if he has indeed happened upon such a forgotten corner of the land. In fact, he attempts to produce a garden that is indistinguishable from the surrounding landscape, and so carves out, or occupies, a different – unclaimed – dimension of the land, which he refers to as "the earth."

A significant part of Coetzee's non-fiction and fiction is concerned with the colonial appropriation not only of South African land, but of the very image of the South African landscape.[33] The colonial enterprise in South Africa was a combination of property claims inscribed on maps and marked by fences and a far more exhaustive aesthetic project, as Coetzee observes in *White Writing*, "to read out and articulate the meaning of the landscape."[34] K's gardening has nothing to do with the articulation of the earth, which does not speak to him. It becomes neither the locus of the assertion of a counter-claim of rights or priority on his part nor, although he understands his attachment in terms of taste, occasion for eloquence in an aesthetic register.[35] And once the voices of his upbringing and of the colonial history of the land have receded into silence, no inner or transcendent voice announces itself. On the contrary, he finds to his relief that he is "one of the fortunate ones who escape being called" (143).

"Gardening" then is not a calling but a relation to the earth, carried on in the absence of more exalted or providential orders of significance. The garden is not

---

[33] For a critical reflection on the presentation of the earth in *Life & Times*, see Derek Wright, "Black Earth, White Myth: Coetzee's *Life & Times of Michael K*," MFS 38.2 (1992): 435–444. For more sympathetic eco-political readings, see Dominic Head, "The (Im)possibility of Ecocriticism," in *Writing the Environment: Ecocriticism and Literature*, ed. Richard Kerridge and Neil Sammells (London: Zed Books, 1998); and Anthony Vital, "Toward an African Ecocriticism: Postcolonialism, Ecology and *Life & Times of Michael K*," Research in African Literatures 39.1 (March 2008): 87–106.

[34] J. M. Coetzee, *White Writing: On the Culture of Letters in South Africa* (New Haven: Yale University Press, 1988), 166.

[35] Coetzee was anxious to emphasize this in his notes: "In a sense, what the narrative continually calls for is a statement of attachment to the earth, a kind of final patriotism. This K– steadfastly refuses to give (he thematizes his refusal in his last declaration): he is not going to be absorbed into a South African rhetoric of liefde vir die bodem," J. M. Coetzee Papers, Container 33.5: 48.

K's property [*Eigentum*]; it is his peculiarity [*Eigentümlichkeit*]. While he finds himself attached by a "cord of tenderness" to this particular "patch of earth," such attachment does not appear to be a function of his work, as in another tradition of European political and colonial thought indebted to Locke. Rather the relation of tenderness appears to be a function of the tenderness of the relation, of its precarious, finite, and potentially irreparable quality: "It seemed to him that one could cut a cord like that only so many times before it would not grow again" (90). It is the tenderness of the relation to the earth that solicits tending, rather than a tenderness that emerges on account of his attention. "All that remains is to be a tender of the soil," he thinks as he tastes its first fruits: "He lifted the first strip to his mouth. Beneath the crisply charred skin the flesh was soft and juicy. He chewed with tears of joy in his eyes. The best, he thought, the very best pumpkin I have tasted. For the first time since he had arrived in the country he found pleasure in eating" (156).

The "passion for justice" that drove the righteous Kohlhaas to partisan revolt is here reinterpreted as the feeling of tenderness to the earth, which discloses a place in which "one can live" without reference to laws and rights. Politically, if one can speak of politics here, K's attitude to the earth has nothing to do with a global outlook that precedes or transcends local and historical determinations in the name of cosmopolitan ideals. He is no "citizen of the world" as Coetzee, reading Kleist, seems initially to have envisaged. Nor does he count, like the partisan or guerrilla in the definition Schmitt developed in the face of the worldwide paramilitary upheavals and liberation movements in the decades following World War II, as one of "the last sentinels of the earth," standing for "a piece of authentic ground."[36] On the contrary, the cord of tenderness draws attention to the contingency, tenuousness, and vulnerability of the affection it occasions and so refuses the defiant self-assertion of authenticity. K is a gardener – on account of its tenderness, he tends the earth that is not his. And this tenderness is disfigured or obliterated in relations of property and exploitation but also in ostensibly liberatory projects of self-determination and autonomy.

At one point on the journey, reflecting on his seeming lack of social and political conviction, K thinks: "Perhaps I am the stony ground" (65). The phrase recited out of a religious education textbook is characteristic of K's attempts to give an account of himself in a second-hand language he never seems able to command, nor wants to. K has no interest in good grounds, nor even fertile ones; these belong, along with its place names, property claims, and articulations of

---

[36] Carl Schmitt, *Theory of the Partisan: Intermediate Commentary on the Concept of the Political*, trans. G. L. Ulman (New York: Telos, 2007), 71, translation modified.

the landscape, to a "natural history" of South Africa that he prefers to abandon. "I have lost my love for that kind of earth," he thinks at one point referring to the "green lawns and oak trees" of Wynberg Park in Cape Town (92). Tending to a garden in the desert, K undergoes what might be described as a political geological transformation: "It is no longer the green and the brown that I want but the yellow and the red; not the wet but the dry; not the dark but the light; not the soft but the hard. I am becoming a different kind of man, he thought, if there are two kinds of man" (92–93).

## Sparrows

Shortly after K begins cultivating his garden for the first time, a Visagie, the grandson of the former owners of the farm, arrives. He is a deserter and plans to go into hiding on the farm. He immediately assumes, despite all protestations to the contrary – "Michael, I am speaking to you as one human being to another" (88) – the old colonial relationship, treating K as his servant. K finds his disability reasserting itself: "Trying to bring out words, K stumbled. The stranger did not shift his gaze from K's bad mouth" (83). The Visagie's manner of speaking recalls the institutional violence K had sought to escape: "The words, whatever they stood for, accusation, threat, reprimand, seemed to K to smother him [...] he felt stupidity creep over him like a fog again. He no longer knew what to do with his face" (88). Before he abandons his budding garden, K kills three sparrows and a dove for the Visagie, who promptly tells him to clean them: "K held up the four dead birds, their feet together in a tangle of claws. There was a pearl of blood at the beak of one of the sparrows. 'So small you don't taste it as it goes down,' he said. 'You wouldn't get yourself dirty, not even your little finger'" (87). This is as close as K comes to a confrontation. Certainly his words spark an infuriated response from the Visagie grandson: "What the hell does that mean? [...] What the fuck do you mean? If you want to say something, say it! Put those things down, I'll take care of them!" (87).

Although sparrows, along with doves and swallows, accompany K inconspicuously throughout the novel, there is one instance in Part II of the book in which the significance of the sparrow is insisted upon. Part II is narrated by the medical officer who tends K at the "rehabilitation camp" to which he is sent after being caught at the end of Part I and suspected of being an *Opgaarder*, holding a stockpile for the guerillas. Representative of the more "humane" side of the regime and facing its inevitable demise (the war is being lost), the medical officer becomes fascinated by the figure of K and what he represents. When K expresses suspicion of the care and attention being afforded him – he does not want to be cared for and will not eat – the officer responds: "'You ask why you are important, Michaels. The answer is that you are not important. But that does not mean you are forgotten. No one is forgotten. Remember the sparrows. Five sparrows are sold for a farthing, and even they are not forgotten'" (186). Invoking lines from Matthew that are recalled in *Hamlet*, the medical officer identifies special providence with the practice of the security regime of which, despite his reservations and even revulsion, he remains very much a part.

The medical officer's political-theological construal of the regime resonates with Foucault's more or less contemporary genealogy of governmentality insofar as it inflects the at once totalizing and individualizing attentions of modern political reason. Indeed his remark anticipates Agamben's recent work, which relates Foucault's governmental apparatus to a "providential machinery" that was already at the center of a longer political-theological tradition concerned with the question of divine governance.[37]

Emulating God's providential regime, the security infrastructure as it is described by the medical officer "takes care" – in the ominous ambivalence of the phrase – of each and every one. And while K neither opposes nor submits but simply *does not care* to engage the system either as participant or antagonist, it proves impossible to evade the zealous attentions of institutions, authorities and even individuals who, to K's bemusement, cannot bear, or cannot believe in and ultimately are profoundly suspicious of, a figure who, without property or even personhood (he is for the most part treated as mentally disabled), demands neither help nor protection. No one is allowed to be forgotten. In this respect, it is tempting to read K's inscrutable words to the Visagie as a commentary on the medical officer's remarks regarding the significance of the sparrow.

Pressed by K – "No one was interested before in what I ate […] So I ask myself why?" (203) – the medical officer writes a letter, which is really a kind of confession or self-justification, addressed to Michaels (the surname provided by the authorities in Prince Albert that he uses even after K has corrected him): "You should have hidden, Michaels. You were too careless of yourself. You should have crept away in the darkest reach of the deepest hole and possessed yourself in patience till the troubles were over" (206). Of course, K had sought to do just that. The novel is indeed about how much "care" it takes not to be taken care of at all.[38] The further he attempts to withdraw from the world of political and ev-

---

**37** Giorgio Agamben, *The Kingdom and the Glory: For a Theological Genealogy of Economy and Government*, trans. Lorenzo Chieso with Matteo Mandarini (Stanford: Stanford University Press, 2011), esp. chap. 5.
**38** When K later escapes the camp and makes his way back to Sea Point where the story had begun, he runs into a pimp and his two "sisters." The man, who takes a particular sympathetic

eryday concerns the more he attracts the attentions and "care" of the authorities.[39]

The medical officer's letter continues:

> Did you think you were a spirit invisible, a visitor on our planet, a creature beyond the reach of the laws of nations? Well, the laws of nations have you in their grip now: they have pinned you down in a bed beneath the grandstand of the old Kenilworth racecourse, they will grind you in the dirt if necessary. The laws are made of iron, Michaels, I hope you are learning that. No matter how thin you make yourself, they will not relax. There is no home left for universal souls, except perhaps in Antarctica or on the high seas. (206–207)

In the drafts, this passage is the last from which the mention of K's allusions to a father as a "citizen of the world" (be he the Khamieskroon killer or Michael Kohlhaas) was struck. In its re-written form the passage insists on the futility of what the medical officer is prepared to recognize as K's particular form of resistance. He cannot bear to see K die a pointless and miserable death under his care – one that would not even count as a hunger strike (if the authorities allowed hunger strikes). The letter concludes: "I appeal to you, Michaels: *yield!* A friend" (208).

Why, K wonders, does the medical officer want to keep him alive? The letter opens with a more revealing confession – one, however, that need not contradict the sentiment expressed in its closing: "The answer is: Because I want to know your story" (204). As K sees things, those who want him to live, indeed may make him live, do so only in order to hear the story of a miserable life that was no life at all: "They want me to open my heart and tell them the story of a life lived in cages" (247). But in a manner that infuriates the medical officer – and not a few of Coetzee's readers – K will not or cannot yield. Obliged to interrogate K about his association with the guerillas and anxious to establish his innocence the medical officer eventually cries out:

> "Give yourself some substance, man, otherwise you are going to slide through life absolutely unnoticed. You will be a digit in the units column at the end of the war when they do the big subtraction sum to calculate the difference, nothing more. You don't want to be simply

---

interest in him and makes all sorts of gestures that K receives with his habitual apathy, states, "'It is difficult to be kind [...] to a person who wants nothing. You must not be afraid to say what you want, then you will get it. That is my advice to you, my thin friend'" (244). "I have become an object of charity," K later observes to himself, "Everywhere I go there are people waiting to exercise their forms of charity on me" (246) and later still, "I have escaped the camps; perhaps, if I lie low, I will escape the charity too" (249).

**39** On care in Coetzee's writings in general, see Katherine Hallemeier, "J. M. Coetzee's Literature of Hospice," *MFS* 62.3 (2016): 481–498.

one of the perished, do you? You want to live, don't you? Well then, talk, make your voice heard, tell your story!" (192)

K does not answer; when obliged to, he produces obscure utterances like his response to the Visagie boy.

Literature attends to the singular. This, in any case, has been the perhaps *schwärmerisch* principle of this book. The imperative that the solicitous but exasperated medical officer shouts at K is thus the imperative that governs a certain kind of modern literature whose purpose it is to lend life – and give a voice – to figures who are otherwise condemned to remain "digits in the units column." As such this literature emerges – and the historical development of the institution of literary fiction in the seventeenth and eighteenth centuries indicates this complicity – as a particular branch of the "providential machinery" of modern governmentality. In the introduction he published to the unwritten book, *The Lives of Infamous Men* (1977), Foucault briefly discusses the curiously transgressive complicity of literature with modern government: "literature belongs to the great system of constraint by which the West obligated the quotidian to enter into discourse."[40] At the same time that "an apparatus was being installed for forcing people to tell the 'insignificant' [*l'infime*]" (here Foucault refers to the workings of the police), literature, in the modern, institutional sense, emerged as a discourse that operated under a similar "kind of injunction to ferret out the most nocturnal and most quotidian elements of existence."[41] Taking upon itself "the duty to say what is most resistant to being said" literature thus becomes, among other things to be sure, the site in which the "discourse of 'infamy'" is preserved.[42] Translated into the providential terms suggested by Agamben reading Foucault, literature takes as its task attending to every possible deviation or perversion, every exception or omission. In this respect literature operates on the principle that everything can be assimilated. Literature, in short, is the space where sparrows are not forgotten and may in fact assume decisive significance.

It is not by chance therefore that the peculiar figures treated in this study – starting with the gypsy soothsayer in *Michael Kohlhaas* whose writing is concealed also from the reader's gaze – are striking for the anti-literary, but for that reason all the more literary, character of their comportment. They occasion literary fascination, and have in some particularly infamous cases, those of Bar-

---

40 Michel Foucault, "The Lives of Infamous Men" (1977), in *Power*, ed. James D. Faubion (London: Penguin Books, 1994), 174.
41 Foucault, "Infamous Men," 173.
42 Foucault, "Infamous Men," 174.

tleby and Gregor, come to present the essence of modern literature precisely because they resist its pretensions providentially to elaborate a world of significance.

If Michael K is a frustrating figure to those of a political but literary disposition, it is because in his postcolonial condition, he is allergic to the cares and attentions of institutions, including the institution of literature.[43] This political drama of the literary is played out in Part II of *Life & Times* in which the attentions of the medical officer are gradually transformed into the more pliant but relentless attentions of a literary critic.

K is treated by the medical officer as the figure he seems indeed to be, namely, the figure who does not yield: "He passes through these institutions and camps and hospitals and God knows what else like a stone" (185). Insofar as he comes to see K as someone categorically beyond both his area of expertise and the reach of the regime in which he is an increasingly unwilling accomplice, the medical officer becomes convinced, not without a certain ironic self-awareness, that K inhabits a realm to which he has no access. As he half-jokingly remarks to the camp commander, "people like Michaels are in touch with things you and I don't understand. They hear the call of the great good master and they obey. Haven't you heard of elephants?" (212). K's exceptionality with regard to the existing political dispensation or, to use a phrase the medical officer suggests, the "originality" of his resistance to it, appears to consist in not taking a stand at all. It seems to lend him a different order of significance and threatens to make him into a literary figure who, despite himself, stands for something. He might escape the camps, but can K evade the even more capacious reach of literature?

In the final section of Part II, the medical officer writes about an imagined encounter with K, who had in fact escaped the camp some days previously, in which he begs K to take him with him. The passage amounts to an account of the genesis of meaning. K is not merely an "original" figure. He figures the origin of figuration. And the significance that seems to concentrate itself in K, the medical officer feels certain, is not a projection of his own needs or desires – "a lack in myself, a lack, say, of something to believe in" (226) – but rather an event of signification, the coming-to-be of meaning in the world:

> "This is not my imagination," I would say to myself. "This sense of a gathering meaningfulness is not something like a ray that I project to bathe this or that bed, or a robe in which I

---

[43] On the complicities between the state and the novel, see Timothy Wright, "The Art of Evasion: Writing and the State in J. M. Coetzee's *Life & Times of Michael K*," *Journal of Literary Studies* 28.3 (September 2012): 55–76.

wrap this or that patient according to whim. Michaels means something, and the meaning he has is not private to me." (226)

The staging of the genesis of a literary figure, whose nature is to escape the structures in which it finds itself, is perhaps the very form of the literary insofar as it concerns itself with originality and the significance of singular figures and events.

Although it is easy to dismiss the medical officer, he is in fact the most attentive of readers. If he ends up sounding mad, certainly madder than K, it is because in this imagined address (or letter or dream or hallucination) he indeed attempts, to the degree that it is possible, to let K be what he seems to be. He is even ready to acknowledge, in fact he has already acknowledged, the futility of his own project: if not to escape with K then at least to understand him. All is lost but he does not give up. Instead he writes fiction. What K represents for the medical officer – and what K seems to represent for Coetzee's readers – is a flight from the world of the camps, from the war and its justifications, and from the stories one tells oneself in order to live. The medical officer's fiction chases after K while attempting to provide a final interpretation of this "great escape artist" (228), who he knows wants nothing to do with him:

> At this moment, I suspect, because such is your nature, you would break into a run. So I would have to run after you [...] calling out: "Your stay in the camp was merely an allegory, if you know that word. It was an allegory – speaking at the highest level – of how scandalously, how outrageously a meaning can take up residence in a system without becoming a term in it." (227–228)

If the reader is to avoid ending up in the same position as the medical officer, shouting desperately after a retreating figure: "'Am I right? [...] Have I understood you? If I am right, hold up your right hand; if I am wrong, hold up your left!'" (229), then that reader cannot insist on simply taking K to the letter, if that were possible.

It cannot, for example, be a matter of letting a sparrow be a sparrow. Perhaps one can take a cue rather from K's peculiar use of language when he speaks of sparrows. In his notes for Part II, which introduce the voice of the medical officer, Coetzee writes the following in quotation marks: "Michaels is him that man in New York who said, 'I prefer not to'..."[44] It seems likely that Coetzee was con-

---

[44] J. M. Coetzee Papers, Container 33.5: 61. On K and Bartleby see Hardt and Negri, *Empire*, 203–204. Buelens and Hoens refer to Deleuze's reading of *Bartleby* in their criticism of Attridge's theory of the singularity of literature, which emerges out of the critique of allegory in his work on Coetzee, see Gert Buelens and Dominiek Hoens, "'Above and Beneath Classifica-

sidering the line as one to be spoken or written by the medical officer. It is, in any case, the sort of thing he might say. However, unlike Bartleby, K does not have a "formula." Instead he says things like, "So small you don't taste it as it goes down [...] You wouldn't get yourself dirty, not even your little finger" (87). I am inclined to call this the language of *improvidence* from which no useful meaning can be derived. For the language of improvidence elaborates a "tender" sphere of insecurity absolved of the exigencies of ends and purposes – even the supposedly most unpragmatic and useless ends of literature.

## Justice

If *Life & Times* contributes to a "criticism of 'political reason'" of the sort that Foucault was pursuing in a historical manner during the same years, it is because the figure of K presents an unassimilable fault in the rationality of the prevailing political dispensation. K's problematic relation to reason and to politics is made explicit when a group of guerillas encamps nearby and the question concerning the political significance of his gardening asserts itself. K considers joining them but does not:

> He even knew the reason why: because enough men had gone off to war saying the time for gardening was when the war was over; whereas there must be men to stay behind and keep gardening alive, or at least the idea of gardening; because once that cord was broken, the earth would grow hard and forget her children. That was why. (150)

It did not take Coetzee's critics to point out that this is a poor reason, perhaps even a suspect one. K himself observes: "Between this reason and the truth that he would never announce himself, however, lay a gap" (150).

On account of this gap, K seems incapable of providing a justification for himself. Indeed, to some critics his faulty reasoning seems to embody a resigna-

---

tion': *Bartleby, Life and Times of Michael K*, and Syntagmatic Participation," *diacritics* 37.2–3 (2007): 157–170; Gilles Deleuze, "Bartleby; or, The Formula," in *Essays Critical and Clinical*, trans. Daniel Smith and Michael Greco (Minneapolis: Minnesota University Press, 1997); and Derek Attridge, *The Singularity of Literature* (London: Routledge, 2004) and *J. M. Coetzee and the Ethics of Reading: Literature in the Event* (Chicago: University of Chicago Press, 2004), chap. 2. The issue hinges in Deleuze's terms on whether through such figures literature stages an instance of recognition (*reconnaître*) or encounter (*rencontre*). In my reading, this question of the "literary" encounter/recognition of the singular is itself staged in Part II of the text.

tion concerning the very possibility of justice.[45] For the gap lies in the nonrelation between the project of self-determination, epitomized by the guerillas, and what K calls gardening. It imposes itself as the missing articulation between an explicit demand for justice that calls for political engagement, and gardening, the impetus for which seems not to be construable in recognizable political terms. It is the very gap that was already evident in Kohlhaas's ambivalence between explicit para-political engagement against the establishment in the name of a more just order in the first part of Kleist's novella and, in the second, an immanent "peculiar" resistance to it, in which the kind of justice at stake remains inarticulate if not *vergeblich*. Coetzee's text, then, which sought to translate the guerila moment in *Michael Kohlhaas* from Prussia to South Africa, ends up bringing into focus the very gap that traverses Kleist's text and now seems indeed to indicate an antinomy in modern political life under the "colonizing" conditions of governmentality.

K dwells on/in this gap:

> Always, when he tried to explain himself to himself, there remained a gap, a hole, a darkness before which his understanding baulked, into which it was useless to pour words. The words were eaten up, the gap remained. His was always a story with a hole in it: a wrong story, always wrong. (150–151)

In the story this "hole" is literally the burrow, in which K hides until the guerillas have moved on. When he returns to the Visagie farm after escaping a "resettlement camp," K resolves to re-embark on his gardening with even more painstaking precautions, to hide it in plain sight: from one perspective a deserted farm, from another, a flourishing garden. In order to evade the authorities, he concludes that "a man must live so that he leaves no trace of his living" (135), and sets about the construction of a burrow, which he takes care to do in an offhand manner, ironically echoing a white colonial South African discourse about black African "idleness:"[46]

---

[45] Nadine Gordimer's review, "The Idea of Gardening," in *The New York Review of Books*, February 2, 1984, remains the most compelling criticism of the politics of *Life & Times* and is the starting point for many such considerations of the political implication of the text and of Coetzee's writings more broadly. See especially, David Attwell, *J. M. Coetzee: South Africa and the Politics of Writing* (Oakland: University of California Press, 1993), chap. 4; Dominic Head, *J. M. Coetzee* (Cambridge: Cambridge University Press, 1997), chap. 5; Jane Poyner, *J. M. Coetzee and the Paradox of Postcolonial Authorship* (Farnham: Ashgate, 2009), chap. 4.

[46] See J. M. Coetzee, "Idleness in South Africa," in *White Writing*, 12–35.

> I am not building a house out here by the dam to pass on to other generations. What I make ought to be careless, makeshift, a shelter to be abandoned without a tugging at the heartstrings. So that if ever they find this place or its ruins, and shake their heads and say to each other: What shiftless creatures, how little pride they took in their work!, it will not matter. (138)

Like his speech, K's burrow is a kind of subversion. The makeshift shelter declines to assert claims to permanence and posterity, which in South Africa (but not only there) characterized the always problematic and therefore insecure claim to "home." And it is met, when discovered by K's captors, with the same incomprehension and anxiety.

The references to Kafka's story *The Burrow* (1923), on which Coetzee published an essay in 1981, are unmistakable. But it is important to emphasize that what K carelessly puts together is the very opposite of the project that obsesses the creature in Kafka's story. Focusing on the strange temporality of the narration, Coetzee argues that Kafka's text presents an attempt to narrate a time of crisis, that is to say, an "eschatological time" in which every moment is decisive without relation to any other, and which therefore breaks altogether with "historical time," understood as a continuum of past-present-future. The attempt on the part of the creature to secure its burrow is complemented by – and in fact proves to be the very same project as – the attempt to "domesticate" such temporality by imposing, through narration, a semblance of the continuum of historical time.[47] To use terms that draw on Foucault and Agamben, the political resonance of Coetzee's reading consists in demonstrating the complicity between the security apparatus or providential machinery and the narrative construct of historical time. The anxious creature/narrator, who finds itself in a time of crisis, understands that securing its burrow coincides with the construction of a sense of historical time. The inevitable failure of this project only increases its insecurities and the urgency of its renewed efforts.[48]

---

[47] This securing of temporality by means of narration deploys, Coetzee argues, an ambiguity, available in German as in English, of tense and aspect. The creature-narrator attempts to overcome its critical anxiety by casting the narration in what appears to be an iterative present (a present that binds the past to the present and the future so as to produce the effect of historical time). But narrated time so construed is constantly straining under the pressure of the time of narration, which intermittently, if decisively, erupts to undermine the semblance of continuity over time.

[48] In his notebook, Coetzee asks the following question: "Does K here hear a whistling in his burrow?" J. M. Coetzee Papers, Container 33.5: 74. The reference is to the whistling that the creature in Kafka's burrow hears at a certain point and then indefinitely in the course of his construction. In Coetzee's reading the whistling is a function of the on-going construction (or simply: the

In contrast, the improvident K lives a disjunctive, eschatological time, just as he lives in his makeshift burrow. In his notebook Coetzee considered adding a recollection about the burrow in the closing of the novel:

> At the end of the book: He remembered life in the burrow…It seemed to him that if he were not the only person who had lived in a burrow, he might as well be. For between him and the other burrow-dwellers there was no likeness, he had nothing to learn from their circumstances. There was nothing to learn from the past of creatures like himself, if there had been such creatures, and that was a great pity. Existence was not a web of meaning connecting the stories of various folk. As for me, he thought, I just live.[49]

Burrow-life can be neither rationalized nor allegorized and thus resists integration into a seamless totality of significance. For to live in a burrow is to abandon the political fiction – deconstructed in Kafka's story – that life can be secured within the providential constructs of historical time. The war, as the medical officer observes, is a time of waiting, yet so is so-called historical time. What it defers and obscures in its urgent project of construction is the time in which one might, as K puts it, "just live."

The "passion + urgency" of Michael Kohlhaas, whose passion for justice was the original conception of the "life & times" of Michael K, gives way to the presentation of a figure who, in or despite the state of emergency, just lives. The scene of K's faulty reasoning makes explicit the antinomy already legible in the life and times of Kleist's Kohlhaas, an antinomy that can now be phrased in terms of two apparently contradictory conceptions of justice: the justice of the guerilla as a political project mapped out and articulated as the end, in both senses, of historical time; and the justice of the gardener, which consists in "just living" in an eschatological time that is always at the end and each moment decisive.

In the published ending, or rather in Part III – since it is by no means clear whether the final section of *Life & Times* is an ending or a re-starting of the book – the notion of "just living" does not appear as such. Instead, the question of how "one can live" (184) is thematized. When K looks back, he consoles himself with being outside of the camps: "Perhaps that is enough of an achievement, for the time being" (248–249). He then draws a lesson from his life as a garden-

---

narration) straining to maintain its coherence at the critical moment in which it always already finds itself – "the whistling that comes from its point(s) of rupture" (579). Coetzee's slip of the pen – *here* instead of *hear* – is altogether consistent with his reading of *The Burrow* in which deictics emerge at the points of rupture that unsettle the narrative effort to constitute a (continuous) account of (continuous) time. K's burrow, in contrast, is as silent as the Karoo earth.
**49** J. M. Coetzee Papers, Container 33.5: 74.

er: "Because if there was one thing I discovered out in the country, it was that there is time enough for everything," and wonders parenthetically whether this might be "the moral of the whole story" (249). That this moral too is flawed is betrayed by the space to which K has retreated – not a burrow but the windowless closet (where his mother had lived in Cape Town) with a warning sign on the door reading "Danger" in three languages. Thinking about his inability to tell stories, K finally arrives at an account of himself: "It excited him, he found, to say, recklessly, *the truth, the truth about me. I am a gardener*, he said again, aloud" (247–248). But if the justification of gardening rang hollow in the Karoo, it is even harder to sustain here: "I am more like an earthworm, he thought. Which is also a kind of gardener. Or a mole, also a gardener, that does not tell stories because it lives in silence. But a mole or an earthworm on a cement floor?" (248). Though there may be time enough for everything, a life like that of an earthworm on a cement floor would be, of course, one of desolation. If K's peculiarity consists in his ability to slip through the gaps in the security apparatus and just live, such a life is not immune from the social, political, and environmental conditions in which he finds himself – even outside the camps. The cement floor presents a concrete objection to the argument for a life just lived. Clearly, the idea of gardening is not untouched by the ravages resulting from the projects and constructions of historical time. The antinomy of political life – between a politics of explicit engagement and an inarticulate mode of peculiar resistance, between constituting a just order and the improvidence of just living, between the time of the guerilla and the time of the gardener – imposes itself once again.

# Bibliography

Achebe, Chinua. "An Image of Africa." *Research in African Literatures* 9.1 (Spring 1978): 1–15.
Agamben, Giorgio. "Bartleby, or on Contingency." In *Potentialities: Collected Essays in Philosophy*, edited and translated by Daniel Heller-Roazen, 343–271. Stanford: Stanford University Press, 1999.
Agamben, Giorgio. *Homo Sacer: Sovereign Power and Bare Life*. Translated by Daniel Heller-Roazen. Stanford: Stanford University Press, 1998.
Agamben, Giorgio. *Opus Dei: An Archaeology of Duty*. Translated Adam Kotsko. Stanford: Stanford University Press, 2013.
Agamben, Giorgio. *Potentialities: Collected Essays in Philosophy*. Edited and translated by Daniel Heller-Roazen. Stanford: Stanford University Press, 1999.
Agamben, Giorgio. *The Kingdom and the Glory: For a Theological Genealogy of Economy and Government*. Translated by Lorenzo Chieso with Matteo Mandarini. Stanford: Stanford University Press, 2011.
Ajala, Adekunle. "The Nature of African Boundaries." *Africa Spectrum* 18.2 (1983): 177–89.
Alexander VI, Pope. Bull *Inter Caetera* (1493). In *International and United States Documents on Oceans Law and Policy*, edited by J. N. Moore, vol. 1, 75–78. Buffalo: William S. Hein, 1986.
Anders, Günther. *Kafka, pro & contra: Die Prozess-Unterlagen*. Munich: Beck, 1951.
Anderson, Mark. *Kafka's Clothes: Ornament and Aestheticism in the Habsburg fin de siècle*. Oxford: Clarendon Press, 1992.
Anghie, Antony. *Imperialism, Sovereignty, and the Making of International Law*. Cambridge: Cambridge University Press, 2005.
Arac, Jonathan. "Romanticism, the Self, and the City: *The Secret Agent* in Literary History." *boundary 2* 9.1 (1980): 75–90.
Arendt, Hannah. *Lectures on Kant's Political Philosophy*. Edited by Ronald Beiner. Chicago: University of Chicago Press, 1982.
Arendt, Hannah. *The Origins of Totalitarianism*. New York: Harcourt, 1966.
Arlt, Herbert and Alexandr W. Belobratow, eds. *Interkulturelle Erforschung der österreichischen Literatur*. St. Ingbert: Röhrig Universitätsverlag, 2000.
Arsić, Branka. *Passive Constitutions, or, 7 1/2 Times Bartleby*. Stanford: Stanford University Press, 2007.
Asiwaju, A. I. "The Conceptual Framework." In *Partitioned Africans*, edited by A. I. Asiwaju, 1–18. New York: St. Martin Press, 1985.
Asiwaju, A. I., ed. *Partitioned Africans*. New York: St. Martin Press, 1985.
Attridge, Derek. *J. M. Coetzee and the Ethics of Reading: Literature in the Event*. Chicago: University of Chicago Press, 2004.
Attridge, Derek. *The Singularity of Literature*. London: Routledge, 2004.
Attridge, John. "Two Types of Secret Agency: Conrad, Causation, and Popular Spy Fiction." *Texas Studies in Literature and Language* 55.2 (2013): 125–158.
Attwell, David. *J. M. Coetzee and the Life of Writing: Face to Face with Time*. New York: Viking, 2015.
Attwell, David. *J. M. Coetzee: South Africa and the Politics of Writing*. Oakland: University of California Press, 1993.

Austin, J. L. *How to Do Things with Words*, 2nd edition. Edited by J. O. Urmson and Marina Sbisà. Cambridge, MA: Harvard University Press, 1975.
Babcock, David. "Professional Subjectivity and the Attenuation of Character in J. M. Coetzee's *Life & Times of Michael K*." *PMLA* 127.4 (2012): 890–904.
Bainton, Roland Herbert. *Here I Stand: A Life of Martin Luther*. New York: Abingdon-Cokesbury Press, 1950.
Bakunin, Mikhail Aleksandrovich. *Statism and Anarchy*. In *Bakunin on Anarchy*, edited and translated by Sam Dolgoff with a preface by Paul Avrich, 323–351. London: Allen & Unwin, 1973.
Balke, Friedrich, Joseph Vogl, and Benno Wagner, eds. *Für Alle und Keinen. Lektüre, Schrift und Leben bei Nietzsche und Kafka*. Zurich: Diaphanes, 2008.
Barnet, Arno in collaboration with Roland Reuß and Peter Staengle, "Polizei-Theater-Zensur. Quellen zu Heinrich von Kleists *Berliner Abendblättern*." *Brandenburger Kleist-Blätter* 11 (1997): 29–353.
Benjamin, Walter. *Selected Writings*. Edited by Michael W. Jennings, Howard Eiland, and Gary Smith. Translated by Harry Zohn. Cambridge, MA: Harvard University Press, 1999.
Benjamin, Walter. *Benjamin über Kafka: Texte, Briefzeugnisse, Aufzeichnungen*. Edited by Hermann Schweppenhäuser. Frankfurt am Main: Suhrkamp, 1981.
Bigagli, Francesco. "'And Who art Thou, Boy?': Face-to-Face with Bartleby; Or Levinas and the Other." *Leviathan* 12.3 (2010): 37–53.
Binder, Hartmut. "Else Lasker-Schüler in Prag." *Wirkendes Wort* 3 (1994): 405–438.
Binder, Hartmut. *Kafka in Paris: Historische Spaziergänge mit alten Photographien*. Munich: Langen Müller, 1999.
Blickle, Peter. *Die Revolution von 1525*, 4th expanded edition. Munich: Oldenbourg, 2004.
Bloch, Ernst. *Naturrecht und menschliche Würde. Gesamtausgabe*, vol. 16. Frankfurt am Main: Suhrkamp, 1985.
Bloch, Ernst. *Thomas Münzer als Theologe der Revolution*. Munich: Wolff, 1921.
Boa, Elizabeth. "Creepy-Crawlies: Gilman's *The Yellow Paper* and Kafka's *The Metamorphosis*." *Paragraph* 13.1 (1990): 19–29.
Bogdal, Klaus-Michael. *Europa erfindet die Zigeuner: Eine Geschichte von Faszination und Verachtung*. Berlin: Suhrkamp, 2014.
Bohnert, Joachim. "Positivität des Rechts und Konflikt bei Kleist." *Kleist-Jahrbuch* (1985): 39–55.
Bolin, John. "Modernism, Idiocy, and the Work of Culture: J. M. Coetzee's *Life & Times of Michael K*." *Modernism/modernity* 22.2 (2015): 343–364.
Born, Jürgen, ed., with assistance from Herbert Mühlfeit and Friedemann Spicker. *Franz Kafka: Kritik und Rezeption zu seinen Lebzeiten, 1912–1914*. Frankfurt am Main: Fischer, 1979.
Brantlinger, Patrick. *Rule of Darkness: British Literature and Imperialism, 1830–1914*. Ithaca: Cornell University Press, 1990.
Brecht, Martin. *Martin Luther: Shaping and Defining the Reformation 1521–1532*. Minneapolis: Fortress Press, 1990.
Breger, Claudia. *Ortlosigkeit des Fremden: "Zigeunerinnen" und "Zigeuner" in der deutschsprachigen Literatur um 1800*. Cologne: Böhlau, 1998.
Britzolakis, Christina. "Pathologies of the Imperial Metropolis: Impressionism as Traumatic Afterimage in Conrad and Ford." *Journal of Modern Literature* 29.1 (2005): 1–20.

Brod, Max. *Über Franz Kafka*. Frankfurt am Main: Fischer, 1966.
Bruce, Maurice. *The Coming of the Welfare State: With a Comparative Essay on American and English Welfare Programs*. New York: Schocken Books, 1966.
Buelens, Gert and Dominiek Hoens, "'Above and Beneath Classification': *Bartleby, Life and Times of Michael K*, and Syntagmatic Participation." *diacritics* 37.2–3 (2007): 157–170.
Burkhardt, Johannes. *Das Reformationsjahrhundert: Deutsche Geschichte zwischen Medienrevolution und Institutionenbildung 1517–1617*. Stuttgart: Kohlhammer, 2002.
Campbell, Joseph. *The Hero with a Thousand Faces*. Princeton: Princeton University Press, 2004.
Campe, Rüdiger and Michael Niehaus, eds. *Gesetz. Ironie*. Heidelberg: Synchron Wissenschaftsverlag, 2004.
Campe, Rüdiger. "Kafkas Institutionenroman. *Der Proceß, Das Schloß*." In *Gesetz. Ironie*, edited by Rüdiger Campe and Michael Niehaus, 197–208. Heidelberg: Synchron Wissenschaftsverlag, 2004.
Cassirer, Ernst. *Heinrich von Kleist und die Kantische Philosophie*. Berlin: Reuther & Reichard, 1919.
Cavarero, Adriana. *Horrorism: Naming Contemporary Violence*. Translated by William McCuaig. New York: Columbia University Press, 2008.
Chakrabarty, Dipesh. *Provincializing Europe: Postcolonial Thought and Historical Difference*. Princeton: Princeton University Press, 2012.
Cicero, Marcus Tullius. *Cicero: The Orations Translated by Duncan, the Offices by Cockman, and the Cato and Lælius by Melmoth*. New York: J. & J. Harper, 1833.
Clark, Jill. "A Tale Told by Stevie: From Thermodynamic to Informational Entropy in *The Secret Agent*." *Conradiana* 36.1–2 (2004): 1–31.
Clarke, Bruce. "Allegories of Victorian Thermodynamics." *Configurations* 4.1 (1996): 67–90.
Coetzee, J. M. *J. M. Coetzee Papers*, Harry Ransom Center, The University of Texas at Austin.
Coetzee, J. M. *Dusklands*. New York: Penguin Books, 1996.
Coetzee, J. M. *Life & Times of Michael K*. London: Secker & Warburg, 1983.
Coetzee, J. M. *White Writing: On the Culture of Letters in South Africa*. New Haven: Yale University Press, 1988.
Conrad, Joseph. "An Anarchist." In *A Set of Six*, Dent Collected Edition. London: Dent, 1954.
Conrad, Joseph. "Geography and some Explorers." *National Geographic*, March 1924.
Conrad, Joseph. *Heart of Darkness*. In *Youth, Heart of Darkness, The End of the Tether*. Edited by Owen Knowles, *The Cambridge Edition of the Works of Joseph Conrad*. Cambridge: Cambridge University Press, 2010.
Conrad, Joseph. *Heart of Darkness: Complete, Authoritative Text with Biographical and Historical Contexts, Critical History, and Essays from Five Contemporary Critical Perspectives*. Edited by Ross C. Murfin. Boston: Bedford Books of St. Martin's Press, 1996.
Conrad, Joseph. *The Collected Letters of Joseph Conrad*. Edited by Frederick Karl and Laurence Davies. Cambridge: Cambridge University Press, 1983–2008.
Conrad, Joseph. *The Secret Agent: A Simple Tale*. Edited by Bruce Harkness and S. W. Reid. *The Cambridge Edition of the Works of Joseph Conrad*. Cambridge: Cambridge University Press, 1990.
Corngold, Stanley and Benno Wagner. *Franz Kafka: The Ghosts in the Machine*. Evanston: Northwestern University Press, 2011.

Corngold, Stanley and Ruth Gross, eds. *Kafka for the Twenty–First Century*. Rochester: Camden House, 2011.

Corngold, Stanley. "Allotria and Excreta in 'In the Penal Colony'." *Modernism/modernity* 8.2 (2001): 281–293.

Corngold, Stanley. *Franz Kafka: The Necessity of Form*. Ithaca: Cornell University Press, 1988.

Corngold, Stanley. *The Commentators' Despair: The Interpretation of Kafka's Metamorphosis*. London: Kennikat Press, 1973.

Coroneos, Con. *Space, Conrad, and Modernity*. Oxford: Oxford University Press, 2002.

David-Ménard, Monique. *La folie dans la raison pure: Kant lecteur de Swedenborg*. Paris: Vrin, 1990.

David, Claude, ed. *Franz Kafka: Themen und Probleme*. Göttingen: Vandenhoeck & Ruprecht, 1980.

Davis, Harold. "Conrad's Revisions of *The Secret Agent*: A Study in Literary Impressionism." *Modern Language Quarterly* 19.3 (1958): 244–254.

De la Durantaye, Leland. "The Suspended Substantive: On Animals and Men in Giorgio Agamben's *The Open*." *diacritics* 33.2 (2003): 3–9.

Deleuze, Gilles and Félix Guattari. *Kafka: Toward a Minor Literature*. Translated by Dana Polan. Minneapolis: University of Minnesota Press, 2012.

Deleuze, Gilles. "Bartleby; or, The Formula." In *Essays Critical and Clinical*, translated by Daniel Smith and Michael Greco, 68–90. Minneapolis: Minnesota University Press, 1997.

Deleuze, Gilles. *Essays Critical and Clinical*. Translated by Daniel Smith and Michael Greco. Minneapolis: Minnesota University Press, 1997.

Derrida, Jacques. *Voice and Phenomena: Introduction to the Problem of the Sign in Husserl's Phenomenology*. Translated by Leonard Lawlor. Evanston: Northwestern University Press, 2011.

Dillingham, William. *Melville's Short Fiction, 1853–1856*. Athens: University of Georgia Press, 2008.

Dilworth, Thomas. "Narrator of 'Bartleby': The Christian-Humanist Acquaintance of John Jacob Astor." *Papers on Language and Literature* 38 (2002): 49–75.

DiSanto, Michael. "'Dramas of Fallen Horses': Conrad, Dostoevsky, and Nietzsche." *Conradiana* 42.3 (Fall 2010): 45–68.

Durkheim, Émile. *Le Suicide*. Paris: Presses universitaires de France, 1993.

"Dynamite: The Modern Agent of Revolution." *The Anarchist*, March, 1885.

Eagleton, Terry. *Against the Grain: Essays 1975–1985*. London: Verso, 1986.

Edelman, Lee. "Occupy Wall Street: 'Bartleby' Against the Humanities." *History of the Present* 3.1 (Spring 2013): 99–118.

Edwards, Jonathan. *Freedom of the Will*. London: Thomas Nelson, 1845.

Edwards, Mark. *Printing, Propaganda, and Martin Luther*. Berkeley: University of California Press, 1994.

Eisenzweig, Uri. "Poetique de l'attentat: Anarchisme et littérature fin-de-siècle." In *Anarchisme et création littéraire*, edited by the Société d'histoire littéraire de la France, 439–452. Paris: Presses universitaires de France, 1999.

Ellrich, Lutz. "Diesseits der Scham. Notizen zu Spiel und Kampf bei Plessner und Kafka." In *Textverkehr*, edited by Claudia Liebrand and Franziska Schößler, 243–272. Würzburg: Königshausen & Neumann, 2004.

Emerson, Ralph Waldo. "The Transcendentalist." In *The Essential Writings of Ralph Waldo Emerson*, edited by Brooks Atkinson, 81–98. New York: Modern Library, 2000.
Engel, Manfred. "Das 'Wahre', das 'Gute' und die 'Zauberlaterne der Begeisterten Phantasie.' Legitimationsprobleme der Vernunft in der spätaufklärerischen Schwärmerdebatte." *German Life & Letters* 62.1 (2009): 53–66.
English, James. "Anarchy in the Flesh: Conrad's 'Counterrevolutionary' Modernism and the *Witz* of the Political Unconscious." *MFS* 38.3 (1992): 615–630.
Erdinast-Vulcan, Daphna. "'Sudden Holes in Space and Time': Conrad's Anarchist Aesthetics in *The Secret Agent*." In *Conrad's Cities: Essays for Hans van Marle*, edited by Gene Moore, 207–222. Amsterdam: Rodopi, 1992.
Ewald, François. *L'état providence*. Paris: Grasset, 1986.
Felman, Shoshana. "Madness and Philosophy or Literature's Reason." *Yale French Studies* 52 (1975): 206–228.
Fenves, Peter. "The Scale of Enthusiasm." In *Enthusiasm and Enlightenment in Europe, 1650–1850*, edited by Lawrence Eliot Klein and Anthony La Vopa. *The Huntington Library Quarterly* 60.1 (1998): 122–135.
Fichte, Johann Gottlieb. *Die Bestimmung des Menschen*. In *Fichtes sämmtliche Werke*, edited by I. H. Fichte, vol. 2. Berlin: Veit & Comp., 1845–1846.
Fichte, Johann Gottlieb. *Die Grundzüge des gegenwärtigen Zeitalters*. In *Gesamtausgabe der Bayerischen Akademie der Wissenschaften*, edited by Erich Fuchs, Reinhard Lauth, Hans Jacobs, and Hans Gliwitzky, vol. 1/8. Stuttgart: Frommann-Holzboog, 1964–2012.
Fichte, Johann Gottlieb. *Grundlage des Naturrechts nach Principien der Wissenschaftslehre*. In *Fichtes sämmtliche Werke*, edited by I. H. Fichte, vol. 3. Berlin: Veit & Comp., 1845–1846.
Firchow, Peter. *Envisioning Africa: Racism and Imperialism in Conrad's Heart of Darkness*. Lexington: University Press of Kentucky, 2000.
Fitzmaurice, Andrew. "The Genealogy of Terra Nullius." *Australian Historical Studies* 38.129 (2007): 1–15.
Fitzmaurice, Andrew. "The Justification of King Leopold II's Congo enterprise by Sir Travers Twiss." In *Law and Politics in British Colonial Thought*, edited by Ian Hunter and Shaunnagh Dorsett, 109–126. New York: Palgrave, 2010.
Fleishman, Avrom. *Conrad's Politics: Community and Anarchy in the Fiction of Joseph Conrad*. Baltimore: Johns Hopkins Press, 1967.
Fleming, Marie. "Propaganda by the Deed: Terrorism and Anarchist Theory in Late Nineteenth-century Europe." *Terrorism* 4.1–4 (1980): 1–23.
Foley, Barbara. "From Wall Street to Astor Place: Historicizing Melville's 'Bartleby.'" *American Literature* 72.1 (2000): 87–116.
Folts, James. *"Duely & constantly kept": A History of the New York Supreme Court, 1691–1847 and an Inventory of its Records (Albany, Utica, and Geneva Offices), 1797–1847*. Albany: New York State Court of Appeals and New York State Archives, 1991.
Forst, Graham. "Up Wall Street towards Broadway: The Narrator's Pilgrimage in Melville's 'Bartleby, the Scrivener'." *Studies in Short Fiction* 24 (Summer 1987): 263–270.
Foucault, Michel. "About the Concept of the 'Dangerous Individual' in 19th-Century Legal Psychiatry." *International Journal of Law and Psychiatry* 1.1 (1978): 1–18.

Foucault, Michel. "*Omnes et singulatim:* Towards a Criticism of 'Political Reason'." In *The Tanner Lectures on Human Values*, edited by Sterling McMurrin, 223–254. Salt Lake City: University of Utah Press, 1981.
Foucault, Michel. *"Society Must Be Defended": Lectures at the Collège de France 1975–1976*. Edited by Mauro Bertani and Alessandro Fontana. Translated by David Macey. New York: Picador, 2003.
Foucault, Michel. "The Lives of Infamous Men." In *Power*, edited by James D. Faubion, 147–175. London: Penguin Books, 1994.
Foucault, Michel. *Abnormal: Lectures at the Collège de France, 1974–1975*. Edited by Valerio Marchetti and Antonella Salomoni. Translated by Graham Burchell. New York: Picador, 2003.
Foucault, Michel. *Discipline and Punish: The Birth of the Prison*. Translated by Alan Sheridan. New York: Vintage, 1995.
Foucault, Michel. *Power.* Edited by James D. Faubion. London: Penguin Books, 1994.
Foucault, Michel. *Security, Territory, Population: Lectures at the Collège de France, 1977–78*. Edited by Michel Senellart, François Ewald, and Alessandro Fontana. New York: Palgrave Macmillan, 2007.
Freeden, Michael. *The New Liberalism: An Ideology of Social Reform*. Oxford: Clarendon Press, 1978.
Freud, Sigmund. "The Ego and the Id." In *The Standard Edition of the Complete Psychological Works of Sigmund Freud*, vol. 19, translated by James Strachey and Anna Freud. New York: Vintage, 2001.
Freud, Sigmund. *Die Traumdeutung*. Leipzig: F. Deuticke, 1900.
Freud, Sigmund. *Zur Psychopathologie des Alltagslebens*. Berlin: Karger, 1901.
Fried, Michael. "'A Blankness To Run At and Dash Your Head Against': On Conrad's *The Secret Agent*." *ELH* 79.4 (2012): 1039–1071.
Gaderer, Rupert. *Querulanz: Skizze eines exzessiven Rechtsgefühls*. Hamburg: Textem-Verlag, 2012.
Gailus, Andreas. *Passions of the Sign: Revolution and Language in Kant, Goethe, and Kleist*. Baltimore: Johns Hopkins University Press, 2006.
Geisenhanslüke, Achim. *Die Sprache der Infamie: Literatur und Ehrlosigkeit*. Paderborn: Fink, 2014.
Geulen, Christian, Anne von der Heiden, and Burkhard Liebsch, eds. *Vom Sinn der Feindschaft*. Berlin: Akademie, 2002.
Gilman, Sander. *Franz Kafka, the Jewish Patient*. New York: Routledge, 1995.
Goebel, Rolf. "Kafka and Postcolonial Critique: *Der Verschollene*, 'In der Strafkolonie,' 'Beim Bau der chinesischen Mauer'." In *A Companion to the Works of Franz Kafka*, edited by James Rollston, 187–212. Rochester: Camden House, 2002.
Goertz, Hans-Jürgen. *Innere und äussere Ordnung in der Theologie Thomas Müntzers*. Leiden: E.J. Brill, 1967.
Goertz, Hans-Jürgen. *Thomas Müntzer: Mystiker, Apokalyptiker, Revolutionär*. Munich: Beck, 1989.
GoGwilt, Christopher. "Joseph Conrad as Guide to Colonial History." In *The Oxford Historical Guide to Joseph Conrad*, edited by John Peters, 166–197. Oxford: Oxford University Press, 2010.
Goldman, Emma. *Anarchism and Other Essays*. New York: Mother Earth Publishing, 1910.

Goldstücker, Eduard. "Kafkas Eckermann?" In *Franz Kafka: Themen und Probleme*, edited by Claude David, 238–255. Göttingen: Vandenhoeck & Ruprecht, 1980.
Gordimer, Nadine. "The Idea of Gardening." *The New York Review of Books*, February 2, 1984.
Guérard, Albert. *Conrad the Novelist*. Cambridge, MA: Harvard University Press, 1958.
Hallemeier, Katherine. "J. M. Coetzee's Literature of Hospice." *MFS* 62.3 (2016): 481–498.
Hamacher, Bernd. "Geschichte und Psychologie der Moderne um 1800 (Schiller, Kleist, Goethe). 'Gegensätzische' Überlegungen zum 'Verbrecher aus Infamie' und zu 'Michael Kohlhaas'." *Kleist-Jahrbuch* (2006): 60–74.
Hamacher, Werner. "Guilt History: Benjamin's Sketch 'Capitalism as Religion'." Translated by Kirk Wetters. *diacritics* 32.3 (2002): 81–106.
Hamilton, John. "Procuratores: On the Limits of Caring for Another." *Telos* 170 (2015): 7–22.
Hamilton, John. *Security: Politics, Humanity, and the Philology of Care*. Princeton: Princeton University Press, 2013.
Hampson, Robert. "Conrad and the Idea of Empire." *L'Epoque Conradienne* (1989): 9–22.
Hardt, Michael and Antonio Negri. *Empire*. Cambridge, MA: Harvard University Press, 2000.
Hargreaves, J. D. "The Making of the Boundaries: Focus on West Africa." In *Partitioned Africans*, edited by A. I. Asiwaju, 19–28. New York: St. Martin Press, 1985.
Head, Dominic. "The (Im)possibility of Ecocriticism." In *Writing the Environment: Ecocriticism and Literature*, edited by Richard Kerridge and Neil Sammells, 27–39. London: Zed Books, 1998.
Head, Dominic. *J. M. Coetzee*. Cambridge: Cambridge University Press, 1997.
Hegel, Georg Wilhelm Friedrich. *Elements of the Philosophy of Right*. Edited by Allen Wood. Translated by H. B. Nisbet. Cambridge: Cambridge University Press, 1991.
Hegel, Georg Wilhelm Friedrich. *The Science of Logic*. Edited and translated by George di Giovanni. Cambridge: Cambridge University Press, 2015.
Hegel, Georg Wilhelm Friedrich. *Werke*. 20 vols. Frankfurt am Main: Suhrkamp, 1979.
Heidegger, Martin. *Being and Time*. Translated by Joan Stambaugh. Albany: SUNY, 2010.
Heidegger, Martin. *Introduction to Metaphysics*. Translated by Gregory Fried and Richard Polt. New Haven: Yale University Press, 2000.
Heidegger, Martin. *Sein und Zeit*. Tübinger: Niemeyer, 1967.
Heiden, Anne von der and Joseph Vogl, eds. *Politische Zoologie*. Zurich: Diaphanes, 2007.
Herbert, Auberon. *The Right and Wrong of Compulsion by the State and Other Essays*. Indianapolis: Liberty Classics, 1978.
Herder, Johann Gottfried. *Herders sämtliche Werke*. Edited by Bernhard Suphan. Berlin: Weidmann, 1877–1913.
Heyd, Michael. *Be Sober and Reasonable: The Critique of Enthusiasm in the Seventeenth and Early Eighteenth Centuries*. New York: Brill, 1995.
Hillis Miller, J. *Versions of Pygmalion*. Cambridge, MA: Harvard University Press, 1990.
Hinske, Norbert. "Die Aufklärung und die Schwärmer – Sinn und Funktion einer Kampfidee." In *Die Aufklärung und die Schwärmer*, edited by Lothar Kreimendahl and Norbert Hinske, 3–6. Hamburg: Meiner, 1988.
Hobbes, Thomas. *Leviathan*. Edited by Richard Tuck. Cambridge: Cambridge University Press, 1996.
Hobhouse, L. T. *Democracy and Reaction*. Edited by P. F Clarke. Brighton: Harvester Press, 1972.
Hobson, J. A. *The Social Problem: Life and Work*. London: J. Nisbet & Co., 1901.

Hochschild, Adam. *King Leopold's Ghost*. London: Pan Macmillan, 1998.
Höcker, Arne and Oliver Simons, eds. *Kafkas Institutionen*. Bielefeld: Transcript, 2007.
Hohmann, Joachim. *Geschichte der Zigeunerverfolgung in Deutschland*, revised edition. Frankfurt am Main: Campus Verlag, 1988.
Honig, Bonnie. "Charged: Debt, Power, and the Politics of the Flesh in Shakespeare's *Merchant*, Melville's *Moby-Dick*, and Eric Santner's *The Weight of All Flesh*." In Eric Santner, *The Weight of All Flesh: On the Subject-Matter of Political Theology*, edited by Kevis Goodman, 131–182. Oxford: Oxford University Press, 2016.
Horn, Peter. "Kafka in der Karoo: John M. Coetzee's *Life & Times of Michael K*." In *Interkulturelle Erforschung der österreichischen Literatur*, edited by Herbert Arlt and Alexandr W. Belobratow, 277–301. St. Ingbert: Röhrig Universitätsverlag, 2000.
Horn, Peter. "Michael K: Pastiche, Parody or the Inversion of Michael Kohlhaas." *Current Writing: Text and Reception in Southern Africa* 17.2 (2005): 56–73.
Horton, David M. and Katherine E. Rich, eds. *The Criminal Anthropological Articles of Cesare Lombroso published in the English Language Periodical Literature during the late 19th and early 20th Centuries*. Lewiston, NY: Edwin Mellen Press, 2004.
Horwarth, Peter. "Auf Den Spuren Teniers, Vouets und Raphaels in Kleists *Michael Kohlhaas*." *Seminar: A Journal of Germanic Studies* 5.2 (September 1969): 102–113.
Houen, Alex. *Terrorism and Modern Literature from Joseph Conrad to Ciaran Carson*. Oxford: Oxford University Press, 2002.
Howe, Irving. *Politics and the Novel*. New York: Horizon Press, 1957.
Huggan, Graham and Stephen Watson, eds. *Critical Perspectives on J. M. Coetzee*. London: Palgrave, 1996.
Hull, Isabel. *Absolute Destruction: Military Culture and the Practices of War in Imperial Germany*. Ithaca: Cornell University Press, 2013.
Hunter, Allan. *Joseph Conrad and the Ethics of Darwinism: The Challenges of Science*. London: Croom Helm, 1983.
Hunter, Ian and Shaunnagh Dorsett, eds. *Law and Politics in British Colonial Thought*. New York: Palgrave, 2010.
Husserl, Edmund. *Introduction to Logical Investigations: A Draft of a Preface to the Logical Investigations*. Edited by E. Fink. Translated by P. J. Bossert and C. H. Peters. The Hague: Martinus Nijhoff, 1975.
Husserl, Edmund. *Logical Investigations*, vol. 2. Edited by Dermot Moran. Translated by J. N. Findlay. New York: Routledge, 2001.
Husserl, Edmund. "Brief an Arnold Metzger, 4 September 1919." *Philosophisches Jahrbuch der Görres-Gesellschaft* 62 (1953): 195–200.
Ingarden, Roman. *On the Motives which led Husserl to Transcendental Idealism*. Dordrecht: Kluwer, 1975.
*International Conference Held at Washington for the Purpose of Fixing a Prime Meridian and a Universal Day: October, 1884. Protocols of the Proceedings*. Washington: Gibson Bros, 1884.
Jacobs, Carol. *Uncontainable Romanticism: Shelley, Brontë, Kleist*. Baltimore: Johns Hopkins University Press, 1989.
Jacobs, Robert. "Comrade Ossipon's Favorite Saint: Lombroso and Conrad." *Nineteenth-Century Fiction* 23.1 (1968): 74–84.

Janouch, Gustav. *Gespräche mit Kafka: Aufzeichnungen und Erinnerungen*. Frankfurt am Main: Fischer, 1951.
Jansen, Markus. *Das Wissen vom Menschen: Franz Kafka und die Biopolitik*. Würzburg: Königshausen & Neumann, 2012.
Jansen, Sarah. *"Schädlinge": Geschichte eines wissenschaftlichen und politischen Konstrukts 1840–1920*. Frankfurt am Main: Campus, 2003.
Johnson, Bruce. "Conrad's Impressionism and Watt's 'Delayed Decoding.'" In *Conrad Revisited: Essays for the Eighties*, edited by Ross Murfin, 51–70. University: University of Alabama Press, 1985.
Kafka, Franz. *Amtliche Schriften*. Edited by Klaus Hermsdorf and Benno Wagner. Frankfurt am Main: Fischer, 2004.
Kafka, Franz. *Briefe 1913–1914*. Edited by Hans-Gerd Koch. Frankfurt am Main: Fischer, 2001.
Kafka, Franz. *Briefe 1914–1917*. Edited by Hans-Gerd Koch. Frankfurt am Main: Fischer, 2005.
Kafka, Franz. *Der Heizer, In der Strafkolonie, Der Bau*. Edited by Malcolm Pasley. Cambridge: Cambridge University Press, 1966.
Kafka, Franz. *Der Verschollene*. Edited by Jost Schillemeit. Frankfurt am Main: Fischer, 1983.
Kafka, Franz. *Die Verwandlung*. Leipzig: Kurt Wolff Verlag, 1915.
Kafka, Franz. *Drucke zu Lebzeiten*. Edited by Wolf Kittler, Hans-Gerd Koch, and Gerhard Neumann. Frankfurt am Main: Fischer, 1996.
Kafka, Franz. *Ein Hungerkünstler. Vier Geschichten*. Berlin: Die Schmiede, 1924.
Kafka, Franz. *Franz Kafka: The Office Writings*. Edited by Stanley Corngold, Jack Greenberg, and Benno Wagner. Translated by Eric Patton and Ruth Hein. Princeton: Princeton University Press, 2009.
Kafka, Franz. *Historisch-Kritische Ausgabe sämtliche Handschriften, Drucke und Typoskripte*. Edited by Roland Reuß and Peter Staengel. Basel: Stroemfeld/Roter Stern, 1995–.
Kafka, Franz. *In der Strafkolonie*. Leipzig: Kurt Wolff Verlag, 1919.
Kafka, Franz. *Kafka's Selected Stories*. Translated by Stanley Corngold. New York: Norton, 2006.
Kafka, Franz. *Letters to Felice*. Edited by Eric Heller and Jürgen Born. Translated by James Stern and Elisabeth Duckworth. New York: Schocken, 1973.
Kafka, Franz. *Letters to Friends, Family, and Editors*. Translated by Clara Winston and Richard Winston. New York: Schocken Books, 1977.
Kafka, Franz. *Nachgelassene Schriften und Fragmente I*. Edited by Malcolm Pasley. Frankfurt am Main: Fischer, 1993.
Kafka, Franz. *Nachgelassene Schriften und Fragmente II*. Edited by Jost Schillemeit. Frankfurt am Main: Fischer, 1992.
Kafka, Franz. *Tagebücher*. Edited by Hans-Gerd Koch, Michael Müller, und Malcolm Pasley. Frankfurt am Main: Fischer, 1990.
Kafka, Franz. *The Diaries of Franz Kafka, 1910–1923*. Translated by Joseph Kresh and Martin Greenberg with the cooperation of Hannah Arendt. New York: Schocken, 1988.
Kafka, Franz. *The Metamorphosis: Translation, Backgrounds and Contexts, Criticism*. Edited and translated by Stanley Corngold. New York: W.W. Norton, 1996.
Kafka, Franz. *The Trial*. Translated by Mike Mitchell. Oxford: Oxford University Press, 2009.
Kaiser, Birgit Mara. *Figures of Simplicity: Sensation and Thinking in Kleist and Melville* Albany: SUNY, 2012.

Kant, Immanuel. *Critique of Practical Reason*. Edited and translated by Mary Gregor. Cambridge: Cambridge University Press, 2015.
Kant, Immanuel. *Critique of the Power of Judgment*. Edited by Paul Guyer. Translated by Paul Guyer and Eric Matthews. Cambridge: Cambridge University Press, 2000.
Kant, Immanuel. *Gesammelte Schriften*. Edited by Preussische (later Deutsche) Akademie der Wissenschaften. 29 vols. Berlin: Reimer (later De Gruyter), 1900–.
Kant, Immanuel. *Groundwork of the Metaphysics of Morals*. Edited and translated by Mary Gregor and Jens Timmermann. Cambridge: Cambridge University Press, 2012.
Kant, Immanuel. *Kant on Swedenborg: Dreams of a Spirit-Seer and other Writings*. Edited by Gregory R Johnson. Translated by Gregory R. Johnson and Glenn Alexander Magee. West Chester: Swedenborg Foundation Publishers, 2007.
Kant, Immanuel. *The Metaphysics of Morals*. Edited and translated by Mary Gregor and Jens Timmermann. Cambridge: Cambridge University Press, 1996.
Kant, Immanuel. *Träume eines Geistersehers: erläutert durch Träume der Metaphysik*. Edited by Rudolf Malter. Stuttgart: Reclam, 1976.
Kern, Stephen. *The Culture of Time and Space 1880–1918*. Cambridge, MA: Harvard University Press, 1983.
Kerridge, Richard and Neil Sammells, eds. *Writing the Environment: Ecocriticism and Literature*. London: Zed Books, 1998.
Kierkegaard, Søren. *Die Wiederholung; Drei Erbauliche Rede 1843*. Translated by Emanuel Hirsch. Düsseldorf: Diederichs, 1955.
Kierkegaard, Søren. *Fear and Trembling; Repetition*. Edited and translated by Edna Hong and Howard Hong. Princeton: Princeton University Press, 1983.
Kierkegaard, Søren. *Søren Kierkegaards Skrifter*. Edited by Niels Jørgen Cappelørn et al. Copenhagen: Forskningscentret and Gads Forlag, 1997.
Kierkegaard, Søren. *The Book on Adler*. Edited and translated by Edna Hong and Howard Hong. Princeton: Princeton University Press, 1998.
Kierkegaard, Søren. *The Point of View*. Edited and translated by Edna Hong and Howard Hong. Princeton: Princeton University Press, 1998.
Kierkegaard, Søren. *The Sickness unto Death: A Christian Psychological Exposition for Upbuilding and Awakening by Anti-Climacus*. Edited and translated by Edna Hong and Howard Hong. Princeton: Princeton University Press, 2011.
Kierkegaard, Søren. *Two Ages: The Age of Revolution and the Present Age: A Literary Review*. Edited and translated by Edna Hong and Howard Hong. Princeton: Princeton University Press, 1978.
Kierkegaard, Søren. *Without Authority*. Edited and translated by Edna Hong and Howard Hong. Princeton: Princeton University Press, 1997.
Kittler, Wolf. "In dubio pro reo. Kafka's Strafkolonie." In *Kafkas Institutionen*, edited by Arne Höcker and Oliver Simons, 33–72. Bielefeld: Transcript, 2007.
Kittler, Wolf. *Die Geburt des Partisanen aus dem Geist der Poesie: Heinrich von Kleist und die Strategie der Befreiungskriege*. Freiburg: Rombach, 1987.
Klein, Lawrence Eliot and Anthony La Vopa, eds. *Enthusiasm and Enlightenment in Europe, 1650–1850*, The Huntington Library Quarterly 60.1 (1998).
Kleist, Heinrich von. *Sämtliche Werke, Brandenburger Kleist-Ausgabe*. Edited by Roland Reuß and Peter Staengle. Basel: Stroemfeld/Roter Stern, 1988–2010.

Knowles, Owen. "The Texts: An Essay." In *Youth, Heart of Darkness, The End of the Tether*, edited by Owen Knowles, 257–324. Cambridge: Cambridge University Press, 2010.
Kohn, Margaret. "Kafka's Critique of Colonialism." *Theory and Event* 8.3 (2005).
Kolani, Ruth. "Secret Agent, Absent Agent? Ethical-Stylistic Aspects of Anarchy in Conrad's *The Secret Agent*." In *The Ethics in Literature*, edited by Dominic Rainsford, Andrew Hadfield, and Tim Woods, 86–100. Basingstoke: Macmillan, 1999.
Kommerell, Max. *Geist und Buchstabe der Dichtung: Goethe, Schiller, Kleist, Hölderlin*. Frankfurt am Main: Vittorio Klostermann, 1991.
Kreimendahl, Lothar and Norbert Hinske, eds. *Die Aufklärung und die Schwärmer*. Hamburg: Meiner, 1988.
La Vopa, Anthony. "The Philosopher and the *Schwärmer*: On the Career of a German Epithet from Luther to Kant." In *Enthusiasm and Enlightenment in Europe, 1650–1850*, edited by Lawrence Eliot Klein and Anthony La Vopa. *The Huntington Library Quarterly* 60.1 (1998): 85–115.
Lackey, Michael. "The Moral Conditions for Genocide in Joseph Conrad's *Heart of Darkness*." *College Literature* 32.1 (2005): 20–41.
Lange, Victor. "Zur Gestalt des Schwärmers im deutschen Roman des 18. Jahrhunderts." In *Festschrift für Richard Alewyn*, edited by Herbert Singer and Benno von Wiese, 151–164. Cologne: Böhlau, 1967.
Larubia-Prado, Francisco. "Horses at the Frontier in Kleist's Michael Kohlhaas." *Seminar: A Journal of Germanic Studies* 46.4 (2010): 330–350.
Lenin, V. I. *What Is to Be Done? Burning Questions of Our Movement*. Peking: Foreign Language Press, 1973.
Leavis, F. R. *The Great Tradition*. New York: New York University Press, 1969.
Levenson, Michael. "The Value of Facts in the Heart of Darkness." *Nineteenth-Century Fiction* 40.3 (1985): 261–280.
Liebrand, Claudia and Franziska Schößler, eds. *Textverkehr*. Würzburg: Königshausen & Neumann, 2004.
Liska, Vivian. *When Kafka Says We: Uncommon Communities in German-Jewish Literature*. Bloomington: Indiana University Press, 2009.
Locke, John. *The Works of John Locke*, vol. 1. London: Thomas Tegg et al., 1823.
Louis, William Roger. *Ends of British Imperialism: The Scramble for Empire, Suez, and Decolonization*. New York: I.B. Tauris, 2006.
Lüdemann, Susanne. "Literarische Fallgeschichten. Schillers 'Verbrecher aus verlorener Ehre' und Kleists 'Michael Kohlhaas.'" In *Das Beispiel. Epistemologie des Exemplarischen*, edited by Jens Ruchatz, Stefan Willer, and Nicolas Pethes, 208–223. Berlin: Kadmos, 2007.
Luther, Martin. *Martin Luthers Werke*. Weimar: Böhlau, 1883–2009.
Mackinder, H. J. "The Geographical Pivot of History." *The Geographical Journal* 23.4 (April 1904): 421–437.
Manning, Patrick. *Slavery and African Life: Occidental, Oriental, and African Slave Trades*. Cambridge: Cambridge University Press, 1990.
Mansell, Darrel. "Trying to Bring Literature Back Alive: The Ivory in Joseph Conrad's 'Heart of Darkness.'" *Criticism* 33.2 (1991): 205–215.
Marcuse, Herbert. "Industrialization and Capitalism." *New Left Review* I.30 (April 1965): 3–17.

Marx, Karl. *Early Political Writings*. Edited and translated by Joseph O'Malley with Richard A. Davis. Cambridge: Cambridge University Press, 1994.
Marx, Karl. *Marx & Engels Collected Works*. New York: International Publishers, 1975–2004.
Marx, Leo. "Melville's Parable of the Walls." *The Sewanee Review* 61.4 (1953): 602–627.
Mbembe, Achille. "Necropolitics." *Public Culture* 15.1 (2003): 11–40.
Mbembe, Achille. *On the Postcolony*. Berkeley: University of California Press, 2001.
McCall, Dan. *The Silence of Bartleby*. Ithaca: Cornell University Press, 1989.
McMurrin, Sterling, ed. *The Tanner Lectures on Human Values*. Salt Lake City: University of Utah Press, 1981.
Mehigan, Tim. *Heinrich von Kleist: Writing after Kant*. Rochester: Camden House, 2011.
Melchiori, Barbara Arnett. *Terrorism in the Late Victorian Novel*. London: Croom Helm, 1985.
Melville, Herman. *Piazza Tales and Other Prose Pieces, 1839–1860. Scholarly Edition*. Edited by Harrison Hayford, Alma A. MacDougall, G. Thomas Tanselle and others. Evanston: Northwestern University Press: 1987.
Mendelssohn, Moses. "Rezension der Träume." In Immanuel Kant, *Träume eines Geistersehers: erläutert durch Träume der Metaphysik*. Edited by Rudolf Malter, 118. Stuttgart: Reclam, 1976.
Menninghaus, Winfried. *Ekel: Theorie und Geschichte einer starken Empfindung*. Frankfurt am Main: Suhrkamp, 2002.
Merivale, Patricia. "Audible Palimpsests: Coetzee's Kafka." In *Critical Perspectives on J. M. Coetzee*, edited by Graham Huggan and Stephen Watson, 152–167. London: Palgrave, 1996.
Mills, Catherine. "Life beyond Law: Biopolitics, Law and Futurity in Coetzee's 'Life and Times of Michael K'." *Griffith Law Review* 15.1 (2006): 177–195.
Miskolcze, Robin. "The Lawyer's Trouble with Cicero." *Leviathan* 15.2 (2013): 43–53.
Möbus, Frank. *Sünden-Fälle: die Geschlechtlichkeit in Erzählungen Franz Kafkas*. Göttingen: Wallstein, 1994.
Monticelli, Daniele. "From Dissensus to Inoperativity: The Strange Case of J. M. Coetzee's Michael K." *English Studies* 97.6 (2016): 618–637.
Moore, Gene, ed. *Conrad's Cities: Essays for Hans van Marle*. Amsterdam: Rodopi, 1992.
Morel, Edmund. *King Leopold's Rule in Africa*. London: William Heinemann, 1904.
Mühlpfordt, Günter. "Luther und die 'Linken.' Eine Untersuchung seiner Schwärmerterminologie." In *Martin Luther: Leben, Werk, Wirkung*, 2nd edition, edited by Günter Vogler, 325–346. Berlin: Akademie, 1986.
Müller-Seidel, Walter. *Die Deportation des Menschen: Kafkas Erzählung "In der Strafkolonie" im europäischen Kontext*. Frankfurt am Main: Fischer, 1989.
Müller-Seidel, Walter. *Versehen und Erkennen: Eine Studie über Heinrich v. Kleist*. Cologne: Böhlau, 1967.
Müller, Gernot. *Kleist und die bildende Kunst*. Tübingen: Francke, 1995.
Muth, Ludwig. *Kleist und Kant. Versuch einer neuen Interpretation*. Cologne: Kölner Universitäts-Verlag, 1954.
Myers, Jeffrey. "The Anxiety of Confluence. Evolution, Ecology, and Imperialism in Conrad's *Heart of Darkness*." *Interdisciplinary Studies in Literature and Environment* 8.2 (2001): 97–108.
Najder, Zdzisław. *Joseph Conrad: A Life*. Rochester: Camden House, 2007.

Nietzsche, Friedrich. *Der Wille zur Macht 1884/88; Versuch einer Umwerthung aller Werthe.* Edited by Elisabeth Förster-Nietzsche. Leipzig: Naumann, 1906.
Nietzsche, Friedrich. *On the Genealogy of Morality.* Edited by Keith Ansell-Pearson. Translated by Carol Diethe. Cambridge: Cambridge University Press, 1994.
Nietzsche, Friedrich. *Sämtliche Werke: Kritische Studienausgabe.* Edited by Giorgio Colli und Mazzino Montinari. Munich: Deutscher Taschenbuch Verlag, 1988.
Nietzsche, Friedrich. *The Gay Science.* Edited by Bernard Williams. Translated by Josefine Nauckhoff. Cambridge: Cambridge University Press, 2001.
Nietzsche, Friedrich. *The Will to Power.* Translated by Walter Kaufmann. New York: Random House, 1968.
Nietzsche, Friedrich. *Thus Spoke Zarathustra: A Book for All and None.* Edited by Adrian Del Caro and Robert B. Pippin. Translated by Adrian Del Caro. Cambridge: Cambridge University Press, 2006.
Nixon, Rob. *London Calling: V.S. Naipaul, Postcolonial Mandarin.* Cambridge, MA: Harvard University Press, 1992.
Nkrumah, Kwame. *Challenge of the Congo.* New York: International Publishers, 1967.
Nohrnberg, Peter. "'I Wish He'd Never Been to School': Stevie, Newspapers and the Reader in *The Secret Agent*." *Conradiana* 35:1–2 (2003): 49–62.
Norris, Margot. *Beasts of the Modern Imagination: Darwin, Nietzsche, Kafka, Ernst and Lawrence.* Baltimore: Johns Hopkins University Press, 1985.
North, Paul. *The Problem of Distraction.* Stanford: Stanford University Press, 2012.
North, Paul. *The Yield: Kafka's Atheological Reformation.* Stanford: Stanford University Press, 2015.
Northey, Anthony. *Kafkas Mischpoche.* Berlin: Wagenbach, 1988.
Nowosadtko, Jutta. *Scharfrichter und Abdecker: der Alltag zweier "unehrlicher Berufe" in der frühen Neuzeit.* Paderborn: Ferdinand Schöningh, 1994.
Ó Donghaile, Deaglán. *Blasted Literature: Victorian Political Fiction and the Shock of Modernism.* Edinburgh: Edinburgh University Press, 2011.
Oliver, Egbert. "A Second Look at Bartleby." *College English* 6 (1945): 431–439.
Oliver, Hermia. *The International Anarchist Movement in Late Victorian London.* London: St. Martin's Press, 1983.
Ott, Michael. *Das ungeschriebene Gesetz: Ehre und Geschlechterdifferenz in der deutschen Literatur um 1800.* Freiburg: Rombach, 2001.
Pakenham, Thomas. *The Scramble for Africa, 1876–1912.* London: Weidenfeld & Nicolson, 1991.
Pasley, Malcolm. "Introduction." In Franz Kafka, *Der Heizer, In der Strafkolonie, Der Bau*, edited by Malcolm Pasley, 17–21. Cambridge: Cambridge University Press, 1966.
Paulus, Jörg. *Der Enthusiast und sein Schatten: literarische Schwärmer- und Philisterkritik um 1800.* Berlin: De Gruyter, 1998.
Pecora, Vincent. "Heart of Darkness and the Phenomenology of Voice." *ELH* 52.4 (1985): 993–1015.
Peters, Paul. "Witness to the Execution: Kafka and Colonialism." *Monatshefte für deutschsprachige Literatur und Kultur* 93 (2001): 401–425.
Pick, Daniel. *Faces of Degeneration: A European Disorder, c.1848–c.1918.* Cambridge: Cambridge University Press, 1989.

Piper, Karen. "The Language of the Machine: A Postcolonial Reading of Kafka." *The Journal of the Kafka Society of America* 20 (1996): 42–54.

Piper, Karen. *Cartographic Fictions: Maps, Race, and Identity.* New Brunswick: Rutgers University Press, 2002.

Poyner, Jane. *J. M. Coetzee and the Paradox of Postcolonial Authorship.* Farnham: Ashgate, 2009.

Pye, Patricia. "Hearing the News in 'The Secret Agent.'" *The Conradian* 34.2 (2009): 51–63.

Reinach, Adolf. "Concerning Phenomenology." Translated by Dallas Willard. *The Personalist* 50 (1969): 194–221.

Reuß, Roland. "Geflügelte Worte. Zwei Notizen zur Redaktion und Konstellation von Artikeln der *Berliner Abendblätter*." *Brandenburger Kleist-Blätter* 11 (1997): 3–9.

Reuß, Roland. "'Michael Kohlhaas' und 'Michael Kohlhaas.' Zwei deutsche Texte, eine Konjektur und das Stigma der Kunst." *Berliner Kleist-Blätter* 3 (1990): 3–43.

Reuß, Roland. "Nachrichten von Hans Kohlhase." *Berliner Kleist-Blätter* 3 (1990): 44–54.

Robertson, Ritchie. "Kafka, Goffman, and the Total Institution." In *Kafka for the Twenty-First Century*, edited by Stanley Corngold and Ruth Gross, 136–150. Rochester: Camden House, 2011.

Rogin, Michael. *Subversive Genealogy: The Politics and Art of Herman Melville.* New York: Knopf, 1983.

Rollston, James, ed. *A Companion to the Works of Franz Kafka.* Rochester: Camden House, 2002.

Ross, Stephen. *Conrad and Empire.* Columbia: University of Missouri Press, 2004.

Rowe, John Carlos. *Through the Custom-House.* Baltimore: Johns Hopkins University Press, 1982.

Rückert, Joachim. "'…der Welt in der Pflicht verfallen….' Kleists 'Kohlhaas' als moral- und rechtsphilosophische Stellungnahme." *Kleist-Jahrbuch* (1988–1989): 375–403.

Ryan, Simon. "Franz Kafka's *Die Verwandlung*: Transformation, Metaphor, and the Perils of Assimilation." *Seminar: A Journal of Germanic Studies* 43.1 (February 2007): 1–18.

Santner, Eric. "Kafka's *Metamorphosis* and the Writing of Abjection." In *The Metamorphosis: Translation, Backgrounds and Contexts, Criticism*, edited and translated by Stanley Corngold, 195–201. New York: W.W. Norton, 1996.

Santner, Eric. *On Creaturely Life: Rilke, Benjamin, Sebald.* Chicago: University of Chicago Press, 2006.

Santner, Eric. *The Weight of All Flesh: On the Subject-Matter of Political Theology.* Edited by Kevis Goodman. Oxford: Oxford University Press, 2016.

Schelling, Friedrich Wilhelm Joseph von. *Darlegung des wahren Verhältnisses der Naturphilosophie zu der verbesserten Fichte'schen Lehre.* In *Schellings sämmtliche Werke*, edited by K.F.A. Schelling, vol. I.7. Stuttgart: Cotta, 1856–1861.

Scherpe, Klaus and Elisabeth Wagner, eds. *Kontinent Kafka: Mosse-Lectures an der Humboldt-Universität zu Berlin.* Berlin: Vorwerk 8, 2006.

Schiller, Friedrich. *Der Verbrecher aus verlorener Ehre. Eine wahre Geschichte.* In *Schillers Werke. Nationalausgabe*, vol. 16, *Erzählungen*, edited by Julius Petersen and Hermann Schneider, 7–32. Weimar: Böhlau: 1954.

Schmidt, Ulrich. "Von der 'Peinlichkeit' der Zeit. Kafkas Erzählung 'In der Strafkolonie'." *Schiller-Jahrbuch* 28 (1984): 407–445.

Schmitt, Carl. *Political Theology*. Translated by George Schwab. Chicago: University of Chicago Press, 2005.
Schmitt, Carl. *Politische Theologie*. Berlin: Duncker & Humblot, 1922.
Schmitt, Carl. *Staat, Großraum, Nomos: Arbeiten aus den Jahren 1916 bis 1969*. Berlin: Duncker & Humblot, 1995.
Schmitt, Carl. *The Concept of the Political*. Translated by George Schwab. Chicago: University of Chicago Press, 2010.
Schmitt, Carl. *The Nomos of the Earth: In the International Law of the Jus Publicum Europaeum*. Translated by G. L. Ulmen. New York: Telos Press, 2006.
Schmitt, Carl. *Theorie des Partisanen. Zwischenbemerkung zum Begriff des Politischen*. Berlin: Duncker & Humblot, 1963.
Schmitt, Carl. *Theory of the Partisan: Intermediate Commentary on the Concept of the Political*. Translated by G. L. Ulman. New York: Telos Press, 2007.
Schnauder, Ludwig. *Free Will and Determinism in Joseph Conrad's Major Novels*. Amsterdam: Rodopi, 2009.
Schopenhauer, Arthur. *Sämtliche Werke*. Edited by Wolfgang Löhneysen. Frankfurt am Main: Suhrkamp, 1989.
Sebald, W. G. *The Rings of Saturn*. Translated by Michael Hulse. London: Vintage, 2002.
Shaftesbury, Earl of. "Letter Concerning Enthusiasm." In *Characteristics of Men, Manners, Opinions, Times*, edited by Lawrence E. Klein, 4–28. Cambridge: Cambridge University Press, 1999.
Shell, Susan Meld. *The Embodiment of Reason: Kant on Spirit, Generation, and Community*. Chicago: University of Chicago Press, 1996.
Sherry, Norman. *Conrad's Western World*. Cambridge: Cambridge University Press, 1980.
Simmel, Georg. "Die Großstädte und das Geistesleben." In *Die Großstadt. Vorträge und Aufsätze zur Städteausstellung*, 185–206. Dresden: v. Zahn & Jaensch, 1903.
Simons, Oliver. "Schuld und Scham. Kafkas episches Theater." In *Kafkas Institutionen*, edited by Arne Höcker and Oliver Simons, 269–294. Bielefeld: Transcript, 2007.
Singh, Frances. "Terror, Terrorism, and Horror in Conrad's *Heart of Darkness*." *Partial Answers: Journal of Literature and the History of Ideas* 5.2 (2007): 199–218.
Smith, Caleb. "Detention without Subjects: Prisons and the Poetics of Living Death." *Texas Studies in Literature and Language* 50.3 (2008): 243–267.
Smock, Ann. "Quiet." *Qui Parle* 2.2 (1988): 68–100.
Sokel, Walter. "Kafka's 'Metamorphosis': Rebellion and Punishment." *Monatshefte* 48.4 (1956): 203–214.
Spector, Scott. *Prague Territories: National Conflict and Cultural Innovation in Franz Kafka's Fin de Siècle*. Berkeley: University of California Press, 2000.
Stach, Rainer. *Kafkas erotischer Mythos: Eine ästhetische Konstruktion des Weiblichen*. Frankfurt am Main: Fischer, 1987.
Stedman Jones, Gareth. *Outcast London: A Study in the Relationship between Classes in Victorian Society*. Oxford: Clarendon Press, 1971.
Steig, Reinhold. *Heinrich von Kleist's Berliner Kämpfe*. Berlin and Stuttgart: Spemann, 1901.
Steinberg, Philip. "Lines of Division, Lines of Connection: Stewardship in the World Ocean." *Geographical Review* 89.2 (1999): 254–264.
Sten, C. W. "Bartleby the Transcendentalist: Melville's Dead Letter to Emerson." *Modern Language Quarterly* 35.1 (1974): 30–44.

Stott, Rebecca. "The Woman in Black: Race and Gender in *The Secret Agent*." *The Conradian* 17.2 (Spring 1993): 39–58.
Strobel, Georg Theodor. *Leben, Schriften und Lehren Thomä Müntzers, des Urhebers des Bauernaufruhrs in Thüringen.* Nürnberg: Monath und Kussler, 1795.
"The Fear of Dynamite," *The Spectator*, April 14, 1883.
Thomson, Matthew. *The Problem of Mental Deficiency: Eugenics, Democracy, and Social Policy in Britain c.1870–1959.* Oxford: Oxford University Press, 1998.
Thoreau, Henry David. *Aesthetic Papers.* Edited by Elizabeth P. Peabody. Boston: The Editor, 1849.
Torgovnick, Marianna. *Gone Primitive: Savage Intellects, Modern Lives.* Chicago: University of Chicago Press, 1990.
Toscano, Alberto. *Fanaticism: On the Uses of an Idea.* London: Verso, 2010.
Trotha, Lothar von. "Politik und Kriegführung." *Berliner Neueste Nachrichten*, February 3, 1909.
Trumpener, Katie. "The Time of the Gypsies: A 'People without History' in the Narratives of the West." *Critical Inquiry* 18.4 (1992): 842–884.
Tugendhat, Ernst. *Der Wahrheitsbegriff bei Husserl und Heidegger.* Berlin: De Gruyter, 1967.
Valente, Joseph. "The Accidental Autist: Neurosensory Disorder in *The Secret Agent*." *Journal of Modern Literature* 38.1 (2014): 20–37.
Valéry, Paul. "Svendenborg." In *Oeuvres*, edited by Jean Hytier, vol 1, 867–883. Paris: Gallimard, 1957.
VanZanten Gallagher, Susan. *A Story of South Africa: J. M. Coetzee's Fiction in Context.* Cambridge, MA: Harvard University Press, 1991.
Vismann, Cornelia. *Files: Law and Media Technology.* Translated by Geoffrey Winthrop-Young. Stanford: Stanford University Press, 2008.
Vital, Anthony. "Toward an African Ecocriticism: Postcolonialism, Ecology and *Life & Times of Michael K*." *Research in African Literatures* 39.1 (March 2008): 87–106.
Vogl, Joseph and Ethel Matala de Mazza. "Bürger und Wölfe. Versuch über politische Zoologie." In *Vom Sinn der Feindschaft*, edited by Christian Geulen, Anne von der Heiden, and Burkhard Liebsch, 207–217. Berlin: Akademie, 2002.
Vogl, Joseph. "Lebende Anstalt." In *Für Alle und Keinen. Lektüre, Schrift und Leben bei Nietzsche und Kafka*, edited by Friedrich Balke, Joseph Vogl, and Benno Wagner, 21–33. Zurich: Diaphanes, 2008.
Vogl, Joseph. *Ort der Gewalt: Kafkas literarische Ethik.* Zurich: Diaphanes, 2010.
Wagner, Benno. "Der Unversicherbare. Kafkas Protokolle." Habilitationsschrift, University of Siegen, 1998.
Wagner, Benno. "Kafkas Phantastisches Büro." In *Kontinent Kafka: Mosse-Lectures an der Humboldt-Universität zu Berlin*, edited by Klaus Scherpe and Elisabeth Wagner, 104–119. Berlin: Vorwerk 8, 2006.
Watt, Ian. *Conrad in the Nineteenth Century.* London: Chatto & Windus, 1980.
Weber, Max. *Economy and Society: An Outline of Interpretive Sociology.* Edited by Guenther Roth and Claus Wittich. Berkeley: University of California Press, 1978.
Weber, Max. *Max Weber Gesamtausgabe.* Edited by Wolfgang Schluchter. Tübingen: J. C. B. Mohr, 1998.
Weber, Max. *The Vocation Lectures.* Edited by David Owen and Tracy B. Strong. Translated by Rodney Livingstone. Indianapolis: Hackett, 2004.

Weinstock, Jeffrey. "Doing Justice to Bartleby." *ATQ* 17.1 (2003): 23–42.
Wesseling, H. L. *Divide and Rule: The Partition of Africa, 1880–1914*. Westport: Praeger Publishers, 1996.
Wright, Derek. "Black Earth, White Myth: Coetzee's *Life & Times of Michael K*." *MFS* 38.2 (1992): 435–444.
Wright, Timothy. "The Art of Evasion: Writing and the State in J. M. Coetzee's *Life & Times of Michael K*." *Journal of Literary Studies* 28.3 (September 2012): 55–76.
Young, Merwin Crawford. *The African Colonial State in Comparative Perspective*. New Haven: Yale University Press, 1994.
Zahavi, Dan. *Husserl's Legacy: Phenomenology, Metaphysics, and Transcendental Philosophy*. Oxford: Oxford University Press, 2017.
Zelnick, Stephen. "Melville's 'Bartleby': History, Ideology, and Literature." *Marxist Perspectives* 2 (Winter 1979–1980): 74–92.
Zieger, Susan Marjorie. *Inventing the Addict: Drugs, Race, and Sexuality in Nineteenth-Century British and American Literature*. Amherst: University of Massachusetts Press, 2008.
Zilcosky, John. "'Samsa war Reisender': Trains, Trauma, and the Unreadable Body." In *Kafka for the Twenty-First Century*, edited by Stanley Corngold and Ruth Gross, 179–206. Rochester: Camden House, 2011.
Zilcosky, John. *Kafka's Travels: Exoticism, Colonialism, and the Traffic of Writing*. New York: Palgrave Macmillan, 2004.
Zimring, Rischona. "Conrad's Pornography Shop." *MFS* 43.2 (1997): 319–348.
Ziolkowski, Theodore. "Kleists Werke im Lichte der zeitgenössischen Rechtskontroverse." *Kleist-Jahrbuch* (1987): 28–51.

# Index

*Abdecker* (knacker)   55–60
abstraction   19–25, 30, 112–114, 118, 123, 129, 136–137, 144
– faculty of   148
– imperial   6
– real   20
– state of   137
  See also Toscano
accident   150–153, 187, 196
accountant (*Heart of Darkness*)   118, 123
*Achtung.*   See under Kant
action   6, 36, 40, 43, 53, 81, 86, 90–94, 130, 135, 151–152
– anarchist   135–146, 150–157
– crisis of   135, 144
– destitute   136, 147, 152, 155–156
– direct   136
– extreme   55
– fanatical   147
– Fichte   45–6, 53
– political   7
– revolutionary   141
– righteous   54
  See also agency; *Fehlleistung*; passivity; speech act
activist   7, 151
– feminist   154
administrative massacre   118
affect   1, 54, 66, 91, 135, 137, 144, 149, 152
– excessive   51
– office   92, 97
– political   226
  See also compassion; passion; *Rechtgefühl*
Agamben, Giorgio   8, 9–12, 202, 237
– "Bartleby"   8, 89, 94, 97
– *Homo Sacer*   9–10, 186
– *Kingdom and the Glory*   235, 242
– *Opus Dei*   76–78, 81–82, 97
– state of exception   9, 11, 157
agency   77, 90, 130, 136, 139–147, 151–156, 196
– crisis of   6, 146, 155
– imperialist   114, 121, 127

– institutional   136
– liberal concept of   130, 134–136
– rationalization of   135
– revolutionary   142
  See also action
agent   6, 30, 90, 135–136, 147–149, 153, 168, 171, 187
– *agent provocateur*   143
– dulling   136–139, 143–144, 147
– grammatical   135
– imperialist   115, 118–119, 121, 126, 158
– moral   139
– of revolution   138, 142
– secret   (see Verloc)
Alexander VI (pope)   102
*Allgemeines Landrecht* (1794)   61–62
anarchism   87, 132, 135–136, 138–139, 141–147, 150–151, 153–156
animals   1–2, 39, 48, 52–58, 188, 199, 202, 209–210, 215. See also bear; dogs; earthworm; horses; human–animal distinction; political zoology; sparrows; *Ungeziefer*; wolves
apathy   137, 142, 144, 149, 152, 190, 221, 236
*Apocalypse Now* (film)   115, 222
apostle   26–29, 32, 112, 143
Arendt, Hannah
– *Lectures on Kant's Political Philosophy*   182
– *Origins of Totalitarianism*   104, 111–112, 118, 120, 122–123, 140
Arsić, Branka   82, 94
Arthur, Chester (U.S. President)   103
*Association Internationale Africaine*   107
Astor, John Jacob   80, 84, 98
Atlantic slave trade   105
Austin, J. L.   127, 145, 179. See also speech act
authenticity/inauthenticity   7, 110, 130, 135, 155, 162, 169, 176, 190–192, 205, 214, 233

## Index

authority   3, 18–19, 24, 62, 67–68, 72, 84, 88, 156, 167, 186, 228
– bureaucratic   6, 18, 157, 168, 170–171, 177–180
– crisis of   157, 167–168
– Kierkegaard on   19, 26–34
authorship   26–29, 38, 177, 241
autonomy   135, 137, 139, 233
avocations   75, 79–81, 84, 86, 96. See also action; office–duty; vocation

Bakunin, Mikhail   141–142, 152
Bartleby (Melville)   4–5, 7–8, 29, 34, 75–101, 149–150, 228, 239, 240
Bauer, Felice   162, 174, 212
bear   47–48, 71. See also under nature
Benjamin, Walter   137, 161, 163, 186, 202
Berlin Conference.   See Congo Conference
Bible   65, 101, 235
biopolitics   2, 6, 12, 30, 38–39, 41 78, 96, 120, 138, 185, 193, 207, 228, 230
Bismarck, Otto von   102, 105
Blanqui, Louis Auguste   139
blasé   137, 139, 149
Bloch, Ernst
– *Natural Law and Human Dignity*   50–53, 60–62
– *Thomas Münzer*   42
Boer War   134, 140
Bohnert, Joachim   56, 62
Bosch, Hieronymus   220
Brod, Max   161, 174, 221
bureaucracy   1–9, 21, 39, 49, 52, 61–63, 71–73, 83, 94–98, 101, 130, 157, 168, 185
– and liberalism   130
– as a vocation   12–18
– as pathology   35
– awe for   54
– case for   35
– contempt for   7, 54
– end of   4, 35, 61
– Hegel   20–25
– history of term   49, 75
– imperialist   109, 111, 118, 122–123
– Kierkegaard   26–28, 31
– Marcuse   17–19

– Marx   23–25
– metaphysics of   16
– political theology of   61–62
– Prussian 21, 23, 49, 61
– Weber   16–19, 157, 168
bureaucracy novella   4, 35, 49, 75, 111, 226. See also *Institutionenroman*
bureaucratic fanatic   2–3, 9, 12, 15, 34, 61, 123
bureaucratic fanaticism   1, 3–4, 6, 9–34, 44, 102, 123, 129, 160, 173, 221
bureaucratization   3–8, 9, 16, 18, 20, 24–25, 35, 44, 49, 54, 61, 75, 78, 82, 118, 146, 164, 185, 223
– fanatical   8, 32, 34, 223
– of authority   157, 167–170
– of philosophy   20–21
– passion for   1, 17

calling.   See vocation
camps   229, 234, 236, 238–239, 241, 243–244
care   7, 75, 85, 90, 96, 98–99, 133, 185–186, 188–199, 201, 205–210, 215–217, 234–236, 238, 242
– *cura*   85
– neglect   36, 56, 59, 150, 186, 201, 214, 229
  See also security
Casement, Roger   120, 148
Cassirer, Ernst   44–45, 47
Cavarero, Adriana   128–129
Chancery, Master in   79–81, 89, 92, 98
charisma   6, 18, 88, 229
Cicero, Marcus Tullius   75–76, 78–79, 85–86, 98, 101
Coetzee, J. M.
– J. M. Coetzee Papers   7, 223–229, 231–233, 239, 243
– *Dusklands*   222–223
– *Life & Times of Michael K*   7, 219–244
– "Time, Tense and Aspect in Kafka's 'The Burrow'"   241–244
– *White Writing*   232, 241
colonialism   6–7, 49, 104–107, 111, 113, 118, 121, 140, 157–158, 180, 182, 188,

222, 224, 227, 231–234, 241–242. *See also* imperialism; Arendt: *Origins*
Columbus, Christopher   102–103
command   52, 77, 167, 170, 180
– commandment   93, 208
– human   69
– of language   178
– of the body   196
   *See also* demand; imperative; office-duty; speech act
communication   3, 9–10, 29, 165, 181, 198
– immediate   42, 66
– indirect   3, 11, 19, 29
– of authority or revelation   3
– official   67
– piecemeal (*allmählich*)   181
compassion/empathy   137, 148–150
confusion   2–3, 5, 9, 16, 29–32, 54, 66, 75, 95, 177, 180, 193, 195, 205, 207–208, 210
confusion of the present age   2, 26–29, 32. *See also* Kierkegaard
Congo   6, 103, 105–108, 113, 117–120, 157–158, 222
Congo Conference   102, 105–106
Congo Free State   107, 111, 120
Conrad, Joseph
– "Geography and some Explorers"   6, 103, 107–108
– *Heart of Darkness*   4, 6, 34, 102–129, 130–132, 157–158, 222
– *Mirror of the Sea*   132
– *Nostromo*   132
– "The Return"   131
– *The Secret Agent*   4, 6, 29, 34, 128, 130–156
conscience   6, 86, 108, 117, 120–121, 123, 157, 205–207
– bite of   172
– Christian   37
   *See also* voice of conscience
conservatism   20, 135, 156. *See also* reaction
corruption   1, 4, 50, 62–63, 219
crimes of passion   146, 153

criminal   35–41, 51, 60–63, 99, 116, 136, 140, 146, 152, 185, 220, 229. *See also under* Schiller
crisis   3, 67, 207
– epistemological   44
– geo-political   105
– Kant crisis   (*see under* Kleist)
– of action   135, 144
– of agency   6, 130, 146, 155
– of authority   6, 157, 167–168
– of Enlightenment reason   2
– of faith   32
– of judgment   176–182
– of masculinity   203
– of conscience   207
– time of   242
   *See also* confusion
Cunninghame Graham, R. B.   123, 131

Dante Alighieri   220
death   56, 96, 101, 150, 179, 236
– apparent   193
– authentic/proper   177, 182
– being-toward-death   176, 215
– civic death   37   (*see also* infamy)
– death sentence   178, 182
– dying from taste   217
– embarrassment   175
– flavour of mortality   127
– "sickness unto death" (*see under* Kierkegaard)
– struggle unto   48, 92
   *see also* mortification
debt   95, 160–161, 163–164, 167, 186, 205–208. *See also* guilt
defiance   34, 60, 70, 139, 222
degenerate   140, 145, 147
Deleuze, Gilles   8, 158, 239–240
demand   75, 91, 93, 95–98, 99, 155, 229
– for action   145
– for justice   241
   *See also* command; imperative; office-duty; speech act
dependence   39, 130, 134, 136, 139, 143, 147, 229
Derrida, Jacques   113

despair 1, 6, 9, 32–34, 64, 75, 83, 96–97, 103, 117–119, 122, 125, 131–136, 152, 154–157, 161, 164, 222
- Bakunin 141–142
See also Kierkegaard: *Sickness unto Death*
destitution. See poverty, the poor
disability 133–134, 147, 229, 234
disaffection 13, 88, 130, 141–142, 156
disenchantment 17, 157
disfigurement 6, 102–108, 114, 121, 129, 157, 229
dogs 35–40, 48–49, 55, 147, 172, 179, 184, 220
- "Like a dog" 40, 179
- rabid 38, 40, 48–49
doubt 15, 29, 33, 47, 63–66, 159–160, 164–173, 176
dream 7, 42, 45, 87, 124, 189–190, 239
- *Dream of Reason produces Monsters* (Goya) 1
- of reason 1, 13–16
See also Kant: *Dreams of a Spirit Seer*
dulden 208–209
duty. See office-duty
Durkheim, Émile 151

earth 3, 4, 6, 16, 101, 102–105, 108, 110–111, 113–117, 126
- attitudes to 222
- rational domination of 6, 44
- tending 219, 233–234
See also gardening; geography; geopolitics
earthworm 244
easiness 83, 86–90. See also snug
ecopolitics 117, 232
Edwards, Jonathan 89–94
*Eigentlichkeit.* See authenticity
*Eigentümlichkeit* 193, 213–215, 233. See also peculiarity
1848 2, 9, 29–31, 84, 140
embarrassment 6, 98, 157–184, 187–188, 190, 192, 194–195, 200. See also *Peinlichkeit, Verlegenheit*
empathy. See compassion
Enlightenment 1, 2, 12–13, 42–43

enthusiasm 2, 14, 30–31, 42–43, 58, 112, 125, 166
- for the abstract 19
*Enthusiasmus* 30, 43
*Entortung* 102–108, 113, 116–119, 123, 125, 158. See also disfigurement; Schmitt
Ewald, François 186–187, 205, 228
exception 9–12, 26–29, 31, 157, 237–238
expansion 2, 16, 49, 104, 122–123

fanatical bureaucrat 3–5, 34, 63, 102, 108, 223
fanaticism 2, 5, 12–13, 16, 19–23, 32, 88, 112, 125, 143, 146, 171
*Fanatismus* 21
*Fehlleistung* (parapraxis) 65, 151
Fichte, Johann Gottlieb 32–33, 43–49, 53, 67, 70–71, 114–115, 117, 123, 221
- *Fundamental Features of the Present Age* 46
- primacy of the practical 5, 77
- *The Determination of the Human* 44–49, 123
- *Wissenschaftslehre* 43–46, 53
See also *Schwärmer*-debates
First World War 6, 16, 19, 157–158, 169, 182
forced labor 107, 115. See also Atlantic slave trade
forgiveness 29, 63–66, 74, 172. See also *Vergebliche*
Foucault, Michel 141, 205, 228, 240
- *Abnormal* 37, 195
- *Discipline and Punish* 196
- *Omnes et Singulatim* 228, 235, 242
- *Security, Territory, Population* 185–186, 204, 208, 228, 235, 242
- "Society Must be Defended" 230
- "The Lives of Infamous Men" 237
French Revolution 2, 12, 41, 42
Fresleven (*Heart of Darkness*) 121–122
Freud, Sigmund 120–121, 124, 151, 196
Freudian slip. See *Fehlleistung*
fulfillment 15, 109–110, 114, 126–128, 131, 155, 193, 207, 213, 216

futility   34, 130, 136, 151–152, 155–156, 193, 236, 239. See also *Vergebliche*
– futile bodily agitation   149, 151, 155, 197

Gaderer, Rupert   49, 61
gardening   219, 229–233, 240–244. See also earth; Michael K
geography   6, 103, 107, 116–117, 157. See also earth
geopolitics   102–104, 108, 113, 116–117, 119, 158. See also earth
Gladstone, William Ewart (British Prime Minister)   134
Gordimer, Nadine   241
governmentality   2, 6, 21, 37, 130, 138, 155, 228, 235, 237, 241
Goya, Francisco
– *Los Caprichos*   1
Greenwich   6, 103, 116, 145–146, 150–151, 155. See also International Meridian Conference
Gregor Samsa (*Metamorphosis*)   4, 6–7, 29, 185–217, 230, 238
Guérard, Albert   124, 130, 144
guerilla.   See partisan
guilt   29, 120, 159–183, 186, 193, 205–208. See also debt
guilt-history (Hamacher)   163, 175
gypsy soothsayer (*Michael Kohlhaas*)   56, 67–73, 238

Hamilton, John   85, 190, 200
Harcourt, William (British Home Secretary)   134
Hegel, Georg Wilhelm Friedrich
– *Differenzschrift*   45, 53, 70
– *Lectures on the Philosophy of History*   19
– *Phenomenology of Spirit*   22
– *Philosophy of Right*   2, 20–26, 32
  See also Marx: "Critique of Hegel's Philosophy of Law"
Hegelianism
– Kierkegaard   26–27
– Schopenhauer   20–21
Heidegger, Martin   159, 202
– *Being and Time*   109, 176, 191–192, 196, 199, 205, 215–216

– *Introduction to Metaphysics*   77
Herbert, Auberon   138
Herder, Johann Gottfried   43
Hobbes, Thomas   36, 55
Hobhouse, L. T.   133–134
Hobson, J. A.   133–134
honor   36–38, 57, 59–61. See also under Schiller
horror   39, 107, 112, 124, 148
– "The horror! The horror!"   29, 116, 127–128
horrorism (Cavarero)   128–129
horses   4, 147–148, 150
– Kohlhaas's *Rappen*   41, 50–66, 70–72, 225
Howe, Irving   132, 135, 150
human, the   4, 16, 21, 25, 31–33, 37, 41, 44–47, 72–73, 90, 104, 119, 144, 188, 192–193, 206, 211, 217
– human–animal distinction   37, 39, 209
– Kierkegaard   31–33
– question of   42, 49–55, 71, 74
  See also Fichte: *Determination of the Human*
humanitarianism   3–4, 35, 39, 54, 96–98, 107, 120, 158, 164, 171, 177, 210
humanity   3, 5, 13, 25, 36–37, 39–43, 50–55, 58, 71–75, 85, 96–98, 131, 180, 201, 209–210
– "Ah Bartleby! Ah humanity!"   5, 95–97
Husserl, Edmund
– *Logical Investigations*   108–115
hunger   88–89, 120, 149, 205–207, 214–215, 230, 236. See also fulfillment

idea   19, 108–112, 124–125, 136, 214, 240–241, 244. See also idealism; Intended; intentionality
idealism   112–115, 126
impassivity   90–91, 96, 150, 229. See also passivity
imperative   5–6, 42, 53–54, 77–78, 81, 82, 85, 87, 95–96, 114
– biopolitical   230
– categorical (Kant)   53, 77
– Fichte   52
– literary   237

– of efficiency 78, 82
– of public safety 85
– of the office 5
– political 42
– security 6
See also command; demand; speech act
imperialism 6, 103, 134
– New Imperialism 102–107, 111–113, 120–123, 158, 222
See also Arendt: Origins; colonialism
impotence 34, 54, 127, 162, 176
improvidence 240, 244
inaction. See impassivity; passivity
indiscretion 7, 189, 192–197, 200–203, 210, 213, 215
infamy 37, 237
inhumanity 42, 84, 164, 171, 176
– of bureaucracy 25, 39, 104
– of reason/rationality 2, 13
injustice 1, 39, 60–63, 139, 149–150, 164, 176, 224, 229
insecurity. See security
institution 142–146, 229
– care 185
– family 6, 185–186, 189–190, 198, 201, 205–210, 214–217
– fatherhood 208, 224, 227, 228, 236
– hospital 185, 202, 229, 231, 238
– literature 7, 56, 70, 156, 237–240
– marriage 64–66, 127, 143, 152, 154
– social security 185
– the Church 26–27, 32, 76–77, 80, 163, 228
*Institutionenroman* (Campe) 187. See also bureaucracy novella
insurance 133, 187, 206
Intended, the (*Heart of Darkness*) 108, 109, 112–113, 116, 124–128, 160. See also idea
intentionality 108–109, 113–115, 127, 136, 197
– imperialist 113, 127
*Inter caetera* (Papal Bull) 102
International Meridian Conference 103
intuition (*Anschauung*) 16, 22, 109, 112
irony 3, 14, 19, 29, 31, 38, 83, 95, 107, 132, 137, 145, 149, 156, 175, 183, 205, 238, 241. *See also* communication: indirect; literature, the literary
ivory 108–115, 126

Jacobs, Carol 67
Janouch, Gustav 192, 197
Job, Book of 101
judgment 16, 31, 170–173, 195–199, 206, 212–213
– crisis of 176–182
– teleological 44, 47–48
See also *Verlegenheit*
justice 4, 7, 35–36, 51, 55, 59–62, 92, 162–165, 220–221, 226–229, 240–244
– divine 42, 62, 220
See also *Rechtgefühl*

Kafka, Franz
– *In the Penal Colony* 4, 6, 29, 34, 157–184, 191, 206
– *The Burrow* 241–243
– *The Hunger Artist* 88, 230
– *The Judgment* 178, 195, 197
– *The Metamorphosis* 4, 6, 34, 185–217, 221
– *The Trial* 161, 179
Kant, Immanuel 2–5, 22, 42–50, 53
– *Achtung* 53, 97
– categorical imperative 53, 77
– *Critique of Practical Reason* 43
– *Critique of Pure Reason* 14, 22, 91
– *Critique of the Power of Judgment* 13, 47, 176, 182
– *Dreams of a Spirit-Seer* 13–16, 19
– primacy of the practical 5, 77
Kant crisis. See *under* Kleist
Kierkegaard, Søren
– *Book on Adler* 2–3, 9, 26–29, 30–33, 177
– *Repetition* 9–12
– *Sickness unto Death* 6, 32–34 119–122
– *Two Ages* 30
See also communication: indirect; confusion
Kleist, Heinrich von
– *Berliner Abendblätter* 37–39, 49

- Kant crisis 44, 47, 62, 219, 221
- *Michael Kohlhaas* 4, 5, 7, 34, 35–74, 91, 115, 121, 149, 193, 220–244
- *On the Marionette Theater* 47–48, 71
knacker See *Abdecker*
Kommerell, Max 63–64
Kurtz (*Heart of Darkness*) 4, 29, 34, 102–129, 160

land appropriation (*Landnahme*) 102, 105, 117, 232. See also Schmitt
law 4, 35–40, 53, 55–57, 66–67, 71–73, 92, 185, 226
- and the legal system 50, 62–63
- copying (*Bartleby*) 82–83, 89, 93–94
- discriminations of 36, 38, 60
- imperialist 116–118, 121
- letter of 50, 64, 73
- of nations 224, 236
- positive 51, 58, 62
  See also natural law
lawyer (*Bartleby*) 4–5, 75–101
Leopold II, King of Belgium 107, 120
Levenson, Michael 109
Liberal welfare reforms (Britain, 1906–1914) 133–134
liberalism 6, 130, 133–134, 185, 200. See also governmentality
liberalization 130, 133–134, 164, 185
lie. See *under* speech act
literalization 6, 178, 197, 205
literature, the literary 3, 7–9, 12, 19, 26, 29, 40, 56, 70, 103, 108, 155–156, 223–227, 237–240. See also bureaucracy novella; communication: indirect; irony; institution: literature; postcolonial literature
Locke, John 89–90, 233
Louvre 218–222
*Lumpenproletariat* 141–142. See also outcast residuum
Luther, Martin 164
- "Against the Robbing and Murdering Hordes of Peasants" 40–41
- in *Michael Kohlhaas* 55–58, 63–67
- *Schwärmerei* 2, 12, 30–31, 40–43, 46

Mackinder, Halford 103–104
madness 16, 39, 119, 136, 142, 152, 154, 156, 228
Matthew, Gospel of 65, 235
Marcuse, Herbert 17–19
Marx, Karl 30, 82, 84–85, 91, 141–142
- "Critique of Hegel's Philosophy of Law" 23–26
measure 77, 117, 119, 121, 146, 209–210
- of the earth 4, 104
- of the human 45
Melville, Herman
- *Bartleby* 4–5, 7–8, 29, 34, 75–101, 149–150, 228, 238, 240
mental deficiency 133, 151. See also disability
Mendelssohn, Moses 13–14
metaphor See literalization
Michael (archangel) 4, 57, 63, 67, 219–220, 222
Michael K (Coetzee) 7, 219–244
Michael Kohlhaas (Kleist) 4, 5, 7, 34, 35–74, 91, 115, 121, 149, 193, 220–244
mood 59, 65, 103, 207, 221. See also despair
- fundamental 6, 32, 119, 130–135, 155
Morel, Edmund 120
mortification 175–179, 182. See also death; embarrassment
Müntzer, Thomas 31, 40–43, 52, 63. See also Luther; *Schwärmerei*
murder 37, 41, 93, 100, 152
Muth, Ludwig 44, 47

natural history 234
natural law 36, 51, 55, 106, 188
nature 35–36, 41, 53–55, 58, 60, 62, 67, 69, 74, 82, 114, 122
- aberration of 188, 197, 210
- bear/bare nature 47–50, 71
- dispositive of 47
- Fichte/Schelling 49, 70, 115
- nature–culture 207
- nemesis of 73
- technique of (Kant) 47
*Naturphilosoph* See gypsy soothsayer (*Michael Kolhaas*); Schelling

*Naturphilosophie* 44–46, 67, 70, 73, 114
necropolitics (Mbembe) 6, 117
Nietzsche, Friedrich
– *Gay Science* 160
– *Genealogy of Morality* 17, 120, 135–136, 152, 159–161, 163–164, 167, 172, 176, 182
– *Will to Power* 190–191
– *Zarathustra* 204
nihilism 190–191
1914 4, 104, 160–162, 164, 174, 177
normalization 186, 195, 198, 213

obligation 64, 75, 77–78, 81, 94–96, 206, 230, 231
– familial 201, 206–209
See also office–duty
office 1, 5, 14–16, 26–27, 57, 62, 75–101, 118, 175–187, 222
office–duty 5, 75, 78, 92–95, 97, 123, 237
office politics 87, 167
officer, the (*Penal Colony*) 4, 6, 29, 34, 159, 164–183, 191
officiousness 2, 67, 84, 166, 200–201
ontology 77–78
– of the office 77, 86, 96–97
outcast residuum 140–142. See also *Lumpenproletariat*; vagrancy
outlaw 4, 35, 37, 40, 51, 61–65, 99, 226
outrage 6, 116, 118, 130, 133, 135, 140, 144, 146–151, 153, 155–156
oversight 5, 22, 39, 49, 56, 60, 69, 84, 86, 97, 143, 185, 187–189

pain 63, 97, 149, 159–164, 174, 182, 208, 217. See also *Peinlichkeit*
parapraxis. See *Fehlleistung*
partisan 4–5, 7, 50–51, 57, 61–62, 65, 68, 224, 226, 229, 233, 241. See also Schmitt
partisan novella 224, 226
passion 13, 85, 93, 137, 144, 146–147, 149, 152, 223–225, 229, 233, 243
– for bureaucratization 1, 17
– negative 142, 152
– for justice (see *Rechtgefühl*)
See also crimes of passion

passivity 34, 52, 69, 71, 88–91, 196, 209, 211. See also impassivity; inaction
pathology 13, 35, 90
peculiar fanatic 3–7, 34, 88, 91, 96, 100, 147–150, 193, 213, 217, 228. See also under resistance
peculiarity 5, 90, 94, 97, 147–151, 193, 213, 217, 226, 233, 238, 241, 244
– taste for 7, 34, 211
*Peinlichkeit* 157, 162–164, 168, 174–176, 182. See also embarrassment; pain; *Verlegenheit*
phenomenology 109, 113–114, 117
police 21, 37–39, 56, 68, 73, 99, 143, 147, 150, 212, 227
political, the 7, 11–12, 29–31, 50, 136, 155, 170, 175, 223, 240
political theology 2, 8–9, 41, 58, 61–62, 76, 123, 235
political theory 3, 7–12, 104
political zoology 2, 38–41, 55, 57, 71, 188, 210
postcolonial literature 157–159, 175, 177, 241
postcolony 171, 174–182
poverty, the poor 36, 57, 91, 99, 130, 133, 135, 137, 138, 140–142, 146–149, 152, 156, 183. See also outcast residuum
power of judgment 177, 180, 195. See also under Kant
prefer not to See under speech act
Prime Meridian See Greenwich; International Meridian Conference
*propaganda par le fait* 145
providence 93, 235. See also governmentality; improvidence
Pufendorf, Samuel 77

querulance 61–63

rabies 38, 40–41, 48–49
racism 111, 114, 119, 123
rage 38–39, 55, 59, 83, 149–150. See also outrage; rabies
Raphael
– *Le grand St Michel* 220
– *Le petit St Michel* 218–223

– *Sistine Madonna* 219
– *Transfiguration* 219
rationality 2–4, 13, 17–19, 43, 49, 54, 69, 240
rationalization 2, 44, 54, 88, 135
– imperative of 54
– of the legal system 80
– Weber 16–20
reaction 133–135, 136, 139, 141, 155. See also conservatism; ressentiment
reason 14, 19, 45, 53–54, 131, 146
– capitalist 18
– dream of 1, 13, 16
– Enlightenment 2, 13, 43
– insanity (*Wahnwitz*) 16
– "political reason" 228, 235, 240
– principle of 94
– relativization of 17
*Rechtgefühl* 4, 35, 41, 51–74, 220
– and bureaucratization 35, 61
– negligible cause 60, 66, 74, 91
– passion for justice 7, 226, 229, 233, 243
– spelling of 51
*Rechtsfanatiker* 61. See also bureaucratic fanatic; partisan
recognition 37, 53, 55, 58, 61, 65, 70
refusal, politics of 7–8, 228
resistance 226, 236, 238
– passive 5, 68, 84, 226
– peculiar 226, 241, 244 (*see also* peculiar fanatic)
*ressentiment* 21, 135, 139. See also reaction
revelation 3, 26–33, 89, 197, 219
revolution 2, 12, 13, 20, 30, 40–42, 50–51, 67, 132, 138, 141–142
Rhodes, Cecil John 104, 122–123
rights 8, 20, 50, 53, 55–56, 58–59, 99, 107, 141, 154, 210, 232–233
Rousseau, Jean-Jacques 219

safety See security; snug
Schelling, Friedrich Wilhelm Joseph 43–49, 67, 70, 114–115, 117. See also *Schwärmer*–debates

Schiller, Friedrich
– *The Criminal out of Lost Honor* 35–41, 59, 68, 70, 185
Schmitt, Carl 9–12, 158
– *Concept of the Political* 170
– *Nomos of the Earth* 102, 104–105, 108, 117, 127
– *Theory of the Partisan* 50–51, 233
Scholem, Gershom 186
Schopenhauer, Arthur 20–21
*Schuld* See guilt
*Schwärmer* 3, 6, 15–16, 19, 31, 35, 41–49, 60, 63, 67, 70, 73, 90, 114, 193, 211. See also bureaucratic fanatic; fanatical bureaucrat; peculiar fanatic
*Schwärmer*–debates 12, 41, 43
– Fichte–Schelling 43–49, 67, 70, 114, 117
*Schwärmerei* 1–8, 17, 31, 43–49, 57, 60, 115, 119, 125, 171, 185, 193, 201, 213, 217, 221, 223, 237
– fallen 6, 193, 215
– Fichte 43–49
– Herder 43
– history of term 2, 12, 41–43
– Kant 13, 16, 43
– Kierkegaard 30
– Luther 30, 41–42
– problem of form 14, 19, 31
– Schelling 43–49
See also bureaucratic fanaticism
science 11, 14, 68, 69, 119, 131, 135, 145–146, 164, 171, 177, 219
– natural 45, 47, 159, 210
– Nietzsche's critique of 17, 160
– political 11
See also Kleist: Kant crisis; Weber: "Science as a Vocation"
security 7, 15, 32, 34, 75, 86, 116, 166, 168, 197–198, 208, 210, 228, 240
– apparatus 230, 242–244
– dispositive 190–191, 218
– regime 235
– *securitas* 85–86, 190
– society of 6, 185–192, 215
– social security 133
See also care; safety; snug
shame 149, 161, 179

silence   84, 113–114, 131, 200, 202, 230–232, 244
slaves   *See* Atlantic slave trade; forced labor
Smock, Anne   95
snug   5, 75, 80, 82, 84–86. *See also* easiness; security
social question   *See* outcast residuum; poverty, the poor
socialism   17–18, 133. *See also* Weber
*Sorge*   189, 191–192, 194, 196–197, 200, 204, 207, 209, 214, 216–217. *See also* care
sovereign   9, 33–34, 60, 63, 66, 103, 106–107, 122, 165, 167, 186, 216
sovereignty   9, 11, 37, 103–107, 122, 157, 164, 186
– postures of   164, 169
sparrows   234–240
speech act
– demand   5
– "Do I not?"   126–127
– failure   180
– infelicitous   65, 145
– lie   112, 116, 127–128
– misexecution   179
– outrage   144
– poor utterances   149
– prefer not to   29, 34, 86–90, 93, 100–101, 150, 239
– question   54
– statement   100
– understatement   81
– *versagen*   201–202
  *See also* Austin; command; *Fehlleistung*; imperative
Stanley, Henry Morton   105, 140
state of emergency   228, 243. *See also* exception
Stevie (*Secret Agent*)   4, 6, 7, 29, 34, 135–137, 143, 144, 147–155, 160
subsumption   21–26, 44, 176
suffering   *See* pain
suffragettes   153–154
suicide   72, 100, 119, 135, 151–152, 154, 215
– bombing   139

superfluous type   111, 120, 122, 140. *See also* Arendt: *Origins*
swarm/swarming   1, 2, 31, 42, 46, 57, 59, 112, 139. See also *Schwärmerei*
Swedenborg, Emanuel   13–16, 26

taking (*nehmen*)   105, 117, 126–127. *See also* land appropriation (*Landnahme*); Schmitt
taste   7, 88, 182, 193, 230–234
– distasteful   107, 208, 211
– dying from taste   210–217
– flavor of mortality   127
– for peculiarity   7, 34, 211
terror, terrorism, terrorist   107, 121, 126, 132, 139, 143–146, 154, 226
Thoreau, Henry David   84
Toscano, Alberto   19–20
traveler, the (*Penal Colony*)   4, 6, 159, 165–166, 169–184
tropical disease   118–119, 132, 180–181
Trotha, Lothar von   158
truth   14, 21, 33, 91, 113, 124, 128, 139, 149, 160, 173–174, 197–198, 239, 244
– as fulfillment   109–110, 127–128, 131, 155
– scientific truth   46, 110
– truth-sayer   69 (*see also* gypsy soothsayer)
  *See also* authenticity/inauthenticity; fulfillment; speech act: lie

*Ungeziefer* (vermin)   57, 185, 187–189, 193–194, 196–199, 203, 206–213, 216, 221

vagrancy   98–101. *See also* outcast residuum
Valéry, Paul   15, 26
*Vergebliche, das*   63–67, 70–71, 74, 90, 115, 147, 193, 241. *See also* forgiveness; futility
*Verlegenheit*   157–158, 161–162, 165, 173, 176, 195. *See also* embarrassment; *Peinlichkeit*
Verloc (*Secret Agent*)   135, 140–156
vermin   *See Ungeziefer*

Vietnam War   115, 222
vocation/calling   12–19, 21, 26, 46, 69, 76, 79, 88, 166, 232. *See also* avocations; Weber
*Vogelfreiheit*   68, 71. *See also* gypsy soothsayer; political zoology
voice   6, 10, 42, 46, 54, 83, 86, 102, 113–117, 121, 123–129, 168, 178, 199, 229–233, 237, 239
– animal   54, 199, 200–201
– narrative   116, 149
  *See also* Kurtz
voice of conscience   45, 48, 53, 115, 121
Voltaire   219

walls   198, 204, 212
– fences   232
– office   25, 82, 84–86, 92, 94, 98, 100
– prison   100
weakness.   *See* impotence; passivity
Weber, Max   1, 3, 16–20, 109, 166
– bureaucratic authority   18, 157, 168
– "Politics as a Vocation"   88–89
– "Science as a Vocation"   16–17
– speech on socialism   17
welfare state   133
"What is to be done?"   91, 95, 97, 145
Winnie (*Secret Agent*)   135–137, 147, 149, 152–156, 160
Wolff, Kurt   162, 164, 173–176, 182, 203
wolves   35–40, 172, 185

www.ingramcontent.com/pod-product-compliance
Lightning Source LLC
Chambersburg PA
CBHW031802220426
**43662CB00007B/496**